Linear Algebra

The Sally Series

Pure and Applied
UNDERGRADUATE TEXTS · 66

Linear Algebra

A Problem-Centered Approach

Róbert Freud

AMERICAN MATHEMATICAL SOCIETY
Providence, Rhode Island USA

2020 *Mathematics Subject Classification.* Primary 15A03, 15-00, 15-01, 15A04, 15A06, 15A15, 15A18, 05C50.

For additional information and updates on this book, visit
www.ams.org/bookpages/amstext-66

Library of Congress Cataloging-in-Publication Data

Names: Freud, Róbert, author.
Title: Linear algebra : A problem-centered approach / Róbert Freud.
Description: Providence, Rhode Island : American Mathematical Society, [2024] | Series: Pure and applied undergraduate texts ; volume 66 | Includes index.
Identifiers: LCCN 2024036316 | ISBN 9781470477363 (paperback) | ISBN 9781470479060 (ebook)
Subjects: LCSH: Algebras, Linear. | AMS: Linear and multilinear algebra; matrix theory – General reference works (handbooks, dictionaries, bibliographies, etc.). | Linear and multilinear algebra; matrix theory – Instructional exposition (textbooks, tutorial papers, etc.). | Linear and multilinear algebra; matrix theory – Basic linear algebra – Vector spaces, linear dependence, rank. | Linear and multilinear algebra; matrix theory – Basic linear algebra – Linear transformations, semilinear transformations. | Linear and multilinear algebra; matrix theory – Basic linear algebra – Linear equations. | Linear and multilinear algebra; matrix theory – Basic linear algebra – Determinants, permanents, other special matrix functions. | Linear and multilinear algebra; matrix theory – Basic linear algebra – Eigenvalues, singular values, and eigenvectors. | Combinatorics – Graph theory – Graphs and linear algebra (matrices, eigenvalues, etc.).
Classification: LCC QA184.2 .F74 2025 | DDC 512/.5–dc23/eng/20240916
LC record available at https://lccn.loc.gov/2024036316

Pure and Applied Undergraduate Texts ISSN: (print)

DOI: https://doi.org/10.1090/amstext/66

∞ The paper used in this book is acid-free and falls within the guidelines established to ensure permanence and durability.
Visit the AMS home page at `https://www.ams.org/`

10 9 8 7 6 5 4 3 2 1 29 28 27 26 25 24

Contents

Introduction

Objectives and target audience

The book is an extended version of an introductory one- or two-semester course in linear algebra with a special emphasis on carefully clarifying not only the "how" but also the "why", i.e., besides covering all notions, theorems, and methods, we try to explain the main points and deeper contexts of the material as well. We kept in mind primarily the needs of students in mathematics and computer science, but the book is suitable for majors in other fields and for individual studies, too.

Prerequisites

No background knowledge beyond high-school material is required, although familiarity with mathematical proofs is assumed. All necessary algebraic notions (complex numbers, polynomials, fields, rings, etc.) and their fundamental properties are summarized in the Appendix "Basic Algebra" at the end of the book. The Appendix, while facilitating the better understanding of the other parts of the book, also provides an insight into an important and useful area of independent interest.

Special features

Compared to the usual linear algebra texts, we highlight three aspects.

(i) We put a strong emphasis on clarifying the material through a problem-centered approach. The notions and statements are thoroughly explained informally beyond their precise formulations and are illustrated by examples. We try to illuminate their many facets, to grasp their essential features, and to point out their connections to previous material. We warn about pitfalls and show how to avoid them. We analyze the reasons for introducing the given notion and always try to proceed from concrete cases towards general observations.

(ii) We show real and entertaining applications to other parts of mathematics based on minimal knowledge. We devote the entire Chapter 9 to combinatorial-type problems, a nice and important area not at all treated in most linear algebra textbooks. Here, linear algebra is a natural (and often the only effective) tool to handle questions in graph theory, geometry, number theory, and other fields of mathematics. Among others, we solve Hilbert's third problem, investigate the topic of Oddtown and Eventown, and study Sidon sets, a favorite of Paul Erdős. Chapter 10 is a short introduction to coding theory, another significant application.

(iii) We introduce determinants at the beginning of the book in a very smooth, detailed, and elementary way. They serve as a powerful and natural tool to handle matrices and systems of linear equations in a unified structure.

We note that the book can also be used effectively by those preferring determinants to occur at a later stage. To do this, begin with Sections 2.1, 3.1, and 3.3. From there, continue to Chapter 4 and possibly Chapter 5, before returning to Chapter 1, and Sections 2.2, 3.2, 3.4, and 3.5.

Exercises

Each section concludes with a generous set of exercises that serve various purposes. Some of them help the better understanding of the actual notions, theorems, and methods; others present additional examples, connections, and applications; again others discuss problems loosely related to the given topic. Some exercises are theorems in "disguise" calling attention to further interesting aspects of the material not treated in detail in the main text.

Accordingly, both the quantity and the difficulty of the exercises vary widely depending on the topic, size, and depth of the actual material. The hard and extra hard exercises (by our judgement) are marked with one and two asterisks, resp. (The difficulty of an exercise is always relative, of course: besides the abilities, interests, and preliminary general knowledge of the solver, it depends strongly, among others, also on the exercises having already been solved previously.)

Supplementary exercises are available electronically to those instructors who have adopted the textbook for classroom use. For more information, please email textbooks@ams.org.

To promote progress in solving the exercises, three extensive online resources are freely accessible on the webpage `bookstore.ams.org/amstext-66`. "Answers" is for checking whether a result is correct. "Hints" give some advice on how to attack a problem. "Solutions" provide a detailed argument for some (mostly harder) problems, their serial numbers are boldface in the text.

The Reader is advised to consult the hint or solution only if the exercise turns out to be absolutely unmanageable. We suggest rather to return to the same problem later or to solve first some special case of it.

When working on an exercise, it is important to unravel its "message" and background, to reveal its position and role in the mathematical landscape. Thinking about a generalization or raising new problems is very useful (even if it is not clear how to solve them).

Instead of just thinking in your head about the exercises, it is important to write down their solutions. Working through the details clarifies the argument, reveals its most important features, and helps to identify any errors or gaps.

Structure of the book

The first three chapters are strongly interrelated dealing with the "most classical" linear algebra material of determinants, matrices, and systems of linear equations. As mentioned before, we changed the usual order by starting with determinants to be able to apply their full power to matrices and systems of linear equations.

Similarly, Chapters 4, 5, and 6 form a tight unit in discussing the basic facts about vector spaces and linear maps and transformations. Most parts of Chapters 4 and 5 can be well understood even independently of the first three chapters.

Chapters 7–10 generally rely strongly on the first six chapters. Chapters 7 and 8 form a block in elaborating bilinear functions and Euclidean spaces. Chapters 9 and 10 deal with applications of combinatorial character and in coding theory, resp. These two chapters do not assume knowledge of each other or of Chapters 7 and 8 except for a few parts of Chapter 9.

We recall that the Appendix is a self-contained detailed summary of the algebraic basics used in the book.

Technical details

The chapters are divided into sections. Definitions, theorems, exercises, and formulas are numbered as $k.m.n$ where k refers to the chapter, m to the section, and n is the serial number within the given section. Definitions and theorems have a common list; thus, e.g., Definition 5.1.4 is followed by Theorem 5.1.5. Examples are numbered with a single number restarting in each section. The phrasing of a definition or theorem is closed by a ♣ sign, the end of a proof is denoted by ■.

The following general notations are applied: vectors are denoted by lower case bold Roman letters (e.g., $\mathbf{a}$); scalars are lower case Greek letters (e.g., α); matrices are uppercase Italic Roman letters (e.g., A); linear maps are uppercase calligraphic Roman letters (e.g., $\mathcal{A}$); bilinear functions are uppercase bold Roman letters (e.g., $\mathbf{A}$). We note that zero can mean quite a lot of things (an integer, the zero element of a ring, a scalar in a field, a vector, a vector space, a subspace, a matrix, a linear map, a bilinear function, etc.), and many of these share the same notation, but the meaning is always clear from the context.

We distinguish the floor and ceiling of (real) numbers; these are denoted by $\lfloor\ \rfloor$ and $\lceil\ \rceil$, resp., thus, e.g., $\lfloor\pi\rfloor = 3$, $\lceil\pi\rceil = 4$ (we do not use the notation $[\pi]$ for the floor). The fractional part is denoted by $\{\ \}$, i.e., $\{c\} = c - \lfloor c\rfloor$. Divisibility, greatest common divisor, and least common multiple are denoted as usual, so, e.g., $7 \mid 42$, $(9, 15) = 3$, $[9, 15] = 45$. The square brackets [ ] can mean a least common multiple, a replacement for (round) parentheses, a closed interval as in Section 9.6. Further, $[\mathcal{A}]$ and $[\mathbf{A}]$ stand for the matrix of the linear map $\mathcal{A}$ and bilinear function $\mathbf{A}$, resp.

The degree of a polynomial f or $f(x)$ is $\deg f$; thus, e.g., $\deg(x^3 + x) = 3$. For the real and imaginary parts of a complex number, we use "Re" and "Im", resp., e.g., $\mathrm{Re}(4 - i) = 4$, $\mathrm{Im}(4 - i) = -1$. As usual, $\mathbf{Q}$, $\mathbf{R}$, and $\mathbf{C}$ stand for the rational, real, and complex numbers, resp.; $\mathbf{Z}$, $\mathbf{Z}_m$, and $F[x]$ mean the integers, the modulo m residue classes, and the polynomials over F. For products and sums we often use the signs $\prod$ and $\sum$.

Style

In the introductory chapters, we present proofs using elementary and less abstract methods, and give detailed explanations instead of overly concise arguments to facilitate the understanding for less experienced Readers. We often recall some assumptions or conditions during the deduction to clarify their point in the argument.

We strive to give a comprehensive illustration of the tight and colorful connection between linear algebra and other branches of mathematics. (Sometimes we may seem to wander too far from the strict limits of linear algebra — though it is very difficult to say where the border is since the various parts of mathematics are interwoven through millions of threads.)

Suggestions

It is self-evident that in mathematics, there is no point in learning notions or theorems without their thorough understanding. We suggest not to skip even the tiniest vague detail, to check all references to parts discussed earlier, and to go meticulously through statements labeled as "easy to see".

Beyond the point-by-point formal understanding, a notion or theorem will not carry a substantial message unless we can find its proper place in the mathematical environment and acquire a clear view of its connections and applications. We can achieve this by going through many illustrative examples and solving several exercises related to the given notion or theorem.

It is better not to stick to the letter of the text in the book, but rather rephrase the given notion or assertion in another way with your own words (but check carefully whether we really say the same). It is highly recommended to analyze where the assumptions or conditions are used in a proof, to formulate the converse of a theorem, and to determine whether it is true or false.

Acknowledgement

The book is an English translation and an improved and extended version of the many Hungarian editions widely used in Hungarian universities.

I am grateful to my colleagues working in algebra and number theory at the Institute of Mathematics at Eötvös Loránd University Budapest for several interesting exercises and important comments. I am also grateful to my students in my over-a-half-a-century career as an instructor not just for their feedback but also for the experience acquired while lecturing to them.

I want to express my special thanks to LÁSZLÓ BABAI for teaching me new aspects of linear algebra that I could incorporate into this book.

I am most grateful to my late wife EDIT GYARMATI since the book's structure, approach, and style bear the brand of her experimenting spirit and substantive innovations both in form and content during her teaching career of many decades. Also, she was my strictest critic and gave lots of professional, didactic, and stylistic recommendations improving the quality of the book.

I express my gratitude to the referees of the Hungarian edition, PÉTER HERMANN and EMIL KISS for their hard and dedicated work. They reviewed the manuscript very carefully, checked the exercises and solutions meticulously, and in addition to detecting errors they made plenty of general, concrete, and stylistic comments most of which I incorporated into the text.

I appreciate the valuable remarks given by ANDRÁS SIMONOVITS who checked the first English version of the manuscript.

Finally, I am very thankful to AMS editors STEPHEN KENNEDY and LORETTA BARTOLINI for all their help in publishing the book, and to KATALIN FRIED who, besides producing a perfect technical realization, corrected many typos in the manuscript and suggested several improvements in the text.

Budapest, June 23, 2024

Róbert Freud
Institute of Mathematics, Eötvös Loránd University, Budapest
freudro8@gmail.com, freud@staff.elte.hu, robert.freud@ext.uni.lu

Chapter 1

Determinants

Determinants were originally invented to describe solutions of systems of linear equations but later they became fundamental in many branches of mathematics. This notion seems to be complicated and artificial at first sight and a certain amount of practising is required to master it properly, but it turned out to be a perfect notational system for a comfortable, elegant, and natural handling of a wide range of problems. We will see several such examples in the next chapters.

1.1. Inversions in Permutations

We shall need the inversions in permutations just for the definition and a few basic properties of the determinant. Therefore, we shall confine ourselves to the simplest "everyday" interpretation of a permutation as an order of distinct elements. For a given integer $n > 0$, there are $n! = n \cdot (n-1) \cdot \ldots \cdot 2 \cdot 1$ such orders of n elements. Indeed, we can place any of the n elements in the first position, and however we fix it, we can put any of the remaining $n-1$ elements in the second position, this gives $n(n-1)$ possibilities for the first two positions, etc.

We shall assume that the elements are numbers and will deal mostly with the permutations of integers $1, 2, \ldots, n$. Our findings remain equally valid for arbitrary elements if we fix a natural order among them and investigate the difference of a permutation from this natural order.

Considering a permutation of $1, 2, \ldots, n$, let $\sigma(1), \sigma(2), \ldots$ denote the numbers standing on the first, second, etc. positions. E.g., taking the permutation 31452 (for $n = 5$), we have $\sigma(1) = 3$, $\sigma(2) = 1$, $\sigma(3) = 4, \sigma(4) = 5$, and $\sigma(5) = 2$. This means that we can interpret a permutation as a *function*: assigning the number $\sigma(i)$ to the ith position, σ acts on the set $\{1, 2, \ldots, n\}$ as a bijection (it is one-to-one and onto).

We note that, in general, it is worth looking at permutations as such bijections. But it will be sufficient for our purposes to think of a permutation just as an order of the numbers $1, 2, \ldots, n$.

Now we turn to the definition of inversion. It means that two elements are positioned conversely compared to the natural order.

Definition 1.1.1. Two elements are in *inversion* in a permutation of $1, 2, \dots, n$ if the bigger number precedes the smaller one. The *number of inversions* in a permutation is the number of pairs standing in inversion. ♣

We denote the number of inversions in permutation σ by $I(\sigma)$. In the example 31452 above, 3 and 1, 3 and 2, 4 and 2, and 5 and 2 are in inversion, so the number of inversions is 4.

Using the notation σ, inversion means that for some $i < j$ we have $\sigma(i) > \sigma(j)$ (that is, element $\sigma(i)$ in position i is greater than element $\sigma(j)$ occurring in the later position j).

In the sequel, only the parity of the number of inversions will play an important role.

Definition 1.1.2. A permutation is *even* or *odd* according to the number of inversions in it being even or odd, respectively. ♣

So, permutation 31452 above is even. The simplest even permutation is the natural order $12 \dots n$ with 0 inversions corresponding to the identity function $\sigma(x) = x$.

Theorem 1.1.3.

I. *Swapping two adjacent elements in a permutation, the number of inversions changes (increases or decreases) by 1.*

II. *Swapping two arbitrary elements in a permutation, the number of inversions changes by an odd integer.* ♣

Proof.

I. The relation of the two swapped elements has changed: if they had been in reverse order originally, then they are not in inversion anymore, and vice versa. As they were adjacent elements, there was no change in their relation to other elements. And the relative position of the other elements has not changed either, of course.

II. Let there be k other elements between the two given elements b and c where the latter stands behind. We first interchange c one-by-one with the elements standing just before it till c gets next to b. This means k swaps of adjacent elements. Now we interchange the (presently adjacent) b and c. Finally, we interchange b one-by-one with the k elements standing behind it. This gives another k swaps of adjacent elements. We performed altogether $2k + 1$ swaps of adjacent elements each yielding a change of 1 in the number of inversions (plus or minus). Therefore the number of inversions has changed by an odd integer (in absolute value not greater than $2k + 1$). ∎

We determine now the number of even and odd permutations.

Theorem 1.1.4. *If $n > 1$, then the number of even and odd permutations of n elements is the same.* ♣

Proof. We swap the first two elements in each odd permutation of $1, 2, \ldots, n$. This results in distinct even permutations. Therefore, there are at least as many even permutations as odd ones. Performing the same procedure for the even permutations, we get that there are at least as many odd permutations as even ones. So, the number of even and odd permutations is the same (namely $n!/2$). ∎

Exercises 1.1

In the exercises, σ is a permutation of the integers $1, 2, \ldots, n$.

1.1.1. (a) Verify $0 \le I(\sigma) \le \binom{n}{2}$.
(b) Let $0 \le k \le \binom{n}{2}$ be an arbitrary integer. Prove that $I(\sigma) = k$ for some σ.

1.1.2. Find the number of inversions ($n = 101$).
(a) $1, 3, 5, \ldots, 99, 101, 100, 98, \ldots, 4, 2$;
(b) $51, 52, 50, 53, 49, \ldots, 101, 1$;
(c) $62, 63, 64, \ldots, 101, 61, 60, 59, \ldots, 1$;
(d) $100, 101, 98, 99, 96, 97, \ldots, 2, 3, 1$.

1.1.3. Pick an arbitrary permutation and write its elements in a reverse order. E.g., starting from 25413 we get the new permutation 31452. Determine the necessary and sufficient condition for the two permutations to have the same parity (that is, either both to be odd, or both to be even).

1.1.4. We carry the first element of a permutation to the last position n (causing the other elements to move forward by one position). What was the original first element if the number of inversions is the same in the new permutation as it was in the original one?

1.1.5. (a) What is the maximal possible change in the number of inversions occurring at swapping two elements and in which case does this occur?
***(b)** P and S play a game: P chooses an arbitrary permutation and then S swaps two elements in it. P wants to maximize the arising change in the number of inversions, whereas S wants to minimize it. What is the change in the number of inversions if both P and S play optimally?

1.1.6. P and S play some new games. P chooses an arbitrary permutation and S has to restore the original order via certain admissible steps. P wants to maximize the number of steps and S wants to minimize it. How many steps are required if they both play optimally and a step means
(a) swapping two elements of adjacent sizes (e.g., 6 and 7, wherever they stand);
*(b) swapping two arbitrary elements;
*(c) swapping 1 and any other element?

1.1.7. Determine all values of n such that there exists a permutation of $1, 2, \dots, n$ where each element is in inversion with exactly

(a) 1; (b) 2; *(c) k

other elements?

*1.1.8. Let $f(n,k)$ denote the number of permutations of $1, 2, \dots, n$ where there are exactly k inversions.

(a) Verify $f(n,k) = \sum_{i=k-n+1}^{k} f(n-1,i)$.

(b) Prove $f(n,k) = f(n-1,k) + f(n,k-1) - f(n-1,k-n)$.

(c) Compute the sum $\sum_k f(n,k)$.

(d) Find a simple formula for $\sum_k k \cdot f(n,k)$.

(e) Show $\max_k f(n,k) \geq 2(n-2)!$ for $n > 2$.

(f) For which k will $f(n,k)$ be maximal (for n fixed)?

1.2. The Definition of a Determinant

Let n be a fixed positive integer. We can think of a determinant in a first approximation as a number computed from n^2 numbers using some complicated rules.

Instead of numbers we shall define the determinant more generally for elements of any commutative *field* F. The exact definition of a commutative field is discussed in Section A.5. We can summarize it as a set where the four basic operations (addition, subtraction, multiplication, and division except by 0) can be performed and the usual operational laws hold. The most important examples of fields are the real numbers **R**, the rational numbers **Q**, the complex numbers **C**, and the modulo p residue classes F_p where p is a prime. We note that the definition of the determinant and its properties discussed in this chapter do not require division, so we can speak about determinants formed from integers and polynomials, as well.

The understanding and proper view of the notions below will not be affected if instead of elements of a commutative field the Reader thinks simply of real numbers for now.

It will be important in the definition of a determinant that the n^2 elements of F are ordered into an $n \times n$ *square* table. This and the more general tables of *rectangular* form are called matrices.

Definition 1.2.1. Let F be a commutative field and let k and n be given positive integers. A $k \times n$ *matrix* over F is a rectangular table having k rows and n columns with elements from F. ♣

Notation. We write the table between delimiters, namely (round) parentheses ( ) or [square] brackets [ ]. In the book we shall use round parentheses ( ).

Example. $\begin{pmatrix} 0 & 7 & 5 \\ 1 & 8 & 4 \end{pmatrix}$ is a 2×3 matrix.

In a general matrix A, we shall denote the jth element of the ith row by α_{ij}. The first index indicates the number of the row in the table where the given element stands, and the second index refers to the corresponding column. In the example above $\alpha_{23} = 4$. The general form of a $k \times n$ matrix is

$$A = \begin{pmatrix} \alpha_{11} & \alpha_{12} & \dots & \alpha_{1n} \\ \alpha_{21} & \alpha_{22} & \dots & \alpha_{2n} \\ \vdots & & & \\ \alpha_{k1} & \alpha_{k2} & \dots & \alpha_{kn} \end{pmatrix}.$$

We shall discuss matrices in the next chapter in more detail.

In the present chapter we shall only deal with $n \times n$ *square* matrices. Their general form is

$$A = \begin{pmatrix} \alpha_{11} & \alpha_{12} & \dots & \alpha_{1n} \\ \alpha_{21} & \alpha_{22} & \dots & \alpha_{2n} \\ \vdots & \vdots & \ddots & \vdots \\ \alpha_{n1} & \alpha_{n2} & \dots & \alpha_{nn} \end{pmatrix}.$$

Now we turn to the definition of the determinant. The determinant $\det A$ of the matrix A is an element in F computed by the following rule. First, we form all products of n factors so that we pick exactly one element from each row and each column of the matrix (there are altogether $n!$ such products). In the next step we attach a + or − sign to each product; this means that either we consider the product itself, or take its negative. The rule for the sign will be given in the next paragraph. Finally we add these signed products (i.e., each term in the sum is either such a product, or the negative of such a product). This sum (of $n!$ terms) is an element in F and is called the determinant of the matrix A. We can also say that it is the determinant formed from the elements α_{ij}. This is an $n \times n$ determinant, or a determinant of order n (instead of order one can say size or dimension, too).

The sign of a product is given by the following rule. We order the factors of the product so that the first factor is from the first row, the second factor is from the second row, etc. As the product contains exactly one element from each column, the indices of columns in this order of the factors yield a permutation of the integers $1, 2, \dots, n$. The product will get a positive sign (i.e., the product itself will occur in the sum) if this permutation is even, and will get a negative sign (i.e., its negative will occur in the sum) if this permutation is odd.

Example. For

$$A = \begin{pmatrix} 1 & 2 & 3 \\ 4 & 5 & 6 \\ 7 & 8 & 9 \end{pmatrix},$$

one of the products is $2 \cdot 6 \cdot 7$. Here the factors are already ordered according to the rows and they are taken from the 2nd, 3rd, and 1st columns, in this order. So, the permutation of the column indices is 231. It contains two inversions; hence it is an even permutation. Therefore the sign of the product is positive (i.e., the product itself and not its negative will appear in the sum providing the determinant).

ATTENTION, please! The sign of the product has nothing to do with the values of the factors themselves, it depends exclusively on the *positions* of the n factors in the matrix. The sign is determined solely by the parity of the permutation of the column indices discussed above, and this permutation always refers to the integers $1, 2, \ldots, n$ independently of the elements of the matrix being (real, complex, etc.) numbers or other mathematical objects (e.g., residue classes).

We denote the determinant by placing the table between two vertical lines (without parentheses or brackets).

We summarize all this in the following definition.

Definition 1.2.2. The *determinant* of a matrix

$$A = \begin{pmatrix} \alpha_{11} & \alpha_{12} & \ldots & \alpha_{1n} \\ \alpha_{21} & \alpha_{22} & \ldots & \alpha_{2n} \\ \vdots & \vdots & \ddots & \vdots \\ \alpha_{n1} & \alpha_{n2} & \ldots & \alpha_{nn} \end{pmatrix}$$

(or, in other words, the determinant formed from the elements α_{ij}, $i, j = 1, 2, \ldots, n$) is

$$\det A = \begin{vmatrix} \alpha_{11} & \alpha_{12} & \ldots & \alpha_{1n} \\ \alpha_{21} & \alpha_{22} & \ldots & \alpha_{2n} \\ \vdots & \vdots & \ddots & \vdots \\ \alpha_{n1} & \alpha_{n2} & \ldots & \alpha_{nn} \end{vmatrix} = \sum_{\sigma} (-1)^{I(\sigma)} \alpha_{1\sigma(1)} \alpha_{2\sigma(2)} \ldots \alpha_{n\sigma(n)}. \clubsuit$$

The sum on the right-hand side refers to all permutations σ of the integers $1, 2, \ldots, n$. Exactly half of the $n!$ terms get a negative sign (i.e., the negative of a product will occur that many times) as the number of odd and even permutations is the same (apart from the trivial case $n = 1$).

Example.

$$\begin{vmatrix} 1 & 2 & 3 \\ 4 & 5 & 6 \\ 7 & 8 & 9 \end{vmatrix} = \begin{cases} (-1)^0\, 1 \cdot 5 \cdot 9 + (-1)^1\, 1 \cdot 6 \cdot 8 + (-1)^1\, 2 \cdot 4 \cdot 9 + (-1)^2\, 2 \cdot 6 \cdot 7 \\ +(-1)^2\, 3 \cdot 4 \cdot 8 + (-1)^3\, 3 \cdot 5 \cdot 7 = 0. \end{cases}$$

The computation of a determinant using the definition is very cumbersome. We generally compute the determinant by the methods discussed in the next sections.

Some further remarks concerning the definition of the determinant. We can interpret the determinant (of order n) as a function assigning elements of F to the $n \times n$ matrices. In this case the determinant of a given matrix is a value of this function. We shall use the word determinant in both senses (i.e., both for the function and its values) hoping that it will not cause a misunderstanding.

If we want to emphasize that we speak about the determinant of a given matrix A, then we use the notation $\det A$. If, for short, we speak about a row, column, or element of a determinant, then we mean the relevant row, column, or element of the corresponding matrix. E.g., saying that we swap two rows in the determinant is a short formulation for swapping two rows in the matrix and considering the resulting determinant.

We stress once again the fundamental difference between a matrix and a determinant. The matrix is a table, i.e., a certain system of elements in F, whereas the determinant is a single element in F. Therefore, we have to be careful about the use of the delimiters; these are (round) parentheses (or square brackets) for matrices and two vertical lines for determinants.

It was important in the definition of the determinant that we have to order the n factors in the product by rows before checking the parity (odd or even) of the permutation of the column indices. If we write the factors in another order, then we get another permutation for the column indices that may have different parity, thus we get no information for the sign. The next theorem makes it possible to find the sign of the product for an arbitrary order of the factors. This method also eliminates the asymmetry between the roles of rows and columns.

Theorem 1.2.3. *Consider a product of n factors occurring in the definition of the determinant. This product contains exactly one element from each row and each column, so it is of the form $\alpha_{\rho(1)\pi(1)}\cdots\alpha_{\rho(n)\pi(n)}$ where ρ and π are the permutations corresponding to the rows and columns, respectively (we wrote the factors in an arbitrary order). Then the sign is determined by $(-1)^{I(\rho)+I(\pi)}$.* ♣

Proof. If ρ is the permutation belonging to the natural order, then this is just the definition of the determinant since $I(\rho) = 0$. We can reach any order of the factors by a sequence of always swapping two suitable factors. Therefore, it is enough to show that the parity of the sum $I(\rho) + I(\pi)$ remains the same in such a swap. Since we swap two elements both in the permutation ρ of the rows and in the permutation π of the columns, the number of inversions changes by an odd number in each permutation. So, the parity of the sum $I(\rho) + I(\pi)$ is preserved. ■

An important corollary of Theorem 1.2.3 is that the roles of rows and columns are interchangeable in the determinant. E.g., we can find the signs by ordering the factors of a product by columns and regarding the parity of the resulting permutation of the row indices. This corresponds to the case in Theorem 1.2.3 when π is the natural order.

Exercises 1.2

In the exercises of this and the next three sections, general determinants are assumed to contain elements from an arbitrary field and concrete determinants are meant to be composed from real numbers, unless stated otherwise. The solution and its difficulty are not affected if we always restrict ourselves to determinants with real numbers.

1.2.1. Compute the coefficient of x^3 in the polynomials.

$$\text{(a)}\ \begin{vmatrix} 3x & 5 & 7 & 1 \\ 2x^2 & 5x & 6 & 2 \\ 1 & x & 0 & 3 \\ 2 & 1 & 4 & 7 \end{vmatrix}; \qquad \text{(b)}\ \begin{vmatrix} 3x^2 & 5 & 7 & 1 \\ 2x^2 & 5x & 6 & 2 \\ 1 & x & 0 & 3 \\ 2 & 1 & 4 & 7 \end{vmatrix}.$$

1.2.2. True or false?

(a) If every element in a matrix is a rational number, then the determinant of the matrix is a rational number.

(b) If every element in a matrix is an irrational number, then the determinant of the matrix is an irrational number.

(c) If exactly one element in a matrix is irrational and all other elements are rational, then the determinant is irrational.
(d) If at least n^2-n+1 elements in an $n\times n$ matrix are 0, then the determinant of the matrix is 0.
(e) If the determinant of a matrix is 0, then the matrix contains an element 0.
(f) If the elements in a matrix are rational numbers and its determinant is 1/27, then the denominator of some element is a power of 3.
(g) If the elements in a matrix are rational numbers and its determinant is 1/27, then the denominator of some element is divisible by 3.

1.2.3. Compute the determinant of order n if
(a) $\alpha_{1j} = 0$ for every j (i.e., every element is 0 in the first row);
(b) $\alpha_{ij} = 0$ for every $i < j$ (i.e., every element is 0 above the main diagonal connecting the upper left and lower right corners of the matrix);
(c) $\alpha_{ij} = 0$, if $i + j > n + 1$ (i.e., every element is 0 under the diagonal connecting the lower left and upper right corners of the matrix).

1.2.4. Compute the determinants of order n (where $n > 1$).
(a) $\alpha_{ij} = \begin{cases} 1, & \text{if } j \equiv i+1 \pmod{n}; \\ 0, & \text{otherwise} \end{cases}$
(i.e., every element just above the main diagonal and in the lower left corner is 1, all other elements are 0);
(b) $\alpha_{ij} = 1$ (i.e., every element is 1);
(c) $\alpha_{ij} = \begin{cases} 1, & \text{if } |j-i| = 1; \\ 0, & \text{otherwise} \end{cases}$
(i.e., every element just above and below the main diagonal is 1, all other elements are 0).

1.2.5. An $n \times n$ matrix contains a rectangular part of k rows and m columns where every element is 0. Prove that the determinant is 0 if $k + m > n$.

1.2.6. We swap two elements in an $n \times n$ matrix, the other elements remain in their previous positions. Write the original and the new determinants using the definition. How many products remain unaltered among the terms if we disregard the signs? What happens if we also consider the signs?

1.2.7. In a 1000×1000 matrix we can modify arbitrary elements as we like. What is the minimal number of changes to achieve a 0 determinant?

1.2.8. Assume that $\begin{vmatrix} \alpha_{11} & \alpha_{12} \\ \alpha_{21} & \alpha_{22} \end{vmatrix} \neq 0$ and investigate the system of linear equations $\alpha_{11}x_1 + \alpha_{12}x_2 = \beta_1$, $\alpha_{21}x_1 + \alpha_{22}x_2 = \beta_2$. Prove that the only solution of the system is

$$x_1 = \frac{\begin{vmatrix} \beta_1 & \alpha_{12} \\ \beta_2 & \alpha_{22} \end{vmatrix}}{\begin{vmatrix} \alpha_{11} & \alpha_{12} \\ \alpha_{21} & \alpha_{22} \end{vmatrix}}, \qquad x_2 = \frac{\begin{vmatrix} \alpha_{11} & \beta_1 \\ \alpha_{21} & \beta_2 \end{vmatrix}}{\begin{vmatrix} \alpha_{11} & \alpha_{12} \\ \alpha_{21} & \alpha_{22} \end{vmatrix}}.$$

1.2.9. Consider a parallelogram with a vertex in the origin and let the two vertices next to the origin be (β_1, β_2) and (δ_1, δ_2). Prove that the area of the parallelogram is the absolute value of $\begin{vmatrix} \beta_1 & \beta_2 \\ \delta_1 & \delta_2 \end{vmatrix}$.

*1.2.10 Consider all $n \times n$ matrices A having at most two non-zero elements in each row. Let $k(A)$ denote the number of non-zero terms (of signed products of n factors) in the sum defining $\det A$. What is the maximal possible value of $k(A)$?

1.2.11. Prove (without calculating the determinant of some remarkable dates from history) that

$$\begin{vmatrix} 1861 & 1492 & 1896 & 1984 \\ 1782 & 1929 & 1918 & 1624 \\ 1776 & 1850 & 1945 & 1756 \\ 1790 & 1914 & 1770 & 1787 \end{vmatrix} \neq 0.$$

1.3. Elementary Properties

The following statements are easy consequences of the definition of the determinant.

Theorem 1.3.1.

I. *If every element below or above the main diagonal (the line connecting the upper left and lower right corners) is 0, then the determinant is the product of the elements in the main diagonal.*

II. *If every element in a row or column is 0, then the determinant is 0.*

III. *Multiplying each element of a row or column by λ, the determinant gets multiplied by λ.* ♣

Proof. I and II were essentially treated in Exercise 1.2.3. As for III, every product in the definition of the determinant is multiplied by λ as there is exactly one factor from the given row or column. The signs of the products remained the same, so factoring out λ from each product we see that the signed sum giving the determinant got multiplied by λ. ■

Note that II is the special case of III for $\lambda = 0$.

We can verify the next property similarly (we set the proof as Exercise 1.3.4).

Theorem 1.3.2. *If every element in a row or column is a two-term sum, then the determinant is the sum of two determinants containing the first and second terms, respectively, of the sums in the given row or column and all other elements are the same in both determinants as in the original one. Expressed in a formula (e.g., for the first row):*

$$\begin{vmatrix} \alpha'_{11}+\alpha''_{11} & \alpha'_{12}+\alpha''_{12} & \dots & \alpha'_{1n}+\alpha''_{1n} \\ \alpha_{21} & \alpha_{22} & \dots & \alpha_{2n} \\ \vdots & \vdots & \ddots & \vdots \\ \alpha_{n1} & \alpha_{n2} & \dots & \alpha_{nn} \end{vmatrix} = \begin{vmatrix} \alpha'_{11} & \alpha'_{12} & \dots & \alpha'_{1n} \\ \alpha_{21} & \alpha_{22} & \dots & \alpha_{2n} \\ \vdots & \vdots & \ddots & \vdots \\ \alpha_{n1} & \alpha_{n2} & \dots & \alpha_{nn} \end{vmatrix} + \begin{vmatrix} \alpha''_{11} & \alpha''_{12} & \dots & \alpha''_{1n} \\ \alpha_{21} & \alpha_{22} & \dots & \alpha_{2n} \\ \vdots & \vdots & \ddots & \vdots \\ \alpha_{n1} & \alpha_{n2} & \dots & \alpha_{nn} \end{vmatrix}. \clubsuit$$

Before moving on to the next properties we make a general observation. We experienced that if a property was true for rows, then it was valid for columns, too, and vice versa. We show that this is always the case. As proved in Theorem 1.2.3, the roles of rows and columns are symmetric. So, swapping the words "row" and "column" everywhere in the definition of the determinant, we arrive at exactly the same notion. (This was not clear from Definition 1.2.2, but it follows from Theorem 1.2.3.) Assume now that we proved a certain property for rows. If we replace "row" by "column" in the proof, we obtain a perfect deduction of this property for columns.

Therefore, any property for rows is valid also for columns. We note that this follows directly from Theorem 1.3.6, as well. In the sequel we shall state the properties only for rows (but they are valid for columns, as well).

Theorem 1.3.3. *If two rows are the same (i.e., the corresponding elements are equal), then the determinant is* 0. ♣

Proof. We show that we can form pairs of the products in the definition of the determinant so that the two products are equal but they have opposite signs. As the sum of each pair is 0, then the sum of the pairs, i.e., the determinant, is 0.

To keep the description simple, we perform the proof for the first two rows, i.e., when $\alpha_{1j} = \alpha_{2j}$ for every j.

We illustrate the idea first on a concrete example. Let $n = 5$ and consider the product

$$S = \alpha_{13}\alpha_{25}\alpha_{34}\alpha_{41}\alpha_{52}$$

in the definition of the determinant. By the assumption, $\alpha_{13} = \alpha_{23}$ and $\alpha_{25} = \alpha_{15}$. Thus the product

$$S' = \alpha_{15}\alpha_{23}\alpha_{34}\alpha_{41}\alpha_{52}$$

is equal to S and appears in the determinant.

Now we investigate the signs of the products S and S'. We have to compare the number of inversions in the permutations of column indices 35412 and 53412. The second permutation arose from the first one by swapping the first two elements. So, the number of inversions changed by an odd number (by 1 as adjacent elements were swapped in our case). This means that the two permutations have opposite parities, thus the products S and S' have opposite signs (yielding terms $-S$ and $+S'$ in the sum in this example).

Exactly the same argument works for the general case. The form of a product in the definition of the determinant is

$$S = \alpha_{1\sigma(1)}\alpha_{2\sigma(2)}\alpha_{3\sigma(3)}\dots\alpha_{n\sigma(n)}.$$

The same product occurs also as

$$S' = \alpha_{1\sigma(2)}\alpha_{2\sigma(1)}\alpha_{3\sigma(3)}\dots\alpha_{n\sigma(n)}$$

since $\alpha_{1\sigma(1)} = \alpha_{2\sigma(1)}$ and $\alpha_{2\sigma(2)} = \alpha_{1\sigma(2)}$. This gives a pairing of the products with $S = S'$.

Finally, we show that S and S' get opposite signs. The sign of S is given by the parity of the permutation $\sigma(1)\sigma(2)\sigma(3)\dots\sigma(n)$, and the sign of S' is determined by the parity of $\sigma(2)\sigma(1)\sigma(3)\dots\sigma(n)$. We obtain the latter permutation from the first one by swapping (the first) two elements, so the two permutations have opposite parity. ∎

Combining Theorem 1.3.3 with part III in Theorem 1.3.1, we obtain the following theorem.

Theorem 1.3.3A. *If a row is λ times some other row, then the determinant is 0.* ♣

The next property is the one we use most often when computing a determinant.

Theorem 1.3.4. *Adding λ times a row to another row, the determinant does not change.* ♣

Proof. For a simpler description we add λ times the second row to the first row. By Theorem 1.3.2

$$\begin{vmatrix} \alpha_{11}+\lambda\alpha_{21} & \alpha_{12}+\lambda\alpha_{22} & \dots & \alpha_{1n}+\lambda\alpha_{2n} \\ \alpha_{21} & \alpha_{22} & \dots & \alpha_{2n} \\ \vdots & \vdots & \ddots & \vdots \\ \alpha_{n1} & \alpha_{n2} & \dots & \alpha_{nn} \end{vmatrix}$$

$$= \begin{vmatrix} \alpha_{11} & \alpha_{12} & \dots & \alpha_{1n} \\ \alpha_{21} & \alpha_{22} & \dots & \alpha_{2n} \\ \vdots & \vdots & \ddots & \vdots \\ \alpha_{n1} & \alpha_{n2} & \dots & \alpha_{nn} \end{vmatrix} + \begin{vmatrix} \lambda\alpha_{21} & \lambda\alpha_{22} & \dots & \lambda\alpha_{2n} \\ \alpha_{21} & \alpha_{22} & \dots & \alpha_{2n} \\ \vdots & \vdots & \ddots & \vdots \\ \alpha_{n1} & \alpha_{n2} & \dots & \alpha_{nn} \end{vmatrix},$$

and the second determinant on the right-hand side is 0 by Theorem 1.3.3A. ∎

Theorem 1.3.5. *Swapping two rows, the determinant changes to its negative.* ♣

Proof. We perform the proof for the first two rows. We shall implement the swap of the two rows by a repeated application of Theorem 1.3.4. First, we subtract the second row from the first one, then we add the (new) first row to the second one, and finally we subtract the (new) second row from the (new) first one. The determinant does not change during the process, and the jth elements of the two rows are modified as follows:

$$\begin{pmatrix}\alpha_{1j}\\ \alpha_{2j}\end{pmatrix} \mapsto \begin{pmatrix}\alpha_{1j}-\alpha_{2j}\\ \alpha_{2j}\end{pmatrix} \mapsto \begin{pmatrix}\alpha_{1j}-\alpha_{2j}\\ \alpha_{1j}\end{pmatrix} \mapsto \begin{pmatrix}-\alpha_{2j}\\ \alpha_{1j}\end{pmatrix}.$$

Factoring out (-1) from the first row, the resulting determinant is (-1) times the original determinant (by Theorem 1.3.1/III). Also, we obtained it by swapping the first two rows in the original one. ■

Theorem 1.3.6. *Reflecting the elements through the main diagonal, the determinant does not change.* ♣

Proof. We show that the factors in the definition are the same in both determinants and their signs are equal, too.

We denote the elements of the original determinant by α_{ij} and the elements of the new one by β_{ij}, so $\beta_{ij} = \alpha_{ji}$ for every i, j.

We use Theorem 1.2.3. Then the form of a product in the original determinant is $\alpha_{\rho(1)\pi(1)} \dots \alpha_{\rho(n)\pi(n)}$ where ρ and π are the permutations of the row and column indices, respectively. The same product will occur in the new determinant, too, as $\beta_{\pi(1)\rho(1)} \dots \beta_{\pi(n)\rho(n)}$. Also the signs of the two products are the same, since they are determined by $(-1)^{I(\rho)+I(\pi)}$ in the original determinant and by $(-1)^{I(\pi)+I(\rho)}$ in the new one. ■

We recall that Theorem 1.3.6 implies that any property of determinants for rows is equally valid for columns, as well.

Applying Theorems 1.3.1–1.3.6, we compute a determinant in general by the following scheme. We try to achieve that all elements below the main diagonal be 0, and then the determinant is the product of the elements in the main diagonal. If we get a 0 row or column, then the determinant is 0. In each step, the determinant remains the same or just changes sign (we have to keep track of the latter, of course).

If $\alpha_{11} \neq 0$, then subtracting a suitable multiple of the first row from the other rows, we achieve that all other elements of the first column are 0. If the element in the upper left corner is 0, then we interchange the first row with another row where the first element differs from 0, and perform the subtractions thereafter. (If there is no such row, then every element in the first column is 0, so the determinant is 0.)

If every element except the first one is 0 in the first column, then it is comfortable to replace the other elements of the first row by 0 automatically. Indeed, subtracting suitable multiples of the first column from the other columns, we obtain that these elements in the first row turn into 0, but all other elements remain the same, and the determinant does not change either. This step, however, can be omitted; we do not really need it.

In the sequel, we must keep the first column intact. Now we move on to (the new) α_{22}. If it is not 0, then subtracting appropriate multiples of the second row from the third, ..., nth rows, we can convert the elements of the second column below the main diagonal into 0. If $\alpha_{22} = 0$, then we interchange the second row with a suitable *later* row. If all elements in the second column are 0, too, then the determinant is 0.

Continuing the process, either the determinant turns out to be 0, or we achieve that all elements below the main diagonal are 0.

Example. We compute the determinant

$$\begin{vmatrix} 1 & 2 & 3 \\ 4 & 5 & 6 \\ 7 & 8 & 9 \end{vmatrix}$$

from the previous section. We subtract 4 and 7 times the first row from the second and third rows, respectively:

$$\begin{vmatrix} 1 & 2 & 3 \\ 0 & -3 & -6 \\ 0 & -6 & -12 \end{vmatrix}.$$

Here the third row is the double of the second row, so the determinant is 0. (The above process yields the same: subtracting the double of the second row from the third one, all elements in the third row will turn into 0.)

We shall also apply a refined version of this method to solve systems of linear equations, which is called Gaussian elimination (see Section 3.1).

Of course, we can work with columns instead of rows, and can even alternate rows and columns in the manipulations. In many cases we can (or have to) use special tricks instead of this general method.

Some words of advice. It is worth writing out many elements even of a large determinant and to follow their changes in each step carefully. A detailed and accurate registration can be a major contribution in itself towards the solution as it often suggests what to do next. It is also important to record the steps just done as otherwise later (e.g., in an eventual debugging) we will not be able to reconstruct our procedure.

It is very dangerous to combine more steps into one. E.g., if we subtract the first row from the second one, then in the next step we shape a new determinant having a new second row. Therefore, a combination works only if the steps in it do not interfere, e.g., we can subtract the first row from every other row.

We have to be very careful with maneuvers like subtracting from each row the one above it. It makes a difference whether we start at the top or at the bottom. If we proceed top-down, then we have to subtract, e.g., a *modified* second row (namely the difference of the original second and first rows) from the third row. If we proceed bottom-up, then we always subtract the original rows. And do not forget that the first row always remains unchanged.

Exercises 1.3

1.3.1. What happens to a determinant if we reflect it through its vertical centerline?

1.3.2. How many complex numbers z satisfy that multiplying all elements of a given $n \times n$ complex matrix by z the determinant of the matrix changes into its negative?

1.3.3. Compute the determinants.

(a) $\begin{vmatrix} 123456 & 123426 \\ 123457 & 123427 \end{vmatrix}$; (b) $\begin{vmatrix} 1111 & 111 & 11 \\ 11111 & 1111 & 111 \\ 12345 & 1234 & 123 \end{vmatrix}$.

1.3.4. Prove Theorem 1.3.2.

1.3.5. Give a direct proof of Theorem 1.3.5 without relying on Theorem 1.3.4.

1.3.6. Show that Theorem 1.3.3 follows directly from Theorem 1.3.5 for determinants of, e.g., real numbers. Is this argument correct for every field?

1.3.7. Let $\alpha \neq 0$ be a fixed complex number. We pick an $n \times n$ matrix (with complex elements) and multiply the jth element in the kth row by (a) α^{j-k}; (b) α^{j+k}, for every j and k. What is the connection between the determinants of the old and new matrices?

1.3.8. Compute the $n \times n$ determinants.

(a) $\alpha_{ij} = \begin{cases} i, & \text{if } i = j; \\ 1, & \text{if } i \neq j. \end{cases}$ (b) $\alpha_{ij} = \min(i, j)$.

(c) $\alpha_{ij} = ij$. (d) $\alpha_{ij} = i + j$. (e) $\alpha_{ij} = i^2 + j^2$.

1.3.9. Compute the determinant if every row is an arithmetic sequence.

1.3.10. Observe that $3 \mid 6969$, $3 \mid 5727$, $3 \mid 1104$, $3 \mid 8625$, and also

$$3 \mid \begin{vmatrix} 6 & 9 & 6 & 9 \\ 5 & 7 & 2 & 7 \\ 1 & 1 & 0 & 4 \\ 8 & 6 & 2 & 5 \end{vmatrix}.$$

Is this accidental? And if 3 is replaced by 23?

1.3.11. Let $\gamma_1, \dots, \gamma_n$ and $\delta_1, \dots, \delta_n$ be arbitrary numbers. Compute the determinant of the $n \times n$ matrix where the jth element in the ith row is $1 + \gamma_i \delta_j$.

1.3.12. Compute the determinant if every element in the main diagonal is γ and all other elements are δ.

1.3.13. A square matrix satisfying $\alpha_{ij} + \alpha_{ji} = 0$ for every i, j is called *skew-symmetric* (or *antisymmetric*). What is the determinant of a skew-symmetric matrix of odd order?

1.3.14. Assume that in a determinant D with complex elements, one can find a the conjugate of every row as a (not necessarily distinct) row (i.e., the corresponding elements in the two rows are complex conjugates of each other). Prove that D^2 is a real number.

1.3.15. Compute the determinant and generalize.

$$\begin{vmatrix} 1 & 1 & 1 & 1 \\ 1 & 2 & 3 & 4 \\ 1 & 3 & 6 & 10 \\ 1 & 4 & 10 & 20 \end{vmatrix}$$

1.3.16. Compute the $n \times n$ determinants. In (b), $\alpha_{ij} = i$, if $j \equiv 1 \pmod{i}$, and all other elements are 1.

$$\text{(a)} \begin{vmatrix} 1 & 1 & 1 & \dots & 1 & 1 \\ 1 & 0 & 0 & \dots & 0 & 1 \\ 0 & 1 & 0 & \dots & 0 & 1 \\ 0 & 0 & 1 & \dots & 0 & 1 \\ \vdots & \vdots & \vdots & \ddots & \vdots & \vdots \\ 0 & 0 & 0 & \dots & 1 & 1 \end{vmatrix}; \qquad \text{(b)} \begin{vmatrix} 1 & 1 & 1 & 1 & 1 & \dots \\ 2 & 1 & 2 & 1 & 2 & \dots \\ 3 & 1 & 1 & 3 & 1 & \dots \\ 4 & 1 & 1 & 1 & 4 & \dots \\ \vdots & \vdots & \vdots & \vdots & \vdots & \\ n & 1 & 1 & 1 & 1 & \dots \end{vmatrix}.$$

1.3.17. Every element in the main diagonal of an $n \times n$ matrix is 1, the $n-1$ elements just below the main diagonal are -1, the $n-1$ elements just above the main diagonal top-down are $1^2, 2^2, \dots, (n-1)^2$, and all other elements are 0. Compute the determinant.

1.3.18. Let $\phi_1, \dots, \phi_n$ be arbitrary angles. Compute the $n \times n$ determinant with elements $\alpha_{ij} = \cos(\phi_i + \phi_j)$.

1.3.19. The elements of an $n \times n$ matrix are integers and no row contains two numbers giving the same remainder when divided by n. Show that the determinant is divisible by n if n is odd. What can we say for n even?

1.3.20. The elements of a $\varphi(n) \times \varphi(n)$ matrix are integers coprime to n and no row contains two numbers giving the same remainder when divided by n. (Here $\varphi(n)$ is Euler's function counting the integers coprime to n in $\{1, 2, \dots, n\}$.) Prove that the determinant is divisible by n if $n > 2$.

*1.3.21 Let $n \geq 7$ and consider the $n \times n$ matrix where α_{ij} is the least *positive* remainder of ij when divided by n (i.e., $a_{ij} = n$, not 0, if n divides ij). Verify that the determinant is 0.

1.4. Expansion

In this section, we get acquainted with another, recursive method for computing a determinant: we evaluate a determinant of order n with the help of n determinants of order $n-1$. This is of theoretical importance primarily but it can be effective also in the practical computation of certain determinants.

We define first the cofactors. We assume $n > 1$ in this section.

Definition 1.4.1. We delete the ith row and jth column from a determinant of order n and multiply the resulting $(n-1) \times (n-1)$ determinant by $(-1)^{i+j}$. The resulting expression is the *cofactor* (or *signed minor*) A_{ij} corresponding to the element α_{ij}. ♣

Example. Considering

$$\begin{vmatrix} 1 & 2 & 3 \\ 4 & 5 & 6 \\ 7 & 8 & 9 \end{vmatrix},$$

the cofactor $A_{23} = (-1)^5 \begin{vmatrix} 1 & 2 \\ 7 & 8 \end{vmatrix} = 6$.

ATTENTION, please! The sign of the cofactor has nothing to do with the sign of the products in the definition of the determinant. No permutation or inversion number occurs in the sign of the cofactor, it only depends on the serial numbers of the deleted row and column: if they are of the same parity, then the sign is + (plus), if they have opposite parity, then the sign is − (minus). Accordingly, the signs follow the so-called chessboard rule:

$$\begin{matrix} + & - & + & - & \dots \\ - & + & - & + & \dots \\ + & - & + & - & \dots \\ - & + & - & + & \dots \\ \vdots & \vdots & \vdots & \vdots & \ddots \end{matrix}$$

Specifically, each cofactor A_{ii} belonging to the main diagonal gets a positive sign.

The significance of the cofactors lies in the following theorem.

Theorem 1.4.2 (Cofactor expansion). *Multiplying each element of a row by the corresponding cofactor, the sum of these products is equal to the determinant:*

$$\det A = \alpha_{i1}A_{i1} + \alpha_{i2}A_{i2} + \cdots + \alpha_{in}A_{in} = \sum_{j=1}^{n} \alpha_{ij}A_{ij}. \clubsuit$$

This is called the expansion of the determinant along the ith row. By the symmetry in the roles of rows and columns, a similar statement holds for columns, too.

Example. Expanding the determinant

$$D = \begin{vmatrix} 1 & 2 & 3 \\ 4 & 5 & 6 \\ 7 & 8 & 9 \end{vmatrix}$$

along its second column, we have

$$D = 2 \cdot (-1)^3 \begin{vmatrix} 4 & 6 \\ 7 & 9 \end{vmatrix} + 5 \cdot (-1)^4 \begin{vmatrix} 1 & 3 \\ 7 & 9 \end{vmatrix} + 8 \cdot (-1)^5 \begin{vmatrix} 1 & 3 \\ 4 & 6 \end{vmatrix} = 0$$

(as we obtained it earlier by other methods).

Proof. We group the products in the definition of D according to which element of the ith row occurs in them. Factoring out this common element, we can write the determinant as $D = \alpha_{i1}\beta_1 + \alpha_{i2}\beta_2 + \cdots + \alpha_{in}\beta_n$. We show that $\beta_j = A_{ij}$ for every j.

We shall make use of Theorem 1.2.3 making it possible to find the sign of a product even if the factors are not (necessarily) ordered by rows.

We write the products containing α_{ij} (for a fixed j) so that the first factor is α_{ij}:

(1.4.1) $$\alpha_{ij}\alpha_{\rho(2)\pi(2)} \dots \alpha_{\rho(n)\pi(n)}.$$

Here ρ and π are the permutations of the row and column indices (so $\rho(1) = i$ and $\pi(1) = j$). By Theorem 1.2.3, the sign of this product in the sum defining D is $(-1)^{I(\rho)+I(\pi)}$ where $I(\rho)$ and $I(\pi)$ denote the number of inversions in the permutations $i, \rho(2), \dots, \rho(n)$ and $j, \pi(2), \dots, \pi(n)$, respectively.

Factoring out α_{ij}, we obtain

$$\beta_j = \sum(-1)^{I(\rho)+I(\pi)}\,\alpha_{\rho(2)\pi(2)}\cdots\alpha_{\rho(n)\pi(n)}. \tag{1.4.2}$$

The summation refers to all products of $n-1$ factors occurring in the definition of the determinant D' obtained by deleting the ith row and jth column of D.

Now we compare the sum on the right-hand side of (1.4.2) (equal to β_j) and the sum in the definition of D'. Both sums contain the same products of $n-1$ factors. The sign of such a product in D' is determined by the number of inversions in the two permutations of row and column indices altogether, by Theorem 1.2.3.

These inversions are not affected by keeping the original numbering of the rows and columns from D, i.e., $1, 2, \ldots, i-1, i+1, \ldots, n$ for the rows and $1, 2, \ldots, j-1, j+1, \ldots, n$ for the columns. Let $\rho' = \rho(2)\rho(3)\ldots\rho(n)$ be the permutation of $1, 2, \ldots, i-1, i+1, \ldots, n$ corresponding to the row indices, and let $\pi' = \pi(2)\pi(3)\ldots\pi(n)$ be the permutation of $1, 2, \ldots, j-1, j+1, \ldots, n$ corresponding to the column indices. We need (the parity of) $I(\rho') + I(\pi')$.

The permutation ρ in (1.4.1) is the completion of $\rho' = \rho(2)\rho(3)\ldots\rho(n)$ by $\rho(1) = i$ from the left. Thus $I(\rho) = I(\rho') + (i-1)$ since $\rho(1) = i$ is in inversion with the integers $1, 2, \ldots, i-1$. Similarly, $I(\pi) = I(\pi') + (j-1)$.

Therefore, $I(\rho) + I(\pi) = i + j - 2 + I(\rho') + I(\pi')$. Substituting back into (1.4.2), we get

$$\beta_j = (-1)^{i+j-2}\sum(-1)^{I(\rho')+I(\pi')}\,\alpha_{\rho(2)\pi(2)}\cdots\alpha_{\rho(n)\pi(n)} = (-1)^{i+j}D' = A_{ij}. \qquad \blacksquare$$

We indicate another proof of Theorem 1.4.2 in Exercise 1.4.4.

Expanding a determinant of order n, we reduce its calculation to computing n determinants of order $(n-1)$. In general, it is worth expanding along a row or column containing many 0 elements as we do not have to compute the cofactors corresponding to them.

Expansion also can be used to compute certain types of general determinants of order n by recursion. This works if we can reduce the determinant to determinants of similar type but of smaller order, e.g., we can express D_n of order n with (say) D_{n-1} and D_{n-2} of order $n-1$ and $n-2$ (we often have to expand the cofactors, too, to achieve this). The resulting recursion formula serves to verify a formula for D_n previously conjectured from the first few values or found systematically.

It is important to write out the elements in detail, to register the steps carefully, and to keep track of the changes precisely when we apply expansion. We should not forget about the proper sign of the cofactor. If we use expansion repeatedly, then we have to take into account the decrease in the size of the determinants and the modified position of the elements in them. Therefore, the structure and the sign of the cofactor belonging to an element changes in each step completely.

It is often useful to combine expansion with elementary properties. There may be a big difference in complexity, time, and computation between the various solutions. So, it can pay off to think a little about the possible approaches, their prospective advantages and disadvantages before rushing into the details. Unfortunately, sometimes it is hard to predict which method is the best for a given determinant, or even, whether a promising idea works at all or not.

Now we verify the so-called skew expansion.

Theorem 1.4.3 (Skew expansion). *If in a matrix A, we multiply the elements of a row with the cofactors corresponding to the elements of another row, then the sum of these products is always 0:*

$$k \neq r \Rightarrow \alpha_{r1}A_{k1} + \alpha_{r2}A_{k2} + \cdots + \alpha_{rn}A_{kn} = \sum_{j=1}^{n} \alpha_{rj}A_{kj} = 0. \clubsuit$$

ATTENTION, please! The word “expansion” is slightly misleading here as the value of the sum has nothing to do with the original determinant; this sum is always 0, independently of the original determinant being 0 or not.

Proof. We construct *another* matrix A' with determinant $D' = 0$ and apply expansion to D'. The only difference between the original matrix A and the new A' is that the kth row of A' is the rth row of A, all other elements are the same in the two matrices:

$$\alpha'_{ij} = \begin{cases} \alpha_{ij}, & \text{if } i \neq k; \\ \alpha_{rj}, & \text{if } i = k. \end{cases}$$

The kth and rth rows in A' are equal (both are the rth row of A), so $D' = 0$. On the other hand, expanding D' along its kth row, we obtain just the sum in the theorem as the cofactors A_{kj} in A and A'_{kj} in A' belonging to the kth row are the same for every j:

$$0 = D' = \alpha'_{k1}A'_{k1} + \alpha'_{k2}A_{k2} + \cdots + \alpha'_{kn}A'_{kn} = \alpha_{r1}A_{k1} + \alpha_{r2}A_{k2} + \cdots + \alpha_{rn}A_{kn}. \quad \blacksquare$$

Exercises 1.4

1.4.1. We multiply every element of an $n \times n$ determinant D with the corresponding cofactor. What will be the sum of these n^2 products?

1.4.2. Compute the determinant if the cofactors belonging to the first two rows are equal: $A_{1j} = A_{2j}$ for every j.

1.4.3. Prove that swapping the first two elements α_{11} and α_{12} in the first row, does not change the determinant if and only if $\alpha_{11} = \alpha_{12}$ or $A_{11} = A_{12}$.

1.4.4. Devise another proof for the cofactor expansion (Theorem 1.4.2) by the following argument:

(i) Verify the statement first for the very special case when the last $n - 1$ elements of the first row are 0 (so only its first element may differ from 0) and we expand the determinant along the first row.

(ii) Consider the still pretty special case when $n - 1$ elements in a row are 0 (i.e., at most one element in it differs from 0) and we expand along that row. By a sequence of swapping rows and swapping columns we can bring this case back to (i).

(iii) Write a general determinant as the sum of determinants of type (ii) using Theorem 1.3.2.

1.4.5. Compute the $n \times n$ determinant if the element in the upper left corner is 1, the other elements of the first row are β, the other elements of the first column are γ, the other elements of the main diagonal are δ, and all other elements are 0.

1.4.6. Compute the $2k \times 2k$ determinant where every element in the main diagonal is γ, every element in the diagonal connecting the lower left and upper right corners is δ, and the other elements are 0.

1.4.7. Compute the $n \times n$ determinant if every element in the main diagonal is $\gamma + \delta$, all the $n-1$ elements just below the main diagonal are 1, all the $n-1$ elements just above the main diagonal are $\gamma\delta$, and the other elements are 0.

1.4.8. We add γ to every element of an $n \times n$ matrix with a non-zero determinant. For how many values of γ is the determinant of the resulting matrix 0?

1.4.9. In an $n \times n$ determinant D the upper left element is $\cos\phi$, the other elements of the main diagonal are $2\cos\phi$, the $2n-2$ elements just above and just below the main diagonal are 1, and the other elements are 0. Verify $D = \cos(n\phi)$.

1.4.10. Let $\beta_1, \dots, \beta_n$ arbitrary. Compute $\det A$ if the elements of the $n \times n$ matrix A are

$$\alpha_{ij} = \begin{cases} 1 - \beta_i^2, & \text{if } i = j; \\ -\beta_i\beta_j, & \text{if } i \neq j. \end{cases}$$

1.4.11. (a) Assume that in a determinant the sum of elements in every row and every column is 0. Prove that all cofactors are equal.

(b) Assume that all cofactors are equal and are not 0. Prove that the sum of elements in every row and every column is 0.

*1.4.12. Compute the $n \times n$ determinant, if every element in the main diagonal is γ, every element above the main diagonal is δ, and every element below the main diagonal is β.

***1.4.13.** (Cf. with Exercise 1.2.7)

(a) P and R play a game. P presents an arbitrary $n \times n$ matrix of real numbers. In each step R replaces any element by any real number until we get a matrix with a non-zero determinant. P wants to maximize the number of steps and R wants to minimize it. How many steps are needed if both P and R play optimally?

(b) Solve the modified game when P selects the positions of the elements to be changed. Namely, after P gives the matrix, R tells how many elements he wants to change, then P selects that many positions and R can replace those elements by other real numbers.

(c) Discuss the games of parts (a) and (b) if R has to create a matrix with 0 determinant.

***1.4.14.** Does there exist for every n an $n \times n$ matrix of real numbers having a 0 determinant, but if exactly one element in it is modified arbitrarily, the determinant is never 0?

*1.4.15. *Laplace expansion* is a generalization of the cofactor expansion Theorem 1.4.2. Let A be an $n \times n$ matrix and let $1 \leq k \leq n-1$. A $k \times k$ *minor* is a determinant of the elements of any k rows and k columns. Its *signed complementary minor* is the $(n-k) \times (n-k)$ minor of the remaining rows and columns multiplied by $(-1)^S$ where S is the sum of row and column indices of the original $k \times k$ minor. We fix k rows and multiply each $k \times k$ minor having these rows with its signed complementary minor. Prove that $\det A$ is the sum of these products. (Cofactor expansion is just the special case $k = 1$.)

1.5. Vandermonde Determinant

In this section we deal with a special type of determinant occurring in many applications:

Definition 1.5.1. Let $\gamma_1, \gamma_2, \ldots, \gamma_n$ be arbitrary. The *Vandermonde determinant* generated by $\gamma_1, \gamma_2, \ldots, \gamma_n$ is

$$V(\gamma_1, \gamma_2, \ldots, \gamma_n) = \begin{vmatrix} 1 & \gamma_1 & \gamma_1^2 & \cdots & \gamma_1^{n-1} \\ 1 & \gamma_2 & \gamma_2^2 & \cdots & \gamma_2^{n-1} \\ 1 & \gamma_3 & \gamma_3^2 & \cdots & \gamma_3^{n-1} \\ \vdots & \vdots & \vdots & & \vdots \\ 1 & \gamma_n & \gamma_n^2 & \cdots & \gamma_n^{n-1} \end{vmatrix}. \clubsuit$$

So, the elements in the ith row of a Vandermonde determinant are the powers of γ_i with exponents $0, 1, \ldots, n-1$ in this order. If two generators are equal, then two rows are identical, so the determinant is 0. The product form below shows that the converse is also true.

Theorem 1.5.2.

$$V(\gamma_1, \gamma_2, \ldots, \gamma_n) = \begin{vmatrix} 1 & \gamma_1 & \gamma_1^2 & \cdots & \gamma_1^{n-1} \\ 1 & \gamma_2 & \gamma_2^2 & \cdots & \gamma_2^{n-1} \\ 1 & \gamma_3 & \gamma_3^2 & \cdots & \gamma_3^{n-1} \\ \vdots & \vdots & \vdots & & \vdots \\ 1 & \gamma_n & \gamma_n^2 & \cdots & \gamma_n^{n-1} \end{vmatrix} = \prod_{1 \leq i < j \leq n} (\gamma_j - \gamma_i). \clubsuit$$

Proof. Proceeding right-to-left, we subtract from each column γ_1 times the preceding column:

$$\begin{vmatrix} 1 & 0 & 0 & \cdots & 0 \\ 1 & \gamma_2 - \gamma_1 & \gamma_2^2 - \gamma_1\gamma_2 & \cdots & \gamma_2^{n-1} - \gamma_1\gamma_2^{n-2} \\ 1 & \gamma_3 - \gamma_1 & \gamma_3^2 - \gamma_1\gamma_3 & \cdots & \gamma_3^{n-1} - \gamma_1\gamma_3^{n-2} \\ \vdots & \vdots & \vdots & & \vdots \\ 1 & \gamma_n - \gamma_1 & \gamma_n^2 - \gamma_1\gamma_n & \cdots & \gamma_n^{n-1} - \gamma_1\gamma_n^{n-2} \end{vmatrix}.$$

Now we subtract the first row from every other row to turn the last $n-1$ elements of the first column into 0 while leaving the other elements unchanged. We can factor out

$\gamma_2-\gamma_1, \gamma_3-\gamma_1, \dots, \gamma_n-\gamma_1$ from the 2nd, 3rd, ..., nth column, respectively. This yields

$$V(\gamma_1,\dots,\gamma_n) = (\gamma_2-\gamma_1)\dots(\gamma_n-\gamma_1)\begin{vmatrix} 1 & 0 & 0 & \dots & 0 \\ 0 & 1 & \gamma_2 & \dots & \gamma_2^{n-2} \\ 0 & 1 & \gamma_3 & \dots & \gamma_3^{n-2} \\ \vdots & \vdots & \vdots & & \vdots \\ 0 & 1 & \gamma_n & \dots & \gamma_n^{n-2} \end{vmatrix}$$

$$= (\gamma_2-\gamma_1)\dots(\gamma_n-\gamma_1)V(\gamma_2,\dots,\gamma_n).$$

So we reduced the problem to a Vandermonde determinant of order one less. Applying the procedure repeatedly (or by induction) we get the formula in the theorem. ∎

Exercises 1.5

1.5.1. Express the products with $V = V(\gamma_1,\dots,\gamma_n)$:
   (a) $\prod_{1\le i<j\le n}(\gamma_i-\gamma_j)$;
   (b) $\prod_{1\le i\ne j\le n}(\gamma_j-\gamma_i)$.

1.5.2. Let $\gamma_2,\dots,\gamma_n$ be fixed. What is the number of solutions (in x) of the equation $V(x,\gamma_2,\dots,\gamma_n) = 0$? Can it happen for some δ that the equation $V(x,\gamma_2,\dots,\gamma_n) = \delta$ has (a) more; (b) fewer solutions?

1.5.3. Compute the determinant without 0 elements where every row is a geometric sequence.

1.5.4. Compute the $n\times n$ determinant if the jth element in the ith row is i^j.

1.5.5. (a) Let $f_0,\dots,f_{n-1}$ be polynomials where $\deg f_k = k$ and let $\gamma_1,\dots,\gamma_n$ be arbitrary. Compute the $n\times n$ determinant where the jth element of the ith row is $f_{i-1}(\gamma_j)$.
   (b) Compute the determinant if we replace the condition on the degrees by $\deg f_k \le n-2$ for every k.

1.5.6. Let $\alpha_1,\dots,\alpha_n,\beta_1,\dots,\beta_n$ satisfy $\alpha_i\beta_j\ne 1$ for every i and j. Compute the $n\times n$ determinant if the jth element in the ith row is $(1-\alpha_i^n\beta_j^n)/(1-\alpha_i\beta_j)$.

1.5.7. Let $\alpha_1,\dots,\alpha_n,\beta_1,\dots,\beta_n$ be arbitrary. Compute the $n\times n$ determinant if the jth element of the ith row is $(\alpha_i+\beta_j)^{n-1}$.

1.5.8. The real numbers $\phi_1,\dots,\phi_n$ have pairwise distinct cosines, in D_1 the jth element of the ith row is $\cos[(j-1)\phi_i]$, and $D_2 = V(\cos\phi_1,\dots,\cos\phi_n)$. Prove that the quotient D_1/D_2 does not depend on the numbers ϕ_i.

1.5.9. Let $\gamma_1,\dots,\gamma_n$ be distinct.
   (a) What happens to the Vandermonde determinant if we swap γ_i and γ_j?
   (b) For which permutations σ is

$$V(\gamma_1,\gamma_2,\dots,\gamma_n) = V(\gamma_{\sigma(1)},\gamma_{\sigma(2)},\dots,\gamma_{\sigma(n)})?$$

1.5.10. The jth element in the ith row is 2^{ij} in an $n\times n$ determinant D. What is the highest power of 2 that divides D?

1.5.11. Let $a_1, \dots, a_n$ be integers. Prove that
(a) the least common multiple of $1, 2, \dots, n-1$; and
*(b) $V(1, 2, \dots, n)$
are divisors of $V(a_1, \dots, a_n)$.

*1.5.12. Let $p > 2$ be a prime and let $V_p = V(1, 2, \dots, p)$. What is the remainder of V_p^2 upon division by p?

*1.5.13. Compute the determinant obtained from $V(\gamma_1, \dots, \gamma_n)$ by replacing the elements γ_i^{n-1} by γ_i^n in the last column.

Chapter 2

Matrices

Matrices are closely related to determinants and to systems of linear equations, but have many other applications, too. Matrix multiplication and some of its weird properties are of special interest and significance. These oddities will obtain a natural interpretation via linear maps to be discussed in Chapter 5.

2.1. Matrix Operations

We introduced matrices in Section 1.2 but for the sake of completeness we repeat the definition and notation:

> **Definition 1.2.1.** Let F be a commutative field and let k and n be given positive integers. A $k \times n$ *matrix* over F is a rectangular table having k rows and n columns with elements from F. ♣
>
> *Notation:* We write the table between delimiters, namely (round) parentheses ( ) or [square] brackets [ ]. In the book we use round parentheses ( ).

We again call attention to the fact that vertical line delimiters | | denote determinants and not matrices.

In a general matrix A, we denote the jth element of the ith row by α_{ij}. The first index indicates the number of the row in the table where the given element stands, and the second index refers to the corresponding column. The general form of a $k \times n$ matrix is

$$A = \begin{pmatrix} \alpha_{11} & \alpha_{12} & \dots & \alpha_{1n} \\ \alpha_{21} & \alpha_{22} & \dots & \alpha_{2n} \\ \vdots & & & \\ \alpha_{k1} & \alpha_{k2} & \dots & \alpha_{kn} \end{pmatrix}.$$

We denote the set of all $k \times n$ matrices over F by $F^{k \times n}$.

We mention again that instead of a general field F one can think in most cases just of the real, rational, or complex numbers. However, finite fields play an important role in applications. Their simplest type is the field F_p of the modulo p residue classes where p is a prime. If the elements of a matrix A are real, rational, or complex numbers, then we shall say for short that A is a real, rational, or complex matrix.

Now we define the addition of two matrices and the multiplication of a matrix by an element of F in a natural way, elementwise.

Definition 2.1.1. Let $A, B \in F^{k \times n}$, $\lambda \in F$. We obtain $A + B$ by adding the elements in the corresponding positions and λA by multiplying every element by λ:

$$A + B = \begin{pmatrix} \alpha_{11} & \alpha_{12} & \dots & \alpha_{1n} \\ \alpha_{21} & \alpha_{22} & \dots & \alpha_{2n} \\ \vdots \\ \alpha_{k1} & \alpha_{k2} & \dots & \alpha_{kn} \end{pmatrix} + \begin{pmatrix} \beta_{11} & \beta_{12} & \dots & \beta_{1n} \\ \beta_{21} & \beta_{22} & \dots & \beta_{2n} \\ \vdots \\ \beta_{k1} & \beta_{k2} & \dots & \beta_{kn} \end{pmatrix}$$

$$= \begin{pmatrix} \alpha_{11} + \beta_{11} & \alpha_{12} + \beta_{12} & \dots & \alpha_{1n} + \beta_{1n} \\ \alpha_{21} + \beta_{21} & \alpha_{22} + \beta_{22} & \dots & \alpha_{2n} + \beta_{2n} \\ \vdots \\ \alpha_{k1} + \beta_{k1} & \alpha_{k2} + \beta_{k2} & \dots & \alpha_{kn} + \beta_{kn} \end{pmatrix}$$

and

$$\lambda A = \lambda \begin{pmatrix} \alpha_{11} & \alpha_{12} & \dots & \alpha_{1n} \\ \alpha_{21} & \alpha_{22} & \dots & \alpha_{2n} \\ \vdots \\ \alpha_{k1} & \alpha_{k2} & \dots & \alpha_{kn} \end{pmatrix} = \begin{pmatrix} \lambda\alpha_{11} & \lambda\alpha_{12} & \dots & \lambda\alpha_{1n} \\ \lambda\alpha_{21} & \lambda\alpha_{22} & \dots & \lambda\alpha_{2n} \\ \vdots \\ \lambda\alpha_{k1} & \lambda\alpha_{k2} & \dots & \lambda\alpha_{kn} \end{pmatrix}. \clubsuit$$

The elements of the field F are called *scalars*, so the operation λA is called the multiplication of a matrix by a scalar.

The above operations obey the usual laws.

Theorem 2.1.2. *The addition in the set $F^{k \times n}$ of the $k \times n$ matrices is associative, commutative, there exists a zero element, and every element has a negative:*

(i) $(A + B) + C = A + (B + C)$ *and* $A + B = B + A$ *for every* A, B, C;

(ii) *there exists a zero matrix* 0 *satisfying* $A + 0 = 0 + A = A$ *for every* A;

(iii) *to every* A *there exists a matrix* $-A$ *satisfying* $A + (-A) = (-A) + A = 0$.

Multiplication by the elements of F *obeys the rules* ($A, B \in F^{k \times n}$, $\lambda, \mu \in F$):

(iv) $(\lambda + \mu)A = \lambda A + \mu A$, $\lambda(A + B) = \lambda A + \lambda B$, $(\lambda\mu)A = \lambda(\mu A)$, $1A = A$, *where 1 is the identity element of* F *(i.e.,* $1\lambda = \lambda 1 = \lambda$ *for every* $\lambda \in F$*).* ♣

To summarize the properties, the $k \times n$ matrices form a commutative group (see Section A.8) under addition and a vector space (see Section 4.1) over F for the two operations.

Proof. Each property follows from the definition of the operations and from the corresponding property in F. E.g., every element of the zero matrix is the zero of F, etc. ■

We shall denote both the zero in F and the zero matrix by 0; this will cause no misunderstanding (hopefully).

Now we turn to the definition of multiplication of two matrices. This is more complicated and seems to be artificial for the time being. The formal definition will be followed by a detailed explanation.

Definition 2.1.3. Let $A \in F^{k\times n}$, $B \in F^{n\times r}$. Then $C = AB \in F^{k\times r}$ and the jth element in the ith row is

$$\gamma_{ij} = \alpha_{i1}\beta_{1j} + \alpha_{i2}\beta_{2j} + \cdots + \alpha_{in}\beta_{nj} = \sum_{s=1}^{n} \alpha_{is}\beta_{sj}. \clubsuit$$

So, matrices A and B can be multiplied (in this order) if and only if the number of *columns* in A equals the number of *rows* in B. Then AB will have as many rows as A and as many columns as B has. The jth element of the ith row in the product AB is the *dot product* (or scalar product, inner product) of the ith row of A and the jth column of B (considered as two vectors with n components), i.e., we add the products of the corresponding components; see Example 2 in Section 7.1.

Example. Let

$$A = \begin{pmatrix} 1 & 2 \\ 3 & 4 \\ 5 & 0 \end{pmatrix}, \quad B = \begin{pmatrix} 6 & 7 & 8 & 9 \\ 10 & 11 & 12 & 13 \end{pmatrix}.$$

The product AB exists as the number of columns in A is the same as the number of rows in B. The product BA, however, makes no sense.

The product AB has 3 rows and 4 columns. The third element of the second row in AB is the dot product of the second row of A and the third column of B: $3\cdot 8+4\cdot 12 = 72$.

Until we acquire skills, the following oblique scheme may be helpful.

$$\begin{array}{cc} & \begin{pmatrix} \beta_{11} & \cdots & \beta_{1r} \\ \beta_{21} & \cdots & \beta_{2r} \\ \beta_{31} & \cdots & \beta_{3r} \\ \vdots & \vdots & \vdots \\ \beta_{n1} & \cdots & \beta_{nr} \end{pmatrix} \\ \begin{pmatrix} \alpha_{11} & \alpha_{12} & \cdots & \alpha_{1n} \\ \alpha_{21} & \alpha_{22} & \cdots & \alpha_{2n} \\ \vdots & & & \\ \alpha_{k1} & \alpha_{k2} & \cdots & \alpha_{kn} \end{pmatrix} & \end{array}$$

Then $C = AB$ will appear between the two matrices (right to A and below B) so that the element γ_{ij} is just in the intersection of the corresponding ith row of A and jth column of B:

$$\begin{array}{cc} & \begin{pmatrix} \beta_{11} & \cdots & \beta_{1r} \\ \beta_{21} & \cdots & \beta_{2r} \\ \beta_{31} & \cdots & \beta_{3r} \\ \vdots & \vdots & \vdots \\ \beta_{n1} & \cdots & \beta_{nr} \end{pmatrix} \\ \begin{pmatrix} \alpha_{11} & \alpha_{12} & \cdots & \alpha_{1n} \\ \alpha_{21} & \alpha_{22} & \cdots & \alpha_{2n} \\ \vdots & & & \\ \alpha_{k1} & \alpha_{k2} & \cdots & \alpha_{kn} \end{pmatrix} & \begin{pmatrix} \gamma_{11} & \cdots & \gamma_{1r} \\ \gamma_{21} & \cdots & \gamma_{2r} \\ \vdots & \vdots & \vdots \\ \gamma_{k1} & \cdots & \gamma_{kr} \end{pmatrix}. \end{array}$$

In the example above:

$$\begin{array}{cc} & \begin{pmatrix} 6 & 7 & 8 & 9 \\ 10 & 11 & 12 & 13 \end{pmatrix} \\ \begin{pmatrix} 1 & 2 \\ 3 & 4 \\ 5 & 0 \end{pmatrix} & \begin{pmatrix} 26 & 29 & 32 & 35 \\ 58 & 65 & 72 & 79 \\ 30 & 35 & 40 & 45 \end{pmatrix}. \end{array}$$

Now we discuss the properties of matrix multiplication. We start with the question of commutativity. Let $A \in F^{k \times n}$, $B \in F^{n \times r}$; then AB makes sense. If $k \neq r$, then BA does not even exist! If $k = r \neq n$, then BA exists, but AB and BA are not of the same form, as AB is a $k \times k$, whereas BA is an $n \times n$ matrix. Finally, if $k = n = r$, i.e., A and B are square matrices of the same size, then $A, B, AB \in F^{n \times n}$. Still, $AB \neq BA$ in general, e.g.,

$$\begin{pmatrix} 1 & 2 \\ 3 & 4 \end{pmatrix}\begin{pmatrix} 5 & 6 \\ 7 & 8 \end{pmatrix} = \begin{pmatrix} 19 & 22 \\ 43 & 50 \end{pmatrix} \neq \begin{pmatrix} 23 & 34 \\ 31 & 46 \end{pmatrix} = \begin{pmatrix} 5 & 6 \\ 7 & 8 \end{pmatrix}\begin{pmatrix} 1 & 2 \\ 3 & 4 \end{pmatrix}.$$

We can feel from this example that two matrices commute very rarely, so matrix multiplication is far from being commutative.

The other usual laws, however, are valid for matrix multiplication.

Theorem 2.1.4. *If $\lambda \in F$ and A, B, C are any matrices such that one side of an equality below makes sense, then the other side makes sense and equality holds.*

I. $A(BC) = (AB)C$ *(associative law);*

II. $A(B + C) = AB + AC$,

$(A + B)C = AC + BC$ *(distributive laws);*

III. $\lambda(AB) = (\lambda A)B = A(\lambda B)$. ♣

As matrix multiplication is not commutative, we have to verify the two distributive laws separately. Also, in III, we can only change the position of the scalar $\lambda \in F$ but must not modify the order of A and B.

Proof. We prove the associative law; the others can be shown by similar computations (see Exercise 2.1.13).

By the definition of matrix multiplication, each of the right– and left-hand sides in I makes sense exactly if the number of columns in A equals the number of rows in B, and the number of columns in B equals the number of rows in C. Let $A \in F^{k\times n}$, $B \in F^{n\times r}$, $C \in F^{r\times t}$, then both $M = A(BC)$ and $N = (AB)C$ are $k\times t$ matrices. We show that the jth elements of the ith row $\mu_{ij} \in M$ and $\nu_{ij} \in N$ are equal. Let $D = BC$. Then

$$\begin{aligned}\mu_{ij} &= \alpha_{i1}\delta_{1j} + \cdots + \alpha_{in}\delta_{nj} \\ &= \alpha_{i1}(\beta_{11}\gamma_{1j} + \cdots + \beta_{1r}\gamma_{rj}) + \cdots + \alpha_{in}(\beta_{n1}\gamma_{1j} + \cdots + \beta_{nr}\gamma_{rj}),\end{aligned}$$

so

$$\mu_{ij} = \sum_{1\le u\le n, 1\le v\le r} \alpha_{iu}(\beta_{uv}\gamma_{vj}).$$

We get similarly

$$\nu_{ij} = \sum_{1\le u\le n, 1\le v\le r} (\alpha_{iu}\beta_{uv})\gamma_{vj}.$$

As multiplication is associative in F, we infer $\mu_{ij} = \nu_{ij}$. ■

Finally we introduce the transpose of a matrix.

Definition 2.1.5. Let $A \in F^{k\times n}$. The *transpose* of A is $B \in F^{n\times k}$ satisfying $\beta_{ij} = \alpha_{ji}$. The transpose of A is denoted by A^T. ♣

Example. The transpose of $\begin{pmatrix} 1 & 2 \\ 3 & 4 \\ 5 & 6 \end{pmatrix}$ is $\begin{pmatrix} 1 & 3 & 5 \\ 2 & 4 & 6 \end{pmatrix}$.

Transposing a matrix swaps the rows and columns, so it means a reflection about the main diagonal, i.e., the line heading from the upper left corner in southwestern direction of 45 degrees downwards and to the right.

For complex matrices another similar notion, the adjoint plays an important role, too (see Section 8.4).

Definition 2.1.6. Let $A \in \mathbf{C}^{k\times n}$. The *adjoint* of A is $B \in \mathbf{C}^{n\times k}$ satisfying $\beta_{ij} = \overline{\alpha_{ji}}$ (where $\overline{z}$ is the conjugate of the complex number z). The adjoint of A is denoted by A^*. ♣

So, the adjoint of a matrix is the conjugate of the transpose. For real matrices $A^* = A^T$.

See Exercise 2.1.20 for the connection of the operations with the transpose and adjoint.

Exercises 2.1

Capital letters A, B, etc. denote matrices.

2.1.1. Compute the sum of all distinct $k \times n$ matrices where every element is one of the real numbers 1, 2, and 3.

2.1.2. Describe all pairs α, β of complex numbers satisfying that every $k\times n$ complex matrix can be represented as $\alpha A + \beta B$ where A and B are real matrices.

2.1.3. In a square matrix I, every element of the main diagonal is 1 and all other elements are 0. Compute the products IA and AI for a general A if multiplication makes sense.

2.1.4. Compute:

$$\text{(a)} \begin{pmatrix} 2 & -4 \\ 1 & -2 \end{pmatrix}^{1111}; \qquad \text{(b)} \begin{pmatrix} 2 & -3 \\ 1 & -2 \end{pmatrix}^{1111}; \qquad \text{(c)} \begin{pmatrix} 1/2 & 1/2 \\ 1/2 & 1/2 \end{pmatrix}^{1111}.$$

2.1.5. Compute:

$$\text{(a)} \begin{pmatrix} 1 & 1 \\ 0 & 1 \end{pmatrix}^{n}; \qquad \text{(b)} \begin{pmatrix} \cos\alpha & -\sin\alpha \\ \sin\alpha & \cos\alpha \end{pmatrix}^{n}; \qquad \text{(c)} \begin{pmatrix} \cos\alpha & \sin\alpha \\ \sin\alpha & -\cos\alpha \end{pmatrix}^{n}.$$

2.1.6. Perform the multiplications.

$$\text{(a)} \begin{pmatrix} a & 1-a \\ a & 1-a \end{pmatrix}\begin{pmatrix} b & 1-b \\ b & 1-b \end{pmatrix}; \qquad \text{(b)} \begin{pmatrix} a & -a \\ a & -a \end{pmatrix}\begin{pmatrix} b & -b \\ b & -b \end{pmatrix}.$$

2.1.7. The sum of elements in each row and in each column of A is 0, and all elements of B are equal. Compute AB and BA if multiplication makes sense.

2.1.8. True or false? (We assume that the products AB and BA exist.)

(a) If A has a 0 row, then this holds for AB, too.
(b) If A has a 0 row, then this holds for BA, too.
(c) If every row of A is an arithmetic sequence, then this holds for AB, too.
(d) If every row of A is an arithmetic sequence, then this holds for BA, too.
(e) If the sum of all elements in A is 0, then this holds for AB, too.
(f) If the sum of all elements in A is 0, then this holds for BA, too.

2.1.9. Comment on the statement and reasoning below.

"If A and B are square matrices of the same size and $A^{100} = B^{100} = 0$, then $(A+B)^{200} = 0$. Indeed,

$$(A+B)^{200} = A^{200} + \binom{200}{1}A^{199}B + \cdots + \binom{200}{k}A^kB^{200-k} + \ldots,$$

and every term is 0 here as $k \geq 100$ implies $A^k = 0$ and $k < 100$ implies $200 - k > 100$, so $B^{200-k} = 0$."

2.1.10. What happens to an $n \times n$ matrix A if we multiply it from the left or right, respectively, by the $n \times n$ matrices

$$B = \begin{pmatrix} 5 & 0 & \ldots & 0 \\ 0 & 1 & \ldots & 0 \\ \vdots & \vdots & \ddots & \vdots \\ 0 & 0 & \ldots & 1 \end{pmatrix}; \qquad C = \begin{pmatrix} 1 & 6 & \ldots & 0 \\ 0 & 1 & \ldots & 0 \\ \vdots & \vdots & \ddots & \vdots \\ 0 & 0 & \ldots & 1 \end{pmatrix}?$$

2.1.11. Let P be an $n \times n$ matrix with exactly one element 1 in each row and column and all other elements are 0. What happens to an $n \times n$ matrix A if we multiply it from the left or right, respectively, by P?

2.1.12. What is the sum of the elements in the main diagonal of $AB - BA$ (where A and B are square matrices of the same size)?

2.1.13. Prove II and III in Theorem 2.1.4.

2.1.14. Assume $A^{100} = A^{72} = A$. How many powers (with positive integer exponents) of A are distinct?

2.1.15. In an $n \times n$ matrix A every element in and below the main diagonal is 0. Prove $A^n = 0$.

*2.1.16. Let p be a prime and let A be a $p \times p$ matrix over the modulo p field where every element in the main diagonal is 1 and every element below it is 0. Compute A^p.

*2.1.17. Let

$$A = \begin{pmatrix} \cos\alpha & -\sin\alpha \\ \sin\alpha & \cos\alpha \end{pmatrix} \text{ and } I = \begin{pmatrix} 1 & 0 \\ 0 & 1 \end{pmatrix}.$$

Prove that there exists an integer $k > 0$ satisfying $I + A + A^2 + \cdots + A^k = 0$ if and only if $\alpha/2\pi$ is a rational number that is not an integer.

***2.1.18.** Find all matrices $A \in F^{n\times n}$ commuting with every $B \in T^{n\times n}$, i.e., $AB = BA$ for every B.

2.1.19. We are given k articles each assembled of n types of components and each component is made of r materials. Let α_{ij} denote the quantity of the jth component in the ith article and let β_{uv} denote the quantity of the vth material necessary for one unit of the uth component. Let A be the $k \times n$ matrix with elements α_{ij} and let B be the $n \times r$ matrix with elements β_{uv}. What is the meaning of the product $C = AB$?

2.1.20. Prove the laws connecting the transpose and the operations: If $\lambda \in F$ and A, B, C are any matrices such that one side of an equality below makes sense, then the other side makes sense and equality holds.

$$(A + B)^T = A^T + B^T, \qquad (\lambda A)^T = \lambda A^T, \qquad (AB)^T = B^T A^T.$$

Formulate and verify the analogous statements for the adjoint, too.

2.1.21. Determine the real matrices A where the sum of elements in the main diagonal of AA^T is 0. Generalize the result for complex matrices.

2.2. The Ring of $n \times n$ Matrices

We apply now the results of the previous section for square matrices:

Theorem 2.2.1. *All $n \times n$ matrices over a field F form a ring under the addition and multiplication of matrices. This ring $F^{n\times n}$ has an identity element but is not commutative for $n > 1$.* ♣

For a precise definition of a ring see Section A.6 (but it is contained implicitly in the proof of this theorem, too).

Proof. Any two such matrices have a sum and a product by Definitions 2.1.1 and 2.1.3. Theorems 2.1.2 and 2.1.4 guarantee that addition is commutative and associative, there exists a zero element, every matrix has a negative, multiplication is associative, and the distributive laws are valid. So, $F^{n\times n}$ is a ring. The identity matrix is I defined in Exercise 2.1.3 with elements 1 in the main diagonal and 0 everywhere else. The lack of the commutative law can be verified with a generalization to every $n > 2$ of the counterexample before Theorem 2.1.4 or with the help of Exercises 2.1.6(a) or 2.1.10. ■

ATTENTION, please! The multiplication is not commutative but some matrices still commute. E.g., the identity matrix I and the zero matrix 0 commute with every matrix, any matrix commutes with its powers, etc.

We denote the identity matrix by I_n if we want to emphasize that it is an $n \times n$ matrix.

Now we turn to further interesting properties of matrix multiplication in the ring $F^{n \times n}$: division does not work in general and the product of two non-zero matrices can be zero.

We define and study the inverse and zero divisor for square matrices. We note that the general properties of inverse and zero divisor in a ring are discussed in Section A.6 but we list them for matrices here, too.

We start with the inverse. In the sequel, we work in the ring $F^{n \times n}$, so a matrix always means an $n \times n$ square matrix and I denotes the $n \times n$ identity matrix. In conformity with the notion of an inverse in a general ring, the *two-sided inverse*, or *inverse*, for short, of a matrix A is a matrix T satisfying $AT = TA = I$. If $LA = I$, then L is a *left inverse* of A, and if $AR = I$, then R is a *right inverse* of A.

It is true in every ring (see Sections A.4 and A.6) that if an element has both a left inverse and a right inverse, then these are equal, and there cannot be more left or right inverses. So, the (two-sided) inverse is unique.

We denote the (two-sided) inverse of a matrix A by A^{-1}.

We can use determinants to characterize which matrices have an inverse.

Theorem 2.2.2.

I. *If* $\det A \neq 0$, *then A has a two-sided inverse.*

II. *If A has a left or right inverse, then* $\det A \neq 0$. ♣

Combining the two statements, we obtain that a left or right inverse of a square matrix A is necessarily its two-sided inverse as the existence of each of the three types of inverses is equivalent to $\det A \neq 0$. If A has a left/right/two-sided inverse, then we say that A is invertible.

The 1×1 matrices and determinants can be identified with the elements of the field F and Theorem 2.2.2 is obvious as exactly the non-zero elements of F have an inverse. So we assume in the sequel that $A \in F^{n \times n}$ where $n > 1$.

We note that the proof of I provides a formula for A^{-1}, too, and we shall make use of it, e.g., in the proof of Theorem 3.2.1.

Proof. The key to I is an identity involving the adjugate of A that is the transpose of the matrix formed from the cofactors of A. The adjugate is sometimes called the classical adjoint of A but it has no connection to the adjoint of a complex matrix introduced in Definition 2.1.6.

Lemma 2.2.3. *Let $\hat{A}$ be the adjugate of A, i.e., its jth element in the ith row is the cofactor A_{ji} (not A_{ij}) corresponding to the element α_{ji} in A:*

$$\hat{A} = \begin{pmatrix} A_{11} & A_{21} & A_{31} & \dots & A_{n1} \\ A_{12} & A_{22} & A_{32} & \dots & A_{n2} \\ \vdots & & & & \\ A_{1n} & A_{2n} & A_{3n} & \dots & A_{nn} \end{pmatrix}.$$

Then

$$A\hat{A} = \hat{A}A = (\det A) \cdot I = \begin{pmatrix} \det A & 0 & \dots & 0 \\ 0 & \det A & \dots & 0 \\ \vdots & \vdots & \ddots & \vdots \\ 0 & 0 & \dots & \det A \end{pmatrix}. \clubsuit$$

Proof of the Lemma. The jth element of the ith row in $A\hat{A}$ is the dot product of the ith row of A and the jth column of $\hat{A}$: $\alpha_{i1}A_{j1} + \cdots + \alpha_{in}A_{jn}$. This sum is $\det A$ if $i = j$, and 0 if $i \neq j$ by the cofactor expansion and skew expansion along rows, respectively (Theorems 1.4.2 and 1.4.3). So, $A\hat{A}$ is $\det A$ multiplied by the identity matrix. Similarly, by expanding along columns, we obtain $\hat{A}A = (\det A) \cdot I$. ∎

Dividing the equalities in Lemma 2.2.3 by $\det A$, we have $A^{-1} = \frac{1}{\det A} \cdot \hat{A}$.

To prove II, we need a theorem stated here without proof:

Theorem 2.2.4 (Product theorem for determinants).

$$\det(AB) = \det A \cdot \det B. \clubsuit$$

This means that multiplying two determinants of the same size by the rules of matrix multiplication, the resulting determinant is the product of the two original determinants. As a determinant does not change by reflecting the elements through the main diagonal, the above theorem remains valid also for row-row, column-column, and column-row multiplication.

Turning to the proof of II, we assume that A has a left inverse, i.e., $LA = I$. By Theorem 2.2.4, $\det L \cdot \det A = \det I = 1$, so $\det A$ cannot be 0. ∎

We give another proof to Theorem 2.2.2 in Section 3.5 using systems of linear equations. A possible proof of Theorem 2.2.4 is sketched in Exercise 2.2.15. Another way is indicated in Exercise 9.8.4 based on the new interpretation of determinants discussed in Section 9.8.

Now we study the zero divisors in $F^{n \times n}$. According to the general notion of zero divisors in rings, we say that a matrix A is a *left zero divisor* if $A \neq 0$, but $AU = 0$ for some $U \neq 0$. We define the right zero divisors analogously.

It is true in every ring (see Theorem A.6.3), that an element having a left inverse can be neither 0 nor a left zero divisor. Similar relation holds for the right inverse and the right zero divisor.

Also the zero divisors among the matrices can be characterized by determinants.

Theorem 2.2.5. *A square matrix $A \neq 0$ is a left/right zero divisor if and only if $\det A = 0$.* $\clubsuit$

Proof. The only if part follows from Theorem 2.2.2 and from the relation of the inverse and zero divisor mentioned above. We prove the if part only for the special case when at least one cofactor A_{ij} in A is not 0, i.e., the adjugate matrix $\hat{A}$ defined in Lemma 2.2.3 is not the zero matrix. By Lemma 2.2.3, $A\hat{A} = \hat{A}A = (\det A) \cdot I = 0 \cdot I = 0$, so $\hat{A} \neq 0$ is a two-sided zero divisor pair to A. ■

We shall give two complete proofs for Theorem 2.2.5 in Section 3.5.

To close this section with a subjective (and non-mathematical) summary, the multiplication in the ring of $n \times n$ matrices has many unusual features, as it is not commutative and there are lots of zero divisors thus having no inverses, so the ring of matrices is very far from being a field.

Exercises 2.2

All exercises refer to matrices in $F^{n\times n}$.

2.2.1. True or false?
(a) If $AB = BA$, then $(A + B)^2 = A^2 + 2AB + B^2$.
(b) If $(A + B)^2 = A^2 + 2AB + B^2$, then $AB = BA$.
*(c) If $(A + B)^3 = A^3 + 3A^2B + 3AB^2 + B^3$, then $AB = BA$.

2.2.2. True or false?
(a) If both A and B are invertible, then AB is invertible.
(b) If AB is invertible, then both A and B are invertible.
(c) If both $A + B$ and $A - B$ are invertible, then $A^2 - B^2$ is invertible.
(d) If both A and B are invertible, then $A + B$ is invertible.
(e) If $A + B$ is invertible, then at least one of A and B is invertible.
(f) If A is invertible, then $A + A^2$ is invertible.
(g) If $A + A^2$ is invertible, then A is invertible.

2.2.3. True or false?
(a) If A is a right zero divisor and $AB \neq 0$, then AB is a right zero divisor.
(b) If AB is a right zero divisor, then both A and B are right zero divisors.
(c) If AB is a right zero divisor, then at least one of A and B is a right zero divisor.
(d) If $A + B$ is a right zero divisor, then at least one of A and B is a right zero divisor.

2.2.4. Check each matrix below whether it is invertible or it is a zero divisor. Compute the inverse if it is invertible and find a pair if it is a zero divisor so that the product of the two non-zero matrices is 0.

(a) $\begin{pmatrix} 1 & 2 \\ 3 & 4 \end{pmatrix}$; (b) $\begin{pmatrix} 1 & 2 \\ 3 & 6 \end{pmatrix}$; (c) $\begin{pmatrix} 1 & 2 & 3 \\ 3 & 4 & 5 \\ 5 & 7 & 9 \end{pmatrix}$; (d) $\begin{pmatrix} 1 & 2 & 3 \\ 3 & 4 & 5 \\ 5 & 7 & 8 \end{pmatrix}$.

2.2.5. Characterize the square matrices where both the matrix and its inverse have integer elements.

2.2.6. We call a square matrix *upper triangular* if every element below the main diagonal is 0. How can we decide easily whether an upper triangular matrix is invertible? Is its inverse necessarily upper triangular?

2.2.7. True or false? ($A, B \in F^{n\times n}$ are given and $X, Y \in F^{n\times n}$ are the variables.)
   (a) If $\det A \neq 0$, then the matrix equation $AX = B$ is solvable.
   (b) If $\det A = 0$, then $AX = B$ is not solvable.
   (c) If $\det A = 0, \det B \neq 0$, then $AX = B$ is not solvable.
   (d) If $\det A = 0, \det B = 0$, then $AX = B$ is solvable.
   (e) $AX = B$ cannot have more than one solution.
   (f) $AX = B$ has exactly one solution if and only if $\det A \neq 0$.
   (g) If $AX = B$ is solvable, then $YA = B$ is also solvable.
   (h) If both $AX = B$ and $YA = B$ have a unique solution, then these solutions are equal.

2.2.8. Give new solutions to Exercises 1.5.5, 1.5.6, and 1.5.7 using the product theorem for determinants (Theorem 2.2.4).

2.2.9. Assume that $n > 1$ is an odd integer, A is an $n \times n$ real matrix with elements α_{ij} and cofactors A_{ij}, and $\det A \neq 0$. Prove that $A^2 = I$ holds if and only if either $\alpha_{ij} = A_{ji}$ for every i and j, or $\alpha_{ij} = -A_{ji}$ for every i and j.

2.2.10. Show that if $\det A \neq 0$, then the adjugate of the adjugate of A is a scalar multiple of A. (This is also true if $\det A = 0$; see Exercise 3.4.17.)

2.2.11. Assume that n is an even integer, A and B are $n \times n$ real matrices, and $\det A \neq 0$. Show that $\hat{A} = \hat{B}$ implies $A = B$. What can we say for complex matrices A and B (for $n > 1$)?

2.2.12. Verify that the 2×2 real matrices of the types below form a ring under the usual matrix operations. Investigate the question of commutativity, determine the left, right, and two-sided identities and zero divisors. If there is a two-sided identity, check which elements have a left, right, or two-sided inverse. Which rings are fields? Show that these fields are basically the same as some well-known fields.
   (a) $\begin{pmatrix} a & 0 \\ 0 & 0 \end{pmatrix}$; (b) $\begin{pmatrix} a & a \\ a & a \end{pmatrix}$; (c) $\begin{pmatrix} a & b \\ b & a \end{pmatrix}$;
   (d) $\begin{pmatrix} a & b \\ -b & a \end{pmatrix}$; (e) $\begin{pmatrix} a & b \\ 0 & 0 \end{pmatrix}$.

***2.2.13.** Let A be a real invertible matrix where all elements in both A and A^{-1} are non-negative numbers. Prove that exactly one element is not zero in every row and every column of A.

*2.2.14. For which integers n does there exist an $n \times n$ rational matrix satisfying $A^3 = 2I$?

*2.2.15. Prove the Product Theorem 2.2.4 for determinants in the steps below. We denote the determinants of matrices A, B, $AB \in F^{n \times n}$ by D_A, D_B, and D_{AB}, respectively.

(a) Verify that the determinant D of the $2n \times 2n$ matrix

$$\begin{pmatrix} A & 0 \\ -I & B \end{pmatrix}$$

is $D_A D_B$ (here 0 and I are the $n \times n$ zero matrix and identity matrix, respectively).

(b) Adding suitable multiples of the first n columns to the last n columns obtain that B in the lower right corner is replaced by 0. Then we get the matrix

$$\begin{pmatrix} A & AB \\ -I & 0 \end{pmatrix}$$

with determinant $D' = D$.

(c) Prove $D' = D_{AB}$.

(d) Parts (a)–(c) imply $D_{AB} = D_A D_B$.

Chapter 3

Systems of Linear Equations

One of the most natural, simple, and practical methods to solve systems of linear equations is Gaussian elimination which has several theoretical implications, too. Cramer's rule refers to special types of systems yielding a formula for the solution via determinants. We introduce the notions of linear independence and the rank of a matrix which play a fundamental role in the sequel. As an application, we revisit the problem of inverses and zero divisors among square matrices.

3.1. Gaussian Elimination

The general form of a *system of linear equations* in k equations and n unknowns (or variables) is

$$\begin{aligned}\alpha_{11}x_1 + \alpha_{12}x_2 + \ldots + \alpha_{1n}x_n &= \beta_1\\ \alpha_{21}x_1 + \alpha_{22}x_2 + \ldots + \alpha_{2n}x_n &= \beta_2\\ &\vdots\\ \alpha_{k1}x_1 + \alpha_{k2}x_2 + \ldots + \alpha_{kn}x_n &= \beta_k\end{aligned}$$

where the coefficients α_{ij} and the constants β_i are from a commutative field F. The number of equations (k) and the number of unknowns (n) are arbitrary and are independent of each other, so we do not restrict ourselves, e.g., to the special case $k = n$.

A *solution* means a sequence $\gamma_1, \ldots, \gamma_n$ of elements in F such that replacing x_i everywhere by γ_i, all equalities hold.

There exist systems with no solution, some others have a unique solution, and again others have more than one solution. The last case does not necessarily mean infinitely many solutions, though we have this experience with systems over the real numbers. We shall see that this is true for all infinite fields. For a finite field F, however, the number of solutions is always finite as the sequence $x_1, \ldots, x_n$ can assume only $|F|^n$ values.

We want to get answers to the following questions: (a) What is the condition for solvability? (b) If solvable, how many solutions do we get? (c) How can we describe all solutions? (d) By what methods can we obtain a solution or all solutions?

In this section, we answer these questions with the help of *Gaussian elimination*.

This algorithm is a repeated application of the following steps that all lead to systems of equations equivalent to the original one, i.e., having exactly the same solutions.

S1 We multiply an equation by a non-zero element of F called a *scalar*.

S2 We add a scalar multiple of an equation to another equation.

S3 We swap two equations.

S4 We delete the equations where all coefficients and the constant on the right-hand side are 0.

We call these steps *elementary equivalent transformations*.

We shall *eliminate* the unknowns one after the other from the system using elementary equivalent transformations.

Assuming $\alpha_{11} \neq 0$, we divide the first equation by α_{11}, i.e., we apply S1 with the scalar $(\alpha_{11})^{-1}$. Then, for every $i > 1$, we subtract α_{i1} times the first equation from the ith equation thus *eliminating* x_1 from all these equations.

Assume that (the new) $\alpha_{22} \neq 0$ in the system obtained after the elimination of x_1. Then we repeat the previous procedure: we divide the second equation by α_{22} and for every $i > 2$, we subtract α_{i2} times the second equation from the ith equation, etc.

If we get stuck, e.g., $\alpha_{22} = 0$ in the previous phase, but (say) $\alpha_{52} \neq 0$, then we swap the 2nd and 5th equations and move on with this system.

If this does not work, i.e., $\alpha_{i2} = 0$ for every $i \geq 2$, then we turn to the third variable and work with α_{23}, etc.

We shall illustrate all these in practice on concrete examples and show how we can obtain the solutions of the systems. But we introduce first a few technical simplifications.

Observe that we only have to trace the changes in the coefficients and the constants on the right-hand side, so there is no need to write the signs x_i, $+$, and $=$ again and again. So we can characterize the system of equations by matrices. The $k \times n$ matrix A formed from the coefficients α_{ij} is called the *coefficient matrix* of the system. Extending A with the $(n+1)$th column $\mathbf{b}$ containing the constants β_i on the right-hand side of the equations, we get the $k \times (n+1)$ *augmented matrix* $A \mid \mathbf{b}$:

$$A = \begin{pmatrix} \alpha_{11} & \alpha_{12} & \dots & \alpha_{1n} \\ \alpha_{21} & \alpha_{22} & \dots & \alpha_{2n} \\ \vdots & & & \\ \alpha_{k1} & \alpha_{k2} & \dots & \alpha_{kn} \end{pmatrix}, \qquad A \mid \mathbf{b} = \left(\begin{array}{cccc|c} \alpha_{11} & \alpha_{12} & \dots & \alpha_{1n} & \beta_1 \\ \alpha_{21} & \alpha_{22} & \dots & \alpha_{2n} & \beta_2 \\ \vdots & & & & \\ \alpha_{k1} & \alpha_{k2} & \dots & \alpha_{kn} & \beta_k \end{array}\right).$$

In the augmented matrix, we separate the coefficients and the constants by a vertical line indicating that the two types of elements play different roles. In the notation $\mathbf{b}$ is a vector with components β_i; see the end of this section for more details.

Clearly, the elementary equivalent transformations S1–S4 in the system of equations correspond to similar operations with the rows of the augmented matrix:

M1 We multiply a row by a non-zero scalar.

M2 We add a scalar multiple of a row to another row.

M3 We swap two rows.

M4 We delete a zero row (where every element is 0).

We call these steps *elementary row-equivalent transformations* of the augmented matrix.

For the sake of the reverse directions of the equivalent steps, we can also allow in S4 and M4 to extend the system or the matrix by a zero equation or row, but clearly, we shall never need to apply these in practice.

We illustrate the elimination on three systems of equations over the real numbers.

Example E1. Solve the system in **R**.

$$\begin{aligned} x_1 + 2x_2 &= 3 \\ 4x_1 + 5x_2 &= 6 \\ 7x_1 + 8x_2 &= 9 \end{aligned}$$

The augmented matrix is

$$\left(\begin{array}{cc|c} 1 & 2 & 3 \\ 4 & 5 & 6 \\ 7 & 8 & 9 \end{array}\right).$$

As indicated in the general scheme, we subtract 4 times the first row from the second row and 7 times the first row from the third row. We get

$$\left(\begin{array}{cc|c} 1 & 2 & 3 \\ 0 & -3 & -6 \\ 0 & -6 & -12 \end{array}\right).$$

We divide the second row by -3 and add 6 times the new second row to the third row. We obtain

$$\left(\begin{array}{cc|c} 1 & 2 & 3 \\ 0 & 1 & 2 \\ 0 & 0 & 0 \end{array}\right).$$

Deleting the zero row, we are left with

$$\left(\begin{array}{cc|c} 1 & 2 & 3 \\ 0 & 1 & 2 \end{array}\right).$$

The second row gives $x_2 = 2$. Substituting back into the equation corresponding to the first row: $x_1 + 2x_2 = x_1 + 2 \cdot 2 = 3$, so $x_1 = -1$. But we can automate even this last step. If we eliminate the second element in the first row, then x_1 is expressed directly. To achieve this, we subtract 2 times the second row from the first row. We get

$$\left(\begin{array}{cc|c} 1 & 0 & -1 \\ 0 & 1 & 2 \end{array}\right),$$

so $x_1 = -1$.

The system has a unique solution: $x_1 = -1$, $x_2 = 2$.

Example E2. Solve the system in **R**.

$$\begin{aligned} x_1 + x_2 + 2x_3 &= 3 \\ 4x_1 + 4x_2 + 5x_3 &= 6 \\ 7x_1 + 7x_2 + 8x_3 &= 10 \end{aligned}$$

The elimination steps of the augmented matrix are

$$\left(\begin{array}{ccc|c} 1 & 1 & 2 & 3 \\ 4 & 4 & 5 & 6 \\ 7 & 7 & 8 & 10 \end{array}\right) \sim \left(\begin{array}{ccc|c} 1 & 1 & 2 & 3 \\ 0 & 0 & -3 & -6 \\ 0 & 0 & -6 & -11 \end{array}\right) \sim \left(\begin{array}{ccc|c} 1 & 1 & 2 & 3 \\ 0 & 0 & 1 & 2 \\ 0 & 0 & 0 & 1 \end{array}\right).$$

Since the third row of the last matrix corresponds to the impossible equation $0x_1 + 0x_2 + 0x_3 = 1$, the system has no solution.

Example E3. Solve the system in **R**.

$$\begin{aligned} x_1 + 2x_2 + 3x_3 &= 4 \\ 5x_1 + 6x_2 + 7x_3 &= 8 \\ 9x_1 + 10x_2 + 11x_3 &= 12 \\ 13x_1 + 14x_2 + 15x_3 &= 16 \end{aligned}$$

The elimination process is now

$$\left(\begin{array}{ccc|c} 1 & 2 & 3 & 4 \\ 5 & 6 & 7 & 8 \\ 9 & 10 & 11 & 12 \\ 13 & 14 & 15 & 16 \end{array}\right) \sim \left(\begin{array}{ccc|c} 1 & 2 & 3 & 4 \\ 0 & -4 & -8 & -12 \\ 0 & -8 & -16 & -24 \\ 0 & -12 & -24 & -36 \end{array}\right) \sim \left(\begin{array}{ccc|c} 1 & 2 & 3 & 4 \\ 0 & 1 & 2 & 3 \\ 0 & 0 & 0 & 0 \\ 0 & 0 & 0 & 0 \end{array}\right)$$

$$\sim \left(\begin{array}{ccc|c} 1 & 2 & 3 & 4 \\ 0 & 1 & 2 & 3 \end{array}\right) \sim \left(\begin{array}{ccc|c} 1 & 0 & -1 & -2 \\ 0 & 1 & 2 & 3 \end{array}\right).$$

We got no restriction on x_3, so it can be an arbitrary real number. Fixing the value of x_3, we obtain unique values for the other unknowns: $x_1 = -2 + x_3$, $x_2 = 3 - 2x_3$. So all solutions are $x_1 = -2 + v$, $x_2 = 3 - 2v$, $x_3 = v$, where v is an arbitrary real number.

The above examples show the general procedure. Working downwards we finally arrive at a matrix where every row except the first one starts with zeros; the first nonzero element in the coefficient part of every top row is 1, called a *leader* (or *pivot*); the leaders stand in distinct columns and follow each other downwards to the right; every element below the leaders is 0; there can be further rows at the bottom containing only 0 in the coefficient part. We say that such a matrix is in *echelon form*. Some echelon forms are the examples

$$\textbf{E4} \left(\begin{array}{ccc|c} 1 & 2 & 3 & 6 \\ 0 & 1 & 7 & 3 \\ 0 & 0 & 1 & 5 \\ 0 & 0 & 0 & 0 \end{array}\right), \qquad \textbf{E5} \left(\begin{array}{cc|c} 1 & 2 & 2 \\ 0 & 1 & 3 \\ 0 & 0 & 2 \end{array}\right), \qquad \textbf{E6} \left(\begin{array}{ccccc|c} 1 & 2 & 3 & 5 & 6 & 7 \\ 0 & 0 & 1 & 2 & 3 & 4 \\ 0 & 0 & 0 & 1 & 2 & 3 \end{array}\right).$$

In example E5 the third row indicates a contradiction, so the system of equations corresponding to E5 has no solution. In E4 the last row is zero, we may delete it.

We can turn the elements above the leaders into 0 by proceeding upwards: we subtract the suitable scalar multiple of the row of the next leader from the rows above it. In Examples E4 and E6 we get the matrices

$$\textbf{E4'}\ \left(\begin{array}{ccc|c} 1 & 0 & 0 & 55 \\ 0 & 1 & 0 & -32 \\ 0 & 0 & 1 & 5 \end{array}\right), \qquad \textbf{E6'}\ \left(\begin{array}{ccccc|c} 1 & 2 & 0 & 0 & -1 & -2 \\ 0 & 0 & 1 & 0 & -1 & -2 \\ 0 & 0 & 0 & 1 & 2 & 3 \end{array}\right).$$

This form is called a *reduced echelon form* or REF, for short. Compared to the echelon form, this has the additional property that the elements above the leaders are 0.

We can easily read the solutions of the system of equations from the REF. In the system corresponding to E4, the unique solution is $x_1 = 55$, $x_2 = -32$, $x_3 = 5$, whereas E6 yields $x_1 = -2 + \nu - 2\mu$, $x_2 = \mu$, $x_3 = -2 + \nu$, $x_4 = 3 - 2\nu$, $x_5 = \nu$ where ν and μ are arbitrary real numbers.

Now we summarize how we can see from the REF whether the system is solvable, and if yes, what the number of solutions is and how to obtain all solutions.

A system of equations is *solvable* if and only if no row occurs in the REF where all elements are 0 in the coefficient part but the constant in the last column differs from 0. We shall call such a row a *forbidden row*. So, a system is solvable if and only if there is no forbidden row in the REF. In fact, we see already in the echelon form if a forbidden row appears, and then there is no need to make further reductions.

The solution is *unique* if and only if (there is no forbidden row and) every column contains a leader, i.e., the number of leaders is the same as the number of unknowns.

ATTENTION, please! This has nothing to do with the occurrence or lack of zero rows, we call these *redundant rows*. A redundant row just means that the corresponding equation follows from the other equations, so it bears no new information, poses no extra requirement, and can be deleted. The third row in the REF corresponding to Example E1 was redundant, and so were the last two rows in Example E3. We can extend every system with redundant equations, e.g., we repeat one of its equations or insert the sum of two equations, etc. Clearly, these cause no change in the solutions of the system, and so do not modify the solvability or the number of solutions either.

If a system has a unique solution, then we get it immediately from the REF. If there is more than one solution, then the unknowns corresponding to the columns without leaders can assume arbitrary values, these are *free parameters*, (or *free variables*) and all other unknowns can be expressed uniquely with these. The variable x_3 in Example E3 and the variables x_2 and x_5 in Example E6 were free parameters. Thus there are infinitely many solutions over an infinite field and $|F|^s$ solutions over a finite field F where s is the number of free parameters.

We summarize all these in the following theorem.

Theorem 3.1.1.

I. *The augmented matrix of a system of linear equations can be converted into a reduced echelon form (REF) by elementary row-equivalent transformations.*

II. *The system is solvable if and only if there is no forbidden row in the (reduced) echelon form.*

III. *The system has a unique solution if and only if (there is no forbidden row and) the number of leaders equals the number of unknowns.*

IV. *If there is more than one solution, then the unknowns corresponding to the leaderfree columns are free parameters (these can be chosen arbitrarily) and the other unknowns can be uniquely expressed by them. So, the number of solutions is infinite over infinite fields and is $|F|^s$ over a finite field F where $s > 0$ is the number of free parameters.*

V. *The solution(s) are readable from the reduced echelon form directly.* ♣

An alternative expression for solvability is that the system of equations is *consistent*.

We note that in the case of more than one solution, these can be parameterized in different ways in general. In Example E3 x_1 or x_2 can be the free parameter, as well, and we obtain the solutions in the form $x_1 = \mu$, $x_2 = -1 - 2\mu$, $x_3 = \mu + 2$, and $x_1 = -1/2 - \tau/2$, $x_2 = \tau$, $x_3 = 3/2 - \tau/2$, respectively. If we want to obtain these via elimination, then first we have to modify the order of the unknowns suitably (to put x_1 or x_2 as the last variable). Of course, one parametrization describes the solutions perfectly, and as swapping the unknowns often causes confusion, it is better not to change their order but rather perform the elimination in the original form of the system.

It is not true that in the case of s free parameters *any* s variables can be chosen freely. E.g., the system $x_1 + 2x_2 + 3x_3 = 1$, $x_1 + 2x_2 + 4x_3 = 1$ determines $x_3 = 0$ uniquely and only one of the two other unknowns can act as a free parameter: the solutions are given by $x_1 = \nu$, $x_2 = 1/2 - \nu/2$, $x_3 = 0$, or $x_1 = 1 - 2\mu$, $x_2 = \mu$, $x_3 = 0$.

Some words of advice. It is recommended to write out the steps of the elimination in detail and to register them carefully as seen at the determinants.

Do not forget that we only work with *rows* during the entire procedure. Do not manipulate the columns as these refer to the unknowns instead of the equations. Swapping two columns means swapping the corresponding variables and it would be very difficult to control such changes in the order of variables all the time. Moreover, a more complicated operation of adding a scalar multiple of a column to another column would produce new unknowns instead of the original ones in a nearly untraceable way.

Finding the (reduced) echelon form is an efficient algorithm to solve the system of equations, but it is by no means an end in itself. If we can get the solutions in another way more easily, then there is no need for elimination. We should be aware, however, that, e.g., guessing a solution does not necessarily mean the general solution, so we have to find the other solutions, too, or *prove* that the system has a unique solution.

ATTENTION, please! We should not infer to the solvability or to the number of solutions from the relation between the numbers of equations and unknowns. It is a fundamentally false approach that "if there are as many equations as unknowns, then there is a unique solution; if there are more unknowns than equations, then there is more than one solution; and if there are more equations than unknowns, then there are no solutions." This argument contains two serious errors. On the one hand, we saw that we can extend any system by redundant equations bearing no new content and so the system will consist of more equations without any change in the solvability or the number of solutions. On the other hand, however many variables occur, we can achieve insolvability with just two equations, e.g., having the same coefficients but a different constant on the right-hand side. In fact, even one equation is sufficient for this with 0 coefficients and a non-zero constant. This might seem to be a degenerate, artificial example, but we should not forget that such a forbidden row appears in every unsolvable system, only in a disguised form before the elimination.

Examples E1 and E3 contained more equations than variables, but E1 had a unique solution and E3 had infinitely many solutions. Example E2 consisted of an equal number of equations and unknowns, but it had no solution. The number of equations and unknowns in itself has almost no effect on whether the system has 0, 1, or more solutions. The only exception is that there cannot be a unique solution if there are more unknowns than equations (but it can happen that there are no solutions!).

Theorem 3.1.2. *If a system of k linear equations with n unknowns has a unique solution, then $n \leq k$.* ♣

Proof. In the case of a unique solution there are n leaders in the REF. These are in distinct rows, so there are at most k of them. Therefore, $n \leq k$. ∎

We shall use a simple but important corollary of this observation many times. We introduce first the notion of a homogeneous system of linear equations:

Definition 3.1.3. A system of linear equations is *homogeneous* if every constant on the right-hand side is 0. ♣

A homogeneous system is solvable always as $x_1 = \cdots = x_n = 0$ is a solution. This is called a *trivial solution*. Therefore, the interesting question is whether there exists a non-trivial solution. We get a sufficient condition for this from the following theorem.

Theorem 3.1.4. *If a homogeneous system of linear equations contains more unknowns than equations, then the system has a non-trivial solution.* ♣

Proof. To prove by contradiction, assume that the only solution is the trivial one. Then the system has a unique solution, so there cannot be more unknowns than equations by Theorem 3.1.2, a contradiction. ∎

We introduce two further useful forms to describe systems of equations. First we shall need the notion of column vectors.

Definition 3.1.5. A matrix consisting of a single column is called a *column vector*. The elements of such a matrix are the *components* or *coordinates* of the vector. The set of vectors with q components over a field F is denoted by F^q (instead of $F^{q\times 1}$). ♣

This is a generalization of the representation of plane or space vectors as pairs or triples of real numbers. Later we shall use the word vector in a much broader sense (see Section 4.1).

We add two vectors in F^q by adding the corresponding components and multiply a vector by a scalar by multiplying the components by the scalar according to the general rules of matrix operations. We shall denote the vectors by boldface lower case Roman letters.

Now we turn to the representation of systems of equations with the help of vectors. Let

$$A = \begin{pmatrix} \alpha_{11} & \alpha_{12} & \dots & \alpha_{1n} \\ \alpha_{21} & \alpha_{22} & \dots & \alpha_{2n} \\ \vdots & & & \\ \alpha_{k1} & \alpha_{k2} & \dots & \alpha_{kn} \end{pmatrix} \in F^{k\times n},$$

$$\mathbf{x} = \begin{pmatrix} x_1 \\ x_2 \\ \vdots \\ x_n \end{pmatrix} \in F^n, \qquad \mathbf{b} = \begin{pmatrix} \beta_1 \\ \beta_2 \\ \vdots \\ \beta_k \end{pmatrix} \in F^k,$$

where A is the coefficient matrix, $\mathbf{x}$ is the vector where the unknowns x_j are the components, and $\mathbf{b}$ is the vector formed from the constants β_i on the right-hand sides of the equations. Then we can rewrite the system using matrix multiplication as $A\mathbf{x} = \mathbf{b}$.

Another representation is based on the column vectors

$$\mathbf{a}_j = \begin{pmatrix} \alpha_{1j} \\ \alpha_{2j} \\ \vdots \\ \alpha_{kj} \end{pmatrix} \in F^k, \quad j = 1, 2, \dots, n,$$

of the coefficient matrix A: we get the system as $x_1\mathbf{a}_1 + x_2\mathbf{a}_2 + \dots + x_n\mathbf{a}_n = \mathbf{b}$.

In homogeneous systems every component of $\mathbf{b}$ is 0; this is called the *zero vector* and is denoted by $\mathbf{0}$. So, the above two forms for a homogeneous system are $A\mathbf{x} = \mathbf{0}$ and $x_1\mathbf{a}_1 + x_2\mathbf{a}_2 + \dots + x_n\mathbf{a}_n = \mathbf{0}$.

Exercises 3.1

3.1.1. Verify that the elementary equivalent transformations S1–S4 lead to systems equivalent to the original one.

3.1.2. (a) Let $F = \mathbf{R}$. Which transformations produce systems equivalent to the original one?

(i) We replace the first equation by the sum of all equations.

(ii) We replace the first equation by the sum of all other equations.

(iii) We replace each of the first two equations by the sum of all equations.

(iv) We replace the first two equations by their sum and difference.

(v) We add 1 to every coefficient and constant in all equations.

(b) Investigate these questions for arbitrary fields instead of **R**.

3.1.3. Solve the systems in **R**.

(a)
$$\begin{aligned} -x + 3y + 3z &= 2 \\ 3x + y + z &= 4 \\ 2x - 2y + 3z &= 10 \end{aligned}$$

(b)
$$\begin{aligned} 2x + 3y + z &= 11 \\ x - y - 2z &= -7 \\ 3x + 2y - z &= 2 \end{aligned}$$

(c)
$$\begin{aligned} 2x + 3y + z &= 11 \\ x - y - 2z &= -7 \\ 3x + 2y - z &= 4 \end{aligned}$$

3.1.4. Determine the number of solutions over the modulo 5 field F_5.

$$\begin{aligned} x_1 + x_2 + x_3 &= 1 \\ x_3 + x_4 + x_5 &= 4 \\ x_1 + x_3 + x_4 &= 2 \\ x_2 + x_3 + x_5 &= 3 \end{aligned}$$

3.1.5. Solve the systems in **C**.

(a)
$$\begin{aligned} x_1 + ix_2 - x_3 &= -i \\ ix_1 - x_2 - ix_3 &= 1 \\ -x_1 - ix_2 + x_3 &= i \\ -ix_1 + x_2 + ix_3 &= -1 \\ x_1 + ix_2 + x_3 &= i \end{aligned}$$

(b)
$$\begin{aligned} x_1 + x_2 + x_3 + x_4 + x_5 &= 0 \\ x_1 + ix_2 + x_3 + x_4 + x_5 &= -1 + i \\ x_1 + x_2 + ix_3 + x_4 + x_5 &= -1 - i \\ x_1 + x_2 + x_3 + ix_4 + x_5 &= 1 - i \\ x_1 + x_2 + x_3 + x_4 + ix_5 &= 1 + i \end{aligned}$$

3.1.6. Solve the system in **R**.

$$x_1 + x_2 = 1, \qquad x_2 + x_3 = 1, \qquad \ldots, \qquad x_n + x_1 = 1.$$

*3.1.7. Let $1 \le m \le n$. When does the $n \times n$ system below have a unique solution in **R**?

$$\begin{aligned} x_1 + x_2 + \ldots + x_m &= 1 \\ x_2 + x_3 + \ldots + x_{m+1} &= 1 \\ &\vdots \\ x_n + x_1 + \ldots + x_{m-1} &= 1 \end{aligned}$$

3.1.8. Let $n > 1$ and solve the system of n equations and n unknowns

$$\begin{array}{cccccc} x_1 + & x_2 + & x_3 + \ldots + & x_n = & n \\ x_1 + & 2x_2 + & 2x_3 + \ldots + & 2x_n = & n-1 \\ x_1 + & 2x_2 + & 3x_3 + \ldots + & 3x_n = & n-2 \\ \vdots & \vdots & \vdots \quad \ddots & \vdots & \vdots \\ x_1 + & 2x_2 + & 3x_3 + \ldots + & nx_n = & 1, \end{array}$$

i.e., $\alpha_{ij} = \min(i, j)$ in the coefficient matrix A.

3.1.9. Let $k, n > 2$ and let p be a prime number greater than kn. We build a $k \times n$ matrix A by writing the numbers from 1 to kn one after the other, so $\alpha_{ij} = (i-1)n + j$. We consider the system of equations $A\mathbf{x} = \mathbf{b}$ over the modulo p field F_p.

(a) If solvable, what is the number of solutions?

(b) For how many vectors $\mathbf{b} \in F_p^k$ is the system solvable?

3.1.10. Show examples of systems of 5 equations and 3 unknowns with

(a) no solution;

(b) a unique solution;

(c) infinitely many solutions;

(d) exactly 7 solutions.

Investigate the situation for systems of 5 unknowns and 3 equations.

3.1.11. Find a relation between the numbers of leaders, free parameters, and unknowns.

3.1.12. Let F be a finite field, $n > k$, and consider a solvable system of k equations and n unknowns over F. Prove that there are at least $|F|^{n-k}$ solutions.

3.1.13. Assume that a system of equations over a(n arbitrary) field F has more than one solution and let H be the set of all values x_1 occurring in the solutions. Show that either $|H| = 1$ or $H = F$.

3.1.14. What is a geometric interpretation of systems of two unknowns (and arbitrarily many linear equations)? How can we read the solvability and the number of solutions? How can we work similarly with three unknowns?

3.1.15. Let $\mathbf{x}'$ be a solution of the system $A\mathbf{x} = \mathbf{b}$. Prove that all solutions are provided by the formula $\mathbf{x}' + \mathbf{x}^*$ where $\mathbf{x}^*$ assumes all solutions of the homogeneous system $A\mathbf{x} = \mathbf{0}$.

3.1.16. Let $A, A_i \in F^{k \times n}$, $\mathbf{x} \in F^n$, $\mathbf{b}, \mathbf{b}_i \in F^k$. True or false?

(a) If both $A\mathbf{x} = \mathbf{b}_1$ and $A\mathbf{x} = \mathbf{b}_2$ are solvable, then so is $A\mathbf{x} = \mathbf{b}_1 + \mathbf{b}_2$.

(b) If both $A_1\mathbf{x} = \mathbf{b}$ and $A_2\mathbf{x} = \mathbf{b}$ are solvable, then so is $(A_1 + A_2)\mathbf{x} = \mathbf{b}$.

(c) If $A\mathbf{x} = \mathbf{b}$ has a unique solution, then $A\mathbf{x} = \mathbf{b}_1$ cannot have more than one solution for any $\mathbf{b}_1$.

(d) If $A\mathbf{x} = \mathbf{b}$ has a unique solution, then $A\mathbf{x} = \mathbf{b}_1$ is solvable for every $\mathbf{b}_1$.

(e) If $k < n$, then for any A there exists a $\mathbf{b}$ such that $A\mathbf{x} = \mathbf{b}$ has no solution.

(f) If $k > n$, then for any A there exists a $\mathbf{b}$ such that $A\mathbf{x} = \mathbf{b}$ has no solution.

3.1.17. Double and Triple each thinks of 5 integers independently. Double can ask the parity of the sum of Triple's any 2 integers, and Triple can ask the parity of the sum of Double's any 3 (distinct) integers. Who can guess the parity of the other person's integers and what is the minimal number of questions needed?

3.1.18. Consider a system of linear equations with integer coefficients and constants. True or false?

(a) If the system is solvable in **Z**, then it is solvable in **C**.

(b) If the system is solvable in **C**, then it is solvable in **Q**.

(c) If the system is solvable in **Q**, then it is solvable in **Z**.

3.1.19. Consider a homogeneous system of linear equations where the coefficients are integers. Interpreting these as elements of the modulo 11 field F_{11} we can also view the system over F_{11}. True or false?

(a) If there exists a non-trivial solution over F_{11}, then there exists a non-trivial solution over **Q**.

(b) If there exists a non-trivial solution over **Q**, then there exists a non-trivial solution over F_{11}.

3.2. Cramer's Rule

This section discusses systems of equations with the same number of variables and equations. Here, the coefficients form a square matrix having a determinant. We show first that if this determinant is not zero, then the system has a unique solution for any right-hand side and we show a formula for this solution.

Theorem 3.2.1 (Cramer's rule). *If $A \in F^{n \times n}$ and $D = \det A \neq 0$, then the system of equations $A\mathbf{x} = \mathbf{b}$ has exactly one solution. In this solution, $x_j = D_j/D$ where the determinant D_j is formed by replacing column j in D by the constants on the right-hand side (i.e., by the components of* **b**). ♣

E.g.,

$$x_2 = \frac{\begin{vmatrix} \alpha_{11} & \beta_1 & \dots & \alpha_{1n} \\ \alpha_{21} & \beta_2 & \dots & \alpha_{2n} \\ \vdots & \vdots & \ddots & \vdots \\ \alpha_{n1} & \beta_n & \dots & \alpha_{nn} \end{vmatrix}}{\begin{vmatrix} \alpha_{11} & \alpha_{12} & \dots & \alpha_{1n} \\ \alpha_{21} & \alpha_{22} & \dots & \alpha_{2n} \\ \vdots & \vdots & \ddots & \vdots \\ \alpha_{n1} & \alpha_{n2} & \dots & \alpha_{nn} \end{vmatrix}}.$$

Proof. The conditions guarantee the existence of A^{-1}. We can "eliminate" A from the left-hand side of $A\mathbf{x} = \mathbf{b}$ via multiplying the system from the left by A^{-1}. This is an equivalent transformation since we can undo it by multiplying by A from the left (and using the properties of matrix multiplication). We obtain $\mathbf{x} = A^{-1}\mathbf{b}$, hence this and only this is the solution of the original system.

We have to bring $\mathbf{x} = A^{-1}\mathbf{b}$ to the form specified in the theorem. By the definition of matrix multiplication, x_j is the product of row j in matrix A^{-1} and the vector $\mathbf{b}$. Using the formula for the inverse of a matrix obtained in the proof of Theorem 2.2.2, $x_j = (1/D)(A_{1j}\beta_1 + \cdots + A_{nj}\beta_n)$ where A_{lm} denotes the corresponding cofactor of determinant D. Since D_j differs only in its jth column from D, therefore the cofactors belonging to the jth column are the same in D and D_j. Thus, expanding D_j along its jth column, we get $D_j = \beta_1 A_{1j} + \cdots + \beta_n A_{nj}$, i.e., $x_j = D_j/D$, indeed. ∎

The primary significance of Cramer's rule is theoretical. It is rarely used for solving a system of equations in practice as it can be applied only in very special cases and even then it requires in general a much larger amount of computation than Gaussian elimination (note that Gaussian elimination providing the complete solution needs just slightly more time than calculating a single determinant of order n, whereas Cramer's rule requires the computation of $n+1$ such determinants).

Obviously, we can rely on Cramer's rule if the determinants D and D_j can be evaluated easily. It is useful also if we can "guess" a solution and verify $D \neq 0$; in this case, the above theorem implies that there cannot be other solutions.

Cramer's rule is interesting from a historic (and high-school didactic) aspect, too. Solving a general system $\alpha_{11}x_1 + \alpha_{12}x_2 = \beta_1$, $\alpha_{21}x_1 + \alpha_{22}x_2 = \beta_2$ of 2 linear equations with 2 variables (over, say, the reals and by any method), we find that it has exactly one solution if and only if $\alpha_{11}\alpha_{22} - \alpha_{12}\alpha_{21} \neq 0$, and then

$$x_1 = \frac{\beta_1\alpha_{22} - \beta_2\alpha_{12}}{\alpha_{11}\alpha_{22} - \alpha_{12}\alpha_{21}}, \qquad x_2 = \frac{-\beta_1\alpha_{21} + \beta_2\alpha_{11}}{\alpha_{11}\alpha_{22} - \alpha_{12}\alpha_{21}}.$$

Observe that this is exactly Cramer's rule for $n = 2$. We can establish the corresponding result also for $n = 3$ using "artisanal" efforts. These and similar observations played a fundamental role in developing the notion of determinants and applying it in handling systems of linear equations.

Cramer's rule requires that the determinant of the coefficient matrix should differ from zero. The formula for x_j makes no sense of course, if $D = 0$, but even the following stronger result holds; if $D = 0$, then the system cannot have exactly one solution.

Theorem 3.2.2. *If $A \in F^{n\times n}$ and $D = \det A = 0$, then the system $A\mathbf{x} = \mathbf{b}$ either has no solution, or has more than one solution.* ♣

Proof. We perform Gaussian elimination on the system $A\mathbf{x} = \mathbf{b}$ but keep the zero rows if they occur. The coefficient matrices arising during the elementary equivalent transformations still have zero determinants by the elementary properties of determinants. Hence, the determinant of the left-hand side in the reduced echelon form is zero. If the system had a unique solution, then every column would have a leader. But then the left-hand side in the reduced echelon form would be the identity matrix having a non-zero determinant. This contradiction guarantees that the system cannot have a unique solution. ∎

Remark. Similar to the proof of Theorem 3.2.2, we get the following result: If the coefficient matrix has a non-zero determinant, then the system has exactly one solution. (In this case, the determinant can change during the Gaussian elimination, but it will never be zero.) This gives a new proof for a part of Cramer's rule (but does not yield the formula for the solution).

By the above, we obtain the following result for homogeneous systems (with the same number of equations and variables).

Theorem 3.2.3. *Let $A \in F^{n\times n}$. The homogeneous system $A\mathbf{x} = \mathbf{0}$ has a non-trivial solution if and only if* $\det A = 0$. ♣

Proof. If $\det A = 0$, then the system cannot have a unique solution by Theorem 3.2.2, so there must be solution(s) besides the trivial (all-zero) one. If $\det A \neq 0$, then the system has exactly one solution by the Remark after Theorem 3.2.2 (or by Theorem 3.2.1), so the trivial solution is the only one. ■

We note that the above arguments provide new (and complete) proofs also for Theorems 2.2.2 and 2.2.5 about inverses and zero divisors of square matrices; we shall discuss these in Section 3.5 in detail.

As an application of Cramer's rule, we now prove a theorem about *interpolation polynomials*.

Theorem 3.2.4. *Let $\gamma_1, \dots, \gamma_n$ be distinct elements in a field F, and let $\beta_1, \dots, \beta_n$ be arbitrary elements in F. Then there exists exactly one polynomial $f \in F[x]$ of degree at most $n-1$ (allowing the zero polynomial having no degree) satisfying $f(\gamma_i) = \beta_i$, $i = 1, 2, \dots, n$.* ♣

Proof. Let $f = \delta_0 + \delta_1 x + \dots + \delta_{n-1}x^{n-1}$, then the condition means the system of equations

$$\begin{array}{ccccccccc} \delta_0 & + & \gamma_1\delta_1 & + & \dots & + & \gamma_1^{n-1}\delta_{n-1} & = & \beta_1 \\ \delta_0 & + & \gamma_2\delta_1 & + & \dots & + & \gamma_2^{n-1}\delta_{n-1} & = & \beta_2 \\ \vdots & & \vdots & & \ddots & & \vdots & & \vdots \\ \delta_0 & + & \gamma_n\delta_1 & + & \dots & + & \gamma_n^{n-1}\delta_{n-1} & = & \beta_n \end{array}$$

where δ_i are the variables. The determinant of the system is the Vandermonde determinant $V(\gamma_1, \dots, \gamma_n)$. Since the values γ_i are all distinct, this determinant is not zero. Therefore, the system has exactly one solution by Cramer's rule. ■

To compute the coefficients δ_i of f we can solve the above system of equations by elimination or applying the formula in Cramer's rule. In certain cases we may be lucky enough to "guess" the polynomial, and then uniqueness guarantees that this is the only one satisfying the condition. Exercises 3.2.9–3.2.11 present two further methods to construct f and an alternative proof to show its uniqueness.

We call f the interpolation polynomial belonging to the given "places" γ_i and "values" β_i. An intuitive meaning of "interpolation" is that the polynomial f realizes the "simplest" function forming its "interposed" values with the help of values β_i assumed at places γ_i.

Exercises 3.2

3.2.1. Solve the system of equations in **C**.

$$\begin{aligned} x_1 + x_2 + x_3 + x_4 &= 2 + 2i \\ x_1 + 2x_2 + 3x_3 + 4x_4 &= 4 + 4i \\ x_1 + 3x_2 + 6x_3 + 10x_4 &= 7 + 6i \\ x_1 + 4x_2 + 10x_3 + 20x_4 &= 11 + 8i \end{aligned}$$

3.2.2. Find the solution(s) in **R**.

$$x_1 + 2x_2 = 3, \qquad x_2 + 3x_3 = 4, \qquad \dots, \qquad x_n + (n+1)x_1 = n + 2.$$

3.2.3. Solve the systems of equations in **R** where $\alpha_i \neq \alpha_j$ if $i \neq j$ in (b).

(a) $$\begin{aligned} x_1 + x_2 + \dots + x_n &= n \\ x_1 + 2x_2 + \dots + nx_n &= n^2 \\ &\vdots \\ x_1 + 2^{n-1}x_2 + \dots + n^{n-1}x_n &= n^n \end{aligned}$$

(b) $$\begin{aligned} x_1 + x_2 + \dots + x_n &= 1 \\ \alpha_1 x_1 + \alpha_2 x_2 + \dots + \alpha_n x_n &= \beta \\ &\vdots \\ \alpha_1^{n-1}x_1 + \alpha_2^{n-1}x_2 + \dots + \alpha_n^{n-1}x_n &= \beta^{n-1} \end{aligned}$$

3.2.4. We use the notation in Theorem 3.2.1. True or false?

(a) If $D = 0$, but $D_j \neq 0$ for some j, then $A\mathbf{x} = \mathbf{b}$ has no solution.

(b) If $D = 0$ and $D_j = 0$ for every j, then $A\mathbf{x} = \mathbf{b}$ has more than one solution.

3.2.5. Let $A_1, A_2 \in F^{n \times n}$ be invertible matrices and consider systems $A_1\mathbf{x} = \mathbf{b}$ and $A_2\mathbf{x} = \mathbf{b}$ (having the same right-hand side). Prove:

(a) The two systems have a common solution for any $\mathbf{b}$ if and only if $A_1 = A_2$.

(b) The two systems have a common solution only for $\mathbf{b} = \mathbf{0}$ if and only if $A_1 - A_2$ is invertible.

3.2.6. (a) Let m_{ij}, $i, j = 1, 2, \dots, n$ be n^2 integers such that the determinant formed from m_{ij} is not a multiple of 7. Assume that for some integers $v_1, \dots, v_n$, the sum $v_1 m_{j1} + \dots + v_n m_{jn}$ is divisible by 7 for every j. Prove that every v_i is a multiple of 7.

*(b) Generalize the problem for any integer instead of 7.

The remaining exercises are related to the interpolation polynomial. We shall use the notations of Theorem 3.2.4.

3.2.7. Determine the polynomial f with complex coefficients and of degree at most three satisfying

(a) $f(-1) = 12$, $f(4) = 7$, $f(6) = 5$, $f(9) = 2$;

(b) $f(1) = f(-1) = 3$, $f(2) = f(-2) = 9$;

(c) $f(1) = i$, $f(i) = -1$, $f(-1) = -i$, $f(-i) = 1$;

(d) $f(-1) = f(i) = f(-i) = 1$, $f(1) = 9$.

3.2.8. How many polynomials f of degree *exactly* (a) $n-1$; (b) n meet the (other) conditions of Theorem 3.2.4?

3.2.9. Give a new proof for the *uniqueness* of f in Theorem 3.2.4 using that a polynomial cannot have more zeroes than its degree.

3.2.10. *Newton's interpolation polynomial.* Give a new proof to Theorem 3.2.4 (i.e., for the *existence* and *uniqueness* of f) by searching it in the form

$$f = \nu_0 + \nu_1(x-\gamma_1) + \nu_2(x-\gamma_1)(x-\gamma_2) + \cdots + \nu_{n-1}(x-\gamma_1)\cdot \cdots \cdot (x-\gamma_{n-1}).$$

3.2.11. *Lagrange's interpolation polynomial.* Let $n > 1$ and $1 \le i, j \le n$.

(a) Find a polynomial L_i of degree at most $n-1$ satisfying $L_i(\gamma_i) = 1$ and $L_i(\gamma_j) = 0$ for every $j \ne i$. (This means that we solve the interpolation problem first in a very special case when one of the prescribed values is 1 and all others are 0.) Polynomials L_i are called the *fundamental Lagrange polynomials* belonging to the points γ_i.

(b) Verify that $f = \sum_{i=1}^{n} \beta_i L_i$ meets the conditions of Theorem 3.2.4.

***3.2.12.** Let $n > 1$.

(a) Which (well-known) polynomial is the sum of the n fundamental Lagrange polynomials L_i?

(b) Let $\gamma_1, \dots, \gamma_n \in \mathbf{R}$ be distinct and let $\nu \in \mathbf{R}$. Compute

(b1) $\sum_{i=1}^{n} \prod_{j\ne i} \frac{\nu-\gamma_j}{\gamma_i-\gamma_j}$;

(b2) $\sum_{i=1}^{n} \prod_{j\ne i} \frac{1}{\gamma_i-\gamma_j}$.

3.2.13. True or false?

(a) If a polynomial f with complex coefficients assumes integer values in every integer point, then all coefficients of f are integers.

(b) If a polynomial f with complex coefficients assumes rational values in every rational point, then all coefficients of f are rational numbers.

*3.2.14. Let F be a finite field. Show that every function $\Phi : F \to F$ is a polynomial function, i.e., there exists a polynomial $f \in F[x]$ satisfying $f(\tau) = \Phi(\tau)$ for every $\tau \in F$.

*3.2.15. Ali Baba wants to leave his treasure to the Forty Thieves, but he is afraid that they will fight for it. Therefore, he wants to find a method assuring that they cannot get through unless 25 thieves have agreed previously how to share the treasure. The road to the treasure is described in a computer defended by a password which is a large integer. Following the advice of Interpol, Ali Baba whispers some words into each thief's ear one by one. If any 25 thieves unite their information, they can find the key number. Any 24 thieves, however, will have absolutely no information about the key even if they join their forces. What was the advice of Interpol(ation) to Ali Baba?

3.3. Linear Independence in F^k

We used the form $A\mathbf{x} = \mathbf{b}$ for systems of linear equations in the previous section. The present investigations are related to the other form introduced at the end of Section 3.1.

To recall this representation, let

$$\mathbf{a}_j = \begin{pmatrix} \alpha_{1j} \\ \alpha_{2j} \\ \vdots \\ \alpha_{kj} \end{pmatrix} \in F^k, \quad j = 1, 2, \ldots, n,$$

be the column vectors of the coefficient matrix A. Then an equivalent form of the system $A\mathbf{x} = \mathbf{b}$ is $x_1\mathbf{a}_1 + x_2\mathbf{a}_2 + \cdots + x_n\mathbf{a}_n = \mathbf{b}$.

Definition 3.3.1. Let $\mathbf{c}_1, \ldots, \mathbf{c}_m \in F^k$ and $\lambda_1, \ldots, \lambda_m \in F$, i.e., take *m vectors* in F^k and *m scalars* in F. Then the vector $\lambda_1\mathbf{c}_1 + \cdots + \lambda_m\mathbf{c}_m \in F^k$ is called a *linear combination* of the vectors $\mathbf{c}_i$ with the scalars λ_i. ♣

So, the solvability of the above system of linear equations means that $\mathbf{b}$ is a linear combination of the column vectors $\mathbf{a}_j$, and the scalars occurring in these linear combinations yield the solutions of the system.

The homogeneous system, i.e., the case $\mathbf{b} = \mathbf{0}$ will play a distinguished role. The trivial solution of the system $x_1\mathbf{a}_1 + x_2\mathbf{a}_2 + \cdots + x_n\mathbf{a}_n = \mathbf{0}$ corresponds to multiplying every $\mathbf{a}_j$ by $\lambda_j = 0$, and the obtained trivial linear combination $0\mathbf{a}_1 + \cdots + 0\mathbf{a}_n$ is clearly $\mathbf{0}$. A non-trivial solution means a *non-trivial* linear combination generating the zero vector so that not all scalars are 0 among the coefficients in the combination.

The next two definitions are of basic importance.

Definition 3.3.2. The vectors $\mathbf{c}_1, \ldots, \mathbf{c}_m \in F^k$ are *linearly dependent* if there exist scalars $\lambda_1, \ldots, \lambda_m \in F$ not all 0 such that $\lambda_1\mathbf{c}_1 + \cdots + \lambda_m\mathbf{c}_m = \mathbf{0}$. ♣

Definition 3.3.3. The vectors $\mathbf{c}_1, \ldots, \mathbf{c}_m \in F^k$ are *linearly independent* if $\lambda_1\mathbf{c}_1 + \cdots + \lambda_m\mathbf{c}_m = \mathbf{0}$ holds **only** when every $\lambda_i = 0$:

$$\lambda_1\mathbf{c}_1 + \cdots + \lambda_m\mathbf{c}_m = \mathbf{0} \Rightarrow \lambda_i = 0, \ i = 1, \ldots, m. \ ♣$$

So, a system of vectors $\mathbf{c}_1, \ldots, \mathbf{c}_m \in F^k$ is linearly independent if and only if it is not linearly dependent. For brevity we sometimes omit the adjective "linear" and say that a system of vectors is dependent or independent.

In a system of vectors the same vector may occur repeatedly (in contrast to a set of vectors). This can influence the question of independence: if a vector occurs more than once, say $\mathbf{c}_1 = \mathbf{c}_2$, then $1\mathbf{c}_1 + (-1)\mathbf{c}_2 + 0\mathbf{c}_3 + \cdots + 0\mathbf{c}_m = \mathbf{0}$, so the system $\mathbf{c}_1, \ldots, \mathbf{c}_m$ is dependent.

Examples.

E1 The vectors

$$\mathbf{c}_1 = \begin{pmatrix}1\\2\\1\end{pmatrix}, \mathbf{c}_2 = \begin{pmatrix}2\\1\\2\end{pmatrix}, \mathbf{c}_3 = \begin{pmatrix}1\\1\\1\end{pmatrix} \in \mathbf{R}^3$$

are linearly dependent as $1\mathbf{c}_1 + 1\mathbf{c}_2 + (-3)\mathbf{c}_3 = \mathbf{0}$.

E2 The system consisting of the previous $\mathbf{c}_1$, $\mathbf{c}_2$, and $\mathbf{c}_4 = \begin{pmatrix}2\\1\\1\end{pmatrix}$ is linearly independent. To verify it, we write out the equality $\lambda_1\mathbf{c}_1 + \lambda_2\mathbf{c}_2 + \lambda_4\mathbf{c}_4 = \mathbf{0}$ in detail:

$$\lambda_1 \begin{pmatrix}1\\2\\1\end{pmatrix} + \lambda_2 \begin{pmatrix}2\\1\\2\end{pmatrix} + \lambda_4 \begin{pmatrix}2\\1\\1\end{pmatrix} = \begin{pmatrix}\lambda_1 + 2\lambda_2 + 2\lambda_4\\ 2\lambda_1 + \lambda_2 + \lambda_4\\ \lambda_1 + 2\lambda_2 + \lambda_4\end{pmatrix} = \begin{pmatrix}0\\0\\0\end{pmatrix}.$$

This is equivalent to

$$\begin{aligned} \lambda_1 + 2\lambda_2 + 2\lambda_4 &= 0\\ 2\lambda_1 + \lambda_2 + \lambda_4 &= 0\\ \lambda_1 + 2\lambda_2 + \lambda_4 &= 0. \end{aligned}$$

We easily get by elimination that this homogeneous system of equations has only a trivial solution, so **only** $\lambda_1 = \lambda_2 = \lambda_4 = 0$ is possible, therefore the vectors $\mathbf{c}_1$, $\mathbf{c}_2$, $\mathbf{c}_4$ are independent.

Of course, to prove the dependence of vectors $\mathbf{c}_1$, $\mathbf{c}_2$, $\mathbf{c}_3$ in Example E1 there is no need to guess the appropriate scalars. In this case we have to investigate the homogeneous system of equations

$$\begin{aligned} \lambda_1 + 2\lambda_2 + \lambda_3 &= 0\\ 2\lambda_1 + \lambda_2 + \lambda_3 &= 0\\ \lambda_1 + 2\lambda_2 + \lambda_3 &= 0. \end{aligned}$$

By elimination, we get the general solution $\lambda_1 = \lambda_2 = -\mu/3$, $\lambda_3 = \mu$ where $\mu \in \mathbf{R}$ is arbitrary. There exist non-trivial solutions, so the vectors are dependent, and any $\mu \neq 0$ provides a suitable non-trivial combination. The combination given above belongs to the value $\mu = -3$ of the free parameter.

All this holds in general, too. To decide the dependence or independence of the vectors $\mathbf{c}_1, \dots, \mathbf{c}_n \in F^k$, consider the homogeneous system $C\mathbf{x} = \mathbf{0}$ of linear equations where $\mathbf{c}_j$ are the column vectors of the matrix $C \in F^{k \times n}$. If this system of equations has only a trivial solution, then the vectors $\mathbf{c}_j$ are independent, and if there exists a non-trivial solution, too, then they are dependent. The elements of a non-trivial solution provide suitable scalars for the coefficients in a non-trivial linear combination representing the zero vector.

We can obtain all this by Gaussian elimination. In the very special case of k vectors, i.e., when the number of vectors is the same as the number of their components, we obtain a square matrix C. So, we can apply Theorem 3.2.3: the vectors are linearly dependent if and only if $\det C = 0$. (But this gives no information about the scalars in a non-trivial combination if the vectors turn out to be dependent.)

If there are more than k vectors, then the homogeneous system contains more unknowns than equations. By Theorem 3.1.4, it always has a non-trivial solution, so the vectors **must be** dependent. We shall use this simple fact many times, so we restate it as the following theorem.

Theorem 3.3.4. *More than k vectors in F^k are always linearly dependent.* ♣

A few simple facts immediately follow from the definitions of linear independence and dependence. A single vector is independent if and only if it is not the zero vector. Two vectors are independent if and only if neither of them is a scalar multiple of the other. This is *false* for more than two vectors; see, e.g., the vectors $\mathbf{c}_1$, $\mathbf{c}_2$, and $\mathbf{c}_3$ in Example E1 where none of them is a scalar multiple of another but they are dependent.

Some further simple consequences of the definitions are summarized in the theorem below. It is strongly recommended checking the proofs only after having tried to prove these statements alone.

Theorem 3.3.5.

I. *Deleting any vector from a linearly independent system of at least two vectors, the remaining system is linearly independent.*

II. *Extending a linearly dependent system by any vector, the new system is linearly dependent.*

III. *A system of at least two vectors is linearly dependent if and only if at least one of the vectors is a linear combination of the others.*

IV. *If extending a linearly independent system by a new vector we get a dependent system, then the new vector is a linear combination of the old ones.*

V. *Assume that a vector is a linear combination of a system of vectors. Then this representation is unique if and only if the system is linearly independent.* ♣

Proof. We prove II first. Assume that $\mathbf{c}_1, \dots, \mathbf{c}_m \in F^k$ are linearly dependent, i.e., there exist scalars $\lambda_1, \dots, \lambda_m \in F$ not all 0 satisfying $\lambda_1\mathbf{c}_1 + \cdots + \lambda_m\mathbf{c}_m = \mathbf{0}$. Extending the system with an arbitrary $\mathbf{c}_{m+1} \in F^k$, we have the non-trivial linear combination $\lambda_1\mathbf{c}_1 + \cdots + \lambda_m\mathbf{c}_m + 0\mathbf{c}_{m+1} = \mathbf{0}$, so the extended system is linearly dependent, as well.

To prove I by contradiction, assume that the remaining vectors are dependent. Adjoining the deleted vector to this system, we get by II that the original system is dependent, a contradiction.

In III we assume first that, say, $\mathbf{c}_m$ is a linear combination of the vectors $\mathbf{c}_1, \dots, \mathbf{c}_{m-1}$: $\mathbf{c}_m = \delta_1\mathbf{c}_1 + \cdots + \delta_{m-1}\mathbf{c}_{m-1}$. After ordering, we get $\mathbf{0} = \delta_1\mathbf{c}_1 + \cdots + \delta_{m-1}\mathbf{c}_{m-1} + (-1)\mathbf{c}_m$. This is a non-trivial representation of the zero vector as the coefficient of $\mathbf{c}_m$ is $-1 \neq 0$. For the converse, assume that the vectors $\mathbf{c}_i$ are dependent and take a non-trivial combination $\lambda_1\mathbf{c}_1 + \cdots + \lambda_m\mathbf{c}_m = \mathbf{0}$ where, say, $\lambda_1 \neq 0$. Expressing $\mathbf{c}_1$, we have $\mathbf{c}_1 = (-\lambda_2/\lambda_1)\mathbf{c}_2 + \cdots + (-\lambda_m/\lambda_1)\mathbf{c}_m$, so $\mathbf{c}_1$ is a linear combination of the other vectors. Observe that this is true for every vector having a non-zero coefficient in the linear combination representing the zero vector.

By the previous sentence, we are done with IV if we check that the coefficient of the new vector $\mathbf{c}_{m+1}$ is not 0 in a non-trivial combination of $\mathbf{c}_1, \dots, \mathbf{c}_m, \mathbf{c}_{m+1}$ expressing the zero vector. If this were not true, then

$$\mathbf{0} = \lambda_1 \mathbf{c}_1 + \cdots + \lambda_m \mathbf{c}_m + 0\mathbf{c}_{m+1} = \lambda_1 \mathbf{c}_1 + \cdots + \lambda_m \mathbf{c}_m$$

would imply that the zero vector is a non-trivial combination of the independent vectors $\mathbf{c}_1, \dots, \mathbf{c}_m$, a contradiction.

Finally, to verify V, observe that

$$\rho_1 \mathbf{c}_1 + \cdots + \rho_m \mathbf{c}_m = \pi_1 \mathbf{c}_1 + \cdots + \pi_m \mathbf{c}_m \iff (\rho_1 - \pi_1)\mathbf{c}_1 + \cdots + (\rho_m - \pi_m)\mathbf{c}_m = \mathbf{0}. \quad \blacksquare$$

To get a right interpretation of III, please note that in a dependent system there are generally *more* vectors that can be expressed as a linear combination of the others. So, in IV, generally, not only is the new vector a linear combination of the old ones, but there exist nearly always some old vectors that are linear combinations of the other old vectors *and* the new one (see Exercises 3.3.4–3.3.5).

Linear independence is a subtle notion and it is not easy to process it properly. Do not forget for example that you cannot grasp it by considering a linear combination with 0 scalars of the given vectors as this always results in the zero vector both for independent and dependent systems.

To get over the initial difficulties concerning linear independence, it is suggested checking every tiny detail based strictly on the definitions and confronting the emerging view with the situation among the vectors in the plane $\mathbf{R}^2$ or space $\mathbf{R}^3$ where we have a clearer geometric vision of the relations.

We shall introduce linear independence in a general vector space in Section 4.4. It will turn out, however, that all its essential features are present already in the special case just defined for F^k; on the one hand, everything can be transferred nearly literally to the general case, and on the other hand, we shall prove that every finite-dimensional vector space is essentially identical with some F^k (see Sections 4.7 and 5.2). Nevertheless or just that's why, linear independence plays a fundamental role in nearly all branches of mathematics.

Exercises 3.3

(See also Exercises 4.4.1–4.4.8.)

3.3.1. Decide whether the vectors in $\mathbf{R}^4$ are linearly dependent or independent. If dependent, express one of them as a linear combination of the others.

(a) $\begin{pmatrix} 1 \\ 2 \\ 3 \\ 4 \end{pmatrix}, \begin{pmatrix} -2 \\ 5 \\ 0 \\ 1 \end{pmatrix}, \begin{pmatrix} -1 \\ 4 \\ 1 \\ 2 \end{pmatrix};$ (b) $\begin{pmatrix} 1 \\ 2 \\ 3 \\ 4 \end{pmatrix}, \begin{pmatrix} -2 \\ 5 \\ 1 \\ 1 \end{pmatrix}, \begin{pmatrix} -1 \\ 4 \\ 1 \\ 2 \end{pmatrix}.$

3.3.2. Decide whether the vectors in $\mathbf{R}^4$ are linearly dependent or independent. Solve the analogous problem if the vectors are considered over the field F_3 instead of $\mathbf{R}$.

(a) $\begin{pmatrix}1\\1\\0\\0\end{pmatrix}, \begin{pmatrix}0\\0\\1\\1\end{pmatrix}, \begin{pmatrix}1\\0\\1\\0\end{pmatrix}, \begin{pmatrix}0\\1\\0\\1\end{pmatrix}$. (b) $\begin{pmatrix}1\\1\\1\\0\end{pmatrix}, \begin{pmatrix}0\\1\\1\\1\end{pmatrix}, \begin{pmatrix}1\\0\\1\\1\end{pmatrix}, \begin{pmatrix}1\\1\\0\\1\end{pmatrix}$.

(c) $\begin{pmatrix}1\\1\\1\\0\end{pmatrix}, \begin{pmatrix}0\\1\\1\\0\end{pmatrix}, \begin{pmatrix}0\\1\\0\\1\end{pmatrix}, \begin{pmatrix}1\\0\\1\\1\end{pmatrix}, \begin{pmatrix}0\\1\\0\\0\end{pmatrix}$.

3.3.3. True or false?

(a) If $\mathbf{c}_1, \dots, \mathbf{c}_5$ are linearly independent but $\mathbf{c}_1, \dots, \mathbf{c}_7$ are linearly dependent, then at least one of $\mathbf{c}_6$ and $\mathbf{c}_7$ is a linear combination of $\mathbf{c}_1, \dots, \mathbf{c}_5$.

(b) If some vector $\mathbf{c} \neq \mathbf{0}$ is a linear combination both of $\mathbf{c}_1, \mathbf{c}_2, \mathbf{c}_3$ and of $\mathbf{c}_4, \mathbf{c}_5, \mathbf{c}_6$, then $\mathbf{c}_1, \dots, \mathbf{c}_6$ are linearly dependent.

(c) If $\mathbf{c}_1, \dots, \mathbf{c}_6$ are non-zero linearly dependent vectors, then some vector $\mathbf{c} \neq \mathbf{0}$ is a linear combination both of $\mathbf{c}_1, \mathbf{c}_2, \mathbf{c}_3$ and of $\mathbf{c}_4, \mathbf{c}_5, \mathbf{c}_6$.

3.3.4. True or false?

(a) If $\lambda_3 \neq 0$ in the non-trivial linear combination $\lambda_1\mathbf{c}_1 + \cdots + \lambda_m\mathbf{c}_m = \mathbf{0}$, then $\mathbf{c}_3$ is a linear combination of the other vectors $\mathbf{c}_i$.

(b) If $\lambda_3 = 0$ in the non-trivial linear combination $\lambda_1\mathbf{c}_1 + \cdots + \lambda_m\mathbf{c}_m = \mathbf{0}$, then $\mathbf{c}_3$ is not a linear combination of the other vectors $\mathbf{c}_i$.

(c) If a system $\mathbf{c}_1, \dots, \mathbf{c}_m$ contains exactly d vectors that can be expressed as a linear combination of the other $m-1$ vectors, then the system contains $m-d$ linearly independent vectors.

(d) If a system $\mathbf{c}_1, \dots, \mathbf{c}_m$ contains exactly d vectors that can be expressed as a linear combination of the other $m-1$ vectors, then the system does not contain more than $m-d$ linearly independent vectors.

3.3.5. Assume that $\mathbf{c}_1, \dots, \mathbf{c}_m$ are linearly independent, $\mathbf{c}_1, \dots, \mathbf{c}_m, \mathbf{v}$ are linearly dependent, and no $\mathbf{c}_i$ is a linear combination of $\mathbf{v}$ and the other $\mathbf{c}_j$ vectors. Determine $\mathbf{v}$.

3.3.6. Assume that $\mathbf{c}_1$, $\mathbf{c}_2$, $\mathbf{c}_3$, $\mathbf{c}_4$ are linearly independent, $\mathbf{c}_5 \neq \mathbf{0}$, and $\mathbf{c}_1$, $\mathbf{c}_2$, $\mathbf{c}_5$ are linearly dependent. What can we say about the independence or dependence of $\mathbf{c}_3$, $\mathbf{c}_4$, $\mathbf{c}_5$? (There are three possible answers: the three vectors are necessarily independent; they are necessarily dependent; they can be independent, but can be dependent, too.)

3.3.7. (a) Do there exist five vectors such that the first three of them are linearly dependent but any other three of them are linearly independent?

(b) Do there exist five non-zero vectors such that the first three of them are linearly independent but any other three of them are linearly dependent?

3.3.8. The vectors $\mathbf{c}_1, \mathbf{c}_2, \mathbf{c}_3, \mathbf{c}_4 \in \mathbf{R}^8$ are linearly independent. Determine whether the systems below are dependent or independent.

(a) $\mathbf{c}_1 - \mathbf{c}_2, \quad \mathbf{c}_2 - \mathbf{c}_3, \quad \mathbf{c}_3 - \mathbf{c}_4;$

(b) $\mathbf{c}_1 - \mathbf{c}_2, \quad \mathbf{c}_2 - \mathbf{c}_3, \quad \mathbf{c}_3 - \mathbf{c}_1;$

(c) $\mathbf{c}_1 + \mathbf{c}_2, \quad \mathbf{c}_2 + \mathbf{c}_3, \quad \mathbf{c}_3 + \mathbf{c}_4, \quad \mathbf{c}_4 + \mathbf{c}_1;$

(d) $\mathbf{c}_1 + \mathbf{c}_2, \quad \mathbf{c}_2 + \mathbf{c}_3, \quad \mathbf{c}_3 + \mathbf{c}_4, \quad \mathbf{c}_4 + \mathbf{c}_2;$

(e) $\mathbf{c}_1 + \pi\mathbf{c}_2 + \sqrt{2}\mathbf{c}_3 + (\sin 1^\circ)\mathbf{c}_4, \qquad 100\mathbf{c}_1 + 77\mathbf{c}_2 + (3/11)\mathbf{c}_3 + \mathbf{c}_4,$

$\mathbf{c}_1 + 5^6\mathbf{c}_2 + \pi^2\mathbf{c}_3 - (1/8)\mathbf{c}_4, \qquad (\lg 3)\mathbf{c}_1 + 9999\mathbf{c}_2 + \mathbf{c}_3 + \mathbf{c}_4,$

$\mathbf{c}_1 + \mathbf{c}_2 + \mathbf{c}_3.$

Solve the analogous problem if $\mathbf{c}_1, \mathbf{c}_2, \mathbf{c}_3, \mathbf{c}_4$ are linearly dependent.

3.3.9. Form q linear combinations of m vectors. What can we say about the independence or dependence of the resulting q vectors if the original m vectors are (a) independent; (b) dependent, and (A) $q = m+1$; (B) $q = m$; (C) $q = m-1$? (This means six questions altogether. There are three possible answers to each: the q vectors are necessarily independent; they are necessarily dependent; they can be independent, but can be dependent, too.)

3.3.10. Let $\mathbf{c}_1, \mathbf{c}_2, \dots, \mathbf{c}_m$ be linearly independent vectors and let λ be a non-zero scalar in F. Verify that both $\lambda\mathbf{c}_1, \mathbf{c}_2, \dots, \mathbf{c}_m$ and $\mathbf{c}_1 + \lambda\mathbf{c}_2, \mathbf{c}_2, \dots, \mathbf{c}_m$ are linearly independent systems.

3.3.11. Prove that the linear independence or dependence of the column vectors in a matrix does not change if we perform elementary (a) column-equivalent; (b) row-equivalent transformations. (The latter mean the steps M1–M4 in Section 3.1 and the former are the corresponding variants for columns.)

*3.3.12. Consider 11 positive integers none of which has a prime divisor greater than 30. Prove that the product of some of these integers is a square. (The product can contain any number of factors from 1 to 11.)

3.4. The Rank of a Matrix

We define three types of ranks for matrices. Two of them are based on linear independence and the third one uses determinants. We show that, for any matrix, they are all equal and their common value equals the number of leaders in the reduced echelon form. So, in general, the simplest way to compute the rank is via elimination. At the end of the section we formulate conditions for solvability and for unique solutions of systems of linear equations with the help of the rank.

A matrix $A \in F^{k\times n}$ has n column vectors, i.e., n vectors in F^k. Similarly, the k row vectors can be interpreted as k vectors in F^n. (The row vectors are formally $1 \times n$ and not $n \times 1$ matrices, but in most cases it is not essential to distinguish $F^{1\times n}$ and $F^{n\times 1}$, so we shall denote both by F^n in general.) The row vectors of A are just the column vectors of the transpose A^T.

The column rank of a matrix is the maximal number of its linearly independent column vectors.

Definition 3.4.1/C. The *column rank* of a matrix A is r if we can select r linearly independent column vectors of A, but cannot select more than r columns with this property. ♣

This means that taking more than r columns arbitrarily, these vectors are necessarily dependent or there are only r columns in A altogether.

A matrix of column rank r often has not just one but several systems of r linearly independent columns, sometimes even any r of its columns can be independent (see Exercise 3.4.13).

Turning to the rows, the row rank means analogously the maximal number of linearly independent row vectors.

Definition 3.4.1/R. The *row rank* of a matrix A is r if we can select r linearly independent row vectors of A, but cannot select more than r rows with this property. ♣

It is easy to check that "more than r" can be replaced by "$r+1$" in both definitions (see Exercise 3.4.1).

As indicated above and to be proven later in this section, the two seemingly different notions of rank are equivalent.

Examples. In

$$\begin{pmatrix} 1 & 2 & 3 & 4 \\ 5 & 6 & 7 & 8 \\ 9 & 10 & 11 & 12 \end{pmatrix}$$

the column rank is 2, since, e.g., the first two columns are independent, but any three columns are dependent. This implies that the row rank is 2, as it can be checked directly, as well. The column or row rank of the zero matrix is 0 as any column or row is the zero vector being dependent in itself. The columns or rows of the $n \times n$ identity matrix I_n are linearly independent, so its rank is n. The column rank of a $k \times n$ matrix is at most n as there are no more columns, and it is at most k since the columns are from F^k where any $k+1$ vectors are dependent. So the rank cannot be greater than the minimum of the numbers of rows and columns.

To introduce the determinant rank we shall need the general notion of a minor (mentioned already in Exercise 1.4.15). A *minor* of order h in a matrix is the determinant of an arbitrary $h \times h$ square submatrix: we select (any) h rows and h columns of the matrix and consider the $h \times h$ determinant of the h^2 elements in the intersections of these rows and columns. The cofactors A_{ij} used in the expansion of an $n \times n$ determinant are the signed minors of order $n-1$.

The determinant rank of a matrix is the maximal order of its non-zero minors.

Definition 3.4.1/D. The *determinant rank* of a matrix A is r if there exists a non-zero minor of order r, but every minor of order greater than r is 0. ♣

We can replace "greater than r" by "$r+1$" here, too (see Exercise 3.4.1).

Similar to the situation with column or row ranks, a matrix of determinant rank r often has not just one but several non-zero $r \times r$ minors (see Exercise 3.4.14).

In the first matrix of the examples above, e.g., the 2×2 minor in the upper left corner is not 0, but every 3×3 minor is 0, so the determinant rank is 2 (conforming with the column and row ranks). Clearly, the determinant rank cannot exceed the number of rows or columns since there exist no minors of greater order. The determinant rank of a matrix and its transpose coincide since the minors of A^T are the transposes of the corresponding minors of A and transposing a determinant does not change its value.

Theorem 3.4.2. *In any matrix, the column, row, and determinant ranks are equal. This common value does not change during the Gaussian elimination and is equal to the number of leaders in the reduced echelon form.* ♣

This common value is called the *rank* of the matrix A and is denoted by $\mathrm{rk}(A)$.

Proof. For a matrix A, let $c(A)$, $r(A)$, and $d(A)$ denote (temporarily) the column, row, and determinant rank.

I. Assume that we have already proved $c(A) = d(A)$. Then $r(A) = d(A)$ follows using the transpose: $r(A) = c(A^T) = d(A^T) = d(A)$.

II. To prove $c(A) = d(A)$ we first show that elementary row-equivalent transformations do not change these ranks. Then it suffices to prove that they are both equal to the number of leaders in the REF obtained by elimination.

III. To determine the column rank we have to investigate the linear independence or dependence of certain columns. This means a homogeneous system of linear equations where the coefficient matrix is a submatrix formed of the given columns in the original matrix. The row-equivalent transformations do not modify the solutions, so there is no change in whether or not this homogeneous system has a non-trivial solution. Therefore the column rank of the matrix does not alter.

IV. We show that the determinant rank does not change by adding a scalar multiple of a row to another row. The verification of the same fact for the other row-equivalent transformations runs easier; we leave the proof to the Reader.

It is sufficient to prove that the determinant rank cannot increase. Then the determinant rank does not increase for the inverse of our transformation either since the inverse means subtracting the same scalar multiple of the row in question from the other row. As we get back to the original matrix and the rank has not increased in either transformation, the determinant rank must have remained the same in both steps.

Assume that (say) we obtained the matrix B by adding λ times the first row of A to its third row. To prove $d(B) \leq d(A)$ we have to show that if every $h \times h$ minor is 0 in A, then the same holds also in B. Consider any minor D of order h in B. If D does not contain the third row of B, then D is a minor of A, too, so it is 0. If D contains both the first and third rows of B, then subtracting λ times the first row from the third row we get a minor D' of A, so $D = D' = 0$. The last case is when D contains the third row of B but does not contain its first row. Assume that (say) D is composed of the first h

columns and of the 2nd, 3rd, ..., $h+1$st rows of B. Then

$$D = \begin{vmatrix} \alpha_{21} & \alpha_{22} & \dots & \alpha_{2h} \\ \alpha_{31}+\lambda\alpha_{11} & \alpha_{32}+\lambda\alpha_{12} & \dots & \alpha_{3h}+\lambda\alpha_{1h} \\ \vdots & \vdots & \ddots & \vdots \\ \alpha_{h+1,1} & \alpha_{h+1,2} & \dots & \alpha_{h+1,h} \end{vmatrix}$$

$$= \begin{vmatrix} \alpha_{21} & \alpha_{22} & \dots & \alpha_{2h} \\ \alpha_{31} & \alpha_{32} & \dots & \alpha_{3h} \\ \vdots & \vdots & \ddots & \vdots \\ \alpha_{h+1,1} & \alpha_{h+1,2} & \dots & \alpha_{h+1,h} \end{vmatrix} + \lambda \begin{vmatrix} \alpha_{21} & \alpha_{22} & \dots & \alpha_{2h} \\ \alpha_{11} & \alpha_{12} & \dots & \alpha_{1h} \\ \vdots & \vdots & \ddots & \vdots \\ \alpha_{h+1,1} & \alpha_{h+1,2} & \dots & \alpha_{h+1,h} \end{vmatrix}$$

$$= D_1 + \lambda D_2.$$

Here D_1 is an $h \times h$ minor in A and so is D_2, too, after swapping rows if necessary. Therefore, both D_1 and D are 0, thus $D = 0$.

V. Finally, we show that both the column and determinant ranks of a matrix A are equal to the number of leaders r in the REF. We can delete the zero rows, so we can assume that A has exactly r rows. Thus neither rank can exceed r. On the other hand, the columns of the leaders form an $r \times r$ identity submatrix I_r. Therefore, these columns are linearly independent and $\det(I_r) = 1 \neq 0$, so both ranks are r. ■

By the theorem just proved, Gaussian elimination is an effective method to compute the rank of a matrix. As rows and columns play symmetric roles concerning the rank, we infer that the rank is not affected by elementary column-equivalent transformations either (this can be verified directly, too). So, similar to determinants, we may also eliminate along columns, or we can even mix the two types of transformations. But we repeatedly call attention to staying away from manipulating columns when solving systems of linear equations.

Now we discuss the relation of the rank and the systems of linear equations.

Theorem 3.4.3. *The system of linear equations $A\mathbf{x} = \mathbf{b}$ is solvable if and only if $\text{rk}(A) = \text{rk}(A \mid \mathbf{b})$, i.e., the rank of the coefficient matrix equals the rank of the augmented matrix. If solvable, the solution is unique exactly if this common rank is equal to the number of unknowns.* ♣

Proof. We use the form $x_1\mathbf{a}_1 + \dots + x_n\mathbf{a}_n = \mathbf{b}$ of the system where $\mathbf{a}_j \in F^k$ are the column vectors of the coefficient matrix $A \in F^{k\times n}$.

I. We assume first that $\text{rk}(A) = \text{rk}(A \mid \mathbf{b}) = r$ and show that the system of equations is solvable. We pick r independent columns from A, let these be (say) $\mathbf{a}_1, \dots, \mathbf{a}_r$. The vectors $\mathbf{a}_1, \dots, \mathbf{a}_r, \mathbf{b}$ are dependent since there are at most $\text{rk}(A \mid \mathbf{b}) = r$ independent columns in the augmented matrix. By Theorem 3.3.5/IV, $\mathbf{b}$ is a linear combination of $\mathbf{a}_1, \dots, \mathbf{a}_r$. Extending this combination by the other column vectors with 0 coefficients, we obtain that $\mathbf{b}$ is a linear combination of all columns of A. This means just the solvability of the system of equations.

II. For the converse, we assume the solvability of the system of equations, so $\mathbf{b}$ is a linear combination of the column vectors $\mathbf{a}_j$:

$$\mathbf{b} = \alpha_1\mathbf{a}_1 + \dots + \alpha_n\mathbf{a}_n. \tag{3.4.1}$$

Let $\text{rk}(A) = r$; we prove that the rank of the augmented matrix $A \mid \mathbf{b}$ is r, too. Clearly $\text{rk}(A \mid \mathbf{b}) \geq \text{rk}(A) = r$, so it suffices to show that any $r+1$ columns of $A \mid \mathbf{b}$ are dependent. If $\mathbf{b}$ is not among the $r+1$ columns, this follows from $\text{rk}(A) = r$. So, we pick $\mathbf{b}$ and r columns of A, say $\mathbf{a}_1, \ldots, \mathbf{a}_r$. If $\mathbf{a}_1, \ldots, \mathbf{a}_r$ are dependent, then so are $\mathbf{a}_1, \ldots, \mathbf{a}_r, \mathbf{b}$ by Theorem 3.3.5/II. Assume now that $\mathbf{a}_1, \ldots, \mathbf{a}_r$ are independent. Then $\mathbf{a}_1, \ldots, \mathbf{a}_r, \mathbf{a}_j$ are dependent for every $j > r$ since $\text{rk}(A) = r$, so $\mathbf{a}_j$ is a linear combination of $\mathbf{a}_1, \ldots, \mathbf{a}_r$ by Theorem 3.3.5/IV. Replacing $\mathbf{a}_j$ in (3.4.1) by this representation, we obtain that $\mathbf{b}$ is a linear combination of just the columns $\mathbf{a}_1, \ldots, \mathbf{a}_r$. But then $\mathbf{a}_1, \ldots, \mathbf{a}_r, \mathbf{b}$ are necessarily dependent, and we are done.

III. If the system of equations is solvable, then a unique solution means that $\mathbf{b}$ is a unique linear combination of the columns in A. By Theorem 3.3.5/V this is equivalent to the linear independence of the columns $\mathbf{a}_j$, so $\text{rk}(A)$ equals the number of columns, i.e., the number of unknowns. (We could have referred to Theorem 3.1.1/III as an alternative justification.) ■

Exercises 3.4

3.4.1. Verify that if any $r+1$ columns in a matrix are linearly dependent, then so are any greater number of columns. Similarly, if every minor of order $r+1$ is 0, then the same holds for every minor of larger order.

3.4.2. How many $h \times h$ minors are there in a $k \times n$ matrix?

3.4.3. Compute the ranks.

$$\text{(a)} \begin{pmatrix} 1 & 2 & 3 \\ 2 & 5 & 6 \\ 3 & 5 & 9 \\ 0 & 1 & 0 \end{pmatrix}; \qquad \text{(b)} \begin{pmatrix} 1 & 3 & 9 \\ 2 & 4 & 8 \\ 9 & 3 & 1 \\ 8 & 4 & 2 \end{pmatrix}; \qquad \text{(c)} \begin{pmatrix} 1 & -2 & 3 \\ -3 & 6 & -9 \\ 2 & -4 & 6 \\ -4 & 8 & -12 \end{pmatrix}.$$

3.4.4. Let $\gamma_1, \ldots, \gamma_k$ and $\delta_1, \ldots, \delta_n$ be arbitrary scalars. Compute the rank of a $k \times n$ matrix where the jth element of the ith row is (a) $\gamma_i \delta_j$; (b) $\gamma_i + \delta_j$.

3.4.5. What is the rank of a matrix where the rows are geometric sequences with distinct non-zero quotients?

3.4.6. (a) Prove that modifying a single element in a matrix changes the rank by at most 1.

(b) Can we modify a suitable element in every matrix so that the rank changes?

(c) Can we modify a suitable element in every 10×20 matrix of rank 5 so that the rank decreases?

(d) Can we modify a suitable element in every 10×20 matrix of rank 5 so that the rank increases?

3.4.7. Prove that the rank of a $k \times n$ matrix is 1 if and only if it is the product of a non-zero column vector in $F^{k \times 1}$ and a non-zero row vector in $F^{1 \times n}$ using matrix multiplication. This is the reason why matrices of rank 1 (and the zero matrix) are called *dyadic products* or simply *dyads*.

3.4.8. If every element of a matrix is 0 or 1, then these can be considered as real numbers, or rational numbers, or the elements of the modulo 2 field F_2. So we get three distinct matrices in $\mathbf{R}^{k\times n}$, $\mathbf{Q}^{k\times n}$, and $F_2^{k\times n}$, respectively. What is the relation of the ranks of these three matrices?

3.4.9. A is a 7×6 matrix and B is the 4×6 submatrix consisting of the first 4 rows of A. True or false?

(a) If the first three columns of B are linearly independent, then so are the first three columns of A.

(b) If the first three columns of A are linearly independent, then so are the first three columns of B.

3.4.10. Prove that if in a matrix both the rows and columns are linearly independent, then it is a square matrix.

3.4.11. Let A be a 6×5 matrix. True or false?

(a) If the first 3 rows are linearly dependent, then the upper left 3×3 minor is 0.

(b) If the upper left 3×3 minor is 0, then the first 3 rows are linearly dependent.

(c) If both the first 3 columns and the last 3 columns are linearly dependent, then the rank of the matrix is at most 3.

(d) If both the first 2 columns and the last 2 columns are linearly dependent, then the rank of the matrix is at most 3.

3.4.12. Verify that two matrices of the same form and over the same field have equal ranks if and only if we can convert one of them into the other by elementary row– and column-equivalent transformations.

3.4.13. (a) Find a 5×7 matrix of rank 4 with exactly 8 possibilities to select 4 independent columns.

(b) Prove that if the rank of a matrix is r and there is only one way to select r independent columns, then the other columns are zero vectors.

*(c) Let k, n be arbitrary and let $1 \leq r \leq \min(n, k)$. Show a $k \times n$ matrix of rank r where any r columns are independent.

3.4.14. (a) Find a 5×8 matrix of rank 3 with exactly 60 non-zero minors of order 3.

(b) Prove that if the rank of a matrix is r and there is only one non-zero minor of order r, then all elements of the matrix outside this minor are 0.

*(c) Let k, n be arbitrary and let $1 \leq r \leq \min(n, k)$. Show a $k \times n$ matrix of rank r where no $r \times r$ minor is 0.

3.4.15. True or false?

(a) If the system of equations $A\mathbf{x} = \mathbf{b}$ is solvable, then the columns of the augmented matrix $A \mid \mathbf{b}$ are linearly dependent.

(b) If the columns of the augmented matrix $A \mid \mathbf{b}$ are linearly dependent, then the system of equations $A\mathbf{x} = \mathbf{b}$ is solvable.

(c) If the columns of the coefficient matrix A are linearly independent, then the system of equations $A\mathbf{x} = \mathbf{b}$ is solvable.

(d) If the rows of the coefficient matrix A are linearly independent, then the system of equations $A\mathbf{x} = \mathbf{b}$ is solvable.

(e) If the system of equations $A\mathbf{x} = \mathbf{b}$ has a unique solution, then the columns of the coefficient matrix A are linearly independent.

(f) If the system of equations $A\mathbf{x} = \mathbf{b}$ has a unique solution, then the rows of the coefficient matrix A are linearly independent.

3.4.16. A matrix A has 10 rows and these are linearly independent. The system of equations $A\mathbf{x} = \mathbf{b}$ has exactly 13 solutions. How many columns are there in A?

*3.4.17. Let $n > 1$ and $A \in F^{n\times n}$ have a 0 determinant.

(a) Verify $\mathrm{rk}(\hat{A}) \leq 1$ where $\hat{A}$ is the adjugate of A (see Lemma 2.2.3).

(b) (Cf. with Exercise 2.2.10.) Prove that the adjugate of the adjugate of A is the zero matrix if $n > 2$ and it is equal to A if $n = 2$.

***3.4.18.** Is every real matrix the sum of such matrices where

(a) every row is an arithmetic sequence;

(b) every row or every column is an arithmetic sequence;

(c) every row is a geometric sequence?

3.4.19. (a) P and R play a game. P presents a system of k linear equations with n unknowns over the real numbers (k and n are fixed positive integers). R has to transform this into a solvable system by repeatedly selecting a coefficient or a constant and replacing it by any other number. P wants to maximize the number of steps and R wants to minimize it. How many steps are needed if they both play optimally?

*(b) Solve the analogous problem if R has to produce a system with no solution.

3.5. Singular and Non-singular Matrices

As an application of the rank and systems of equations, we extend considerably the results of Section 2.2 about the invertibility of matrices.

Definition 3.5.1. A square matrix is *singular* if its determinant is 0, and is *non-singular* if its determinant is not 0. ♣

We have proven several equivalent conditions for a matrix to be singular or non-singular, we summarize these first.

Theorem 3.5.2. *The non-singularity of a matrix $A \in F^{n\times n}$ is equivalent to any of the conditions below:*

(D) $\det A \neq 0$;

(I) *A has a two-sided inverse;*

(LI) *A has a left inverse;*

(RI) *A has a right inverse;*

(NLZ) *A is not zero and is not a left zero divisor;*

(NRZ) *A is not zero and is not a right zero divisor;*

(T) *the homogeneous system of equations $A\mathbf{x} = \mathbf{0}$ has only a trivial solution;*

(EU) *there exists some* $\mathbf{b}(\in F^n)$ *such that the system of equations* $A\mathbf{x} = \mathbf{b}$ *has a unique solution;*

(AU) *for any* $\mathbf{b}(\in F^n)$*, the system of equations* $A\mathbf{x} = \mathbf{b}$ *has a unique solution;*

(RK) $\mathrm{rk}(A) = n$*;*

(CLI) *the columns of A are linearly independent;*

(RLI) *the rows of A are linearly independent.* ♣

Proof. Condition (D) is the definition of non-singularity. Its equivalence to the other conditions was given in the following theorems:

(I), (LI), (RI): Theorem 2.2.2.

(NLZ), (NRZ): Theorem 2.2.5.

(T): Theorem 3.2.3.

(EU), (AU): Theorems 3.2.1 and 3.2.2.

(RK), (CLI), (RLI): Theorem 3.4.2. ■

We suggest the Reader formulate the complementary conditions equivalent to the singularity of a matrix.

Now we discuss a direct connection between the inverse of a matrix and systems of equations. It will lead to a new proof of Theorem 2.2.2 and make possible to compute the inverse of a matrix via elimination. This method is much quicker in general than applying the formula derived in Theorem 2.2.2.

To determine the right inverse X of a matrix $A \in F^{n\times n}$ we have to solve the matrix equation $AX = I_n$ where I_n is the $n \times n$ identity matrix. Let $\mathbf{x}_1, \dots, \mathbf{x}_n$ denote the columns of X and let $\mathbf{u}_1, \dots, \mathbf{u}_n$ denote the columns of I_n, so $\mathbf{u}_j$ is the jth *unit vector* where the jth component is 1 and the other components are 0. Then $AX = I_n$ is equivalent to $A\mathbf{x}_1 = \mathbf{u}_1, \dots, A\mathbf{x}_n = \mathbf{u}_n$. Thus we have to solve these n systems of equations where the coefficient matrix of every system is A.

If $\det A \neq 0$, then each system of equations has a (unique) solution by the Remark after Theorem 3.2.2. (We avoid the reference to Theorem 3.2.1 as its proof uses Theorem 2.2.2.) So, A has a right inverse in this case.

If $\det A = 0$, then we show that $A\mathbf{x} = \mathbf{u}_j$ is not solvable for at least one j, so A cannot have a right inverse. For a proof by contradiction, we assume that each system is solvable and $\mathbf{c}_j$ is a solution of $A\mathbf{x} = \mathbf{u}_j$, $j = 1, 2, \dots, n$. Then $A\mathbf{x} = \mathbf{b}$ is solvable for every $\mathbf{b} \in F^n$ as $\sum_{j=1}^{n} \beta_j \mathbf{c}_j$ is a solution where $\beta_1, \dots, \beta_n$ are the components of $\mathbf{b}$. On the other hand, $\det A = 0$ implies that the rows of A are linearly dependent. Therefore, some row of the coefficient matrix is a linear combination of the other rows. If the system $A\mathbf{x} = \mathbf{b}$ is solvable, then the same relation must hold for the corresponding constants β_j. This cannot hold for every $\mathbf{b}$, a contradiction.

The left inverse Y is a solution of the matrix equation $YA = I_n$. This is equivalent to its transpose $A^T Y^T = I_n$ and we can apply the previous argument. This concludes the new proof of Theorem 2.2.2.

Let us see how we can compute the inverse of a matrix in practice. The systems of equations $A\mathbf{x}_1 = \mathbf{u}_1, \dots, A\mathbf{x}_n = \mathbf{u}_n$ have the same coefficient matrix, so we can handle them simultaneously. We write (a single copy of) A followed by the vectors $\mathbf{u}_1, \dots, \mathbf{u}_n$ after the vertical line. So, we put I_n after A: $A \mid \mathbf{u}_1 \dots \mathbf{u}_n = A \mid I_n$. We now perform the elimination. If $\det A \neq 0$, then the reduced echelon form of A will be I_n and the n unit vectors on the right-hand side will turn into the solutions of the corresponding systems of equations, so we obtain A^{-1}. If $\det A = 0$, then the last row of the reduced echelon form of A is zero producing a forbidden row at least at one of the n systems of equations, so A has no inverse. There is no need to find out in advance whether or not $\det A$ is zero; this gets revealed automatically by the elimination. We summarize the procedure in the following theorem.

Theorem 3.5.3. *Write the identity matrix I_n after $A \in F^{n\times n}$: $A \mid I_n$. A is invertible if and only if elimination transforms $A \mid I_n$ into $I_n|B$ and then $B = A^{-1}$.* ♣

Similar to the above, we can connect zero divisors and systems of equations. A is a left zero divisor if and only if $A \neq 0$ and the matrix equation $AX = 0$ has a solution $X \neq 0$. Denoting the columns of X by $\mathbf{x}_1, \dots, \mathbf{x}_n$ we can rewrite $AX = 0$ as $A\mathbf{x}_1 = \mathbf{0}$, $\dots$, $A\mathbf{x}_n = \mathbf{0}$. This means n identical copies of the homogeneous system of equations $A\mathbf{x} = \mathbf{0}$. Therefore $A \neq 0$ is a left zero divisor exactly if $A\mathbf{x} = \mathbf{0}$ has a non-trivial solution. By Theorem 3.2.3, this is equivalent to $\det A = 0$. For the right zero divisors we can apply the trick of transposing used at the inverse. This yields a new proof to Theorem 2.2.5.

We do not have to compute the determinant in this case either as Gaussian elimination yields whether or not $A\mathbf{x} = \mathbf{0}$ has a non-trivial solution. Entering (trivial and eventual non-trivial) solutions independently into the columns of X, we obtain all solutions of the matrix equation $AX = 0$. These are $X = 0$ and all right zero divisor pairs of A if A is a left zero divisor.

A third proof to Theorem 2.2.5 uses the matrix rank. We know from the previous proof that $A \neq 0$ is a left zero divisor if and only if $A\mathbf{x} = \mathbf{0}$ has a non-trivial solution. This condition means that the columns of A are linearly dependent, so the column rank of A is less than n. But then the same holds for the determinant rank, so this condition is equivalent to $\det A = 0$.

Exercises 3.5

3.5.1. Determine the inverses of the 4×4 real matrices.

$$\text{(a)} \begin{pmatrix} 1 & 0 & 1 & 0 \\ 0 & 1 & 0 & 1 \\ 1 & 0 & 0 & 1 \\ 0 & 1 & 1 & 1 \end{pmatrix}; \qquad \text{(b)} \begin{pmatrix} 1 & 1 & 1 & 1 \\ 1 & 2 & 3 & 4 \\ 1 & 3 & 6 & 10 \\ 1 & 4 & 10 & 20 \end{pmatrix}.$$

3.5.2. Determine the inverses of the $n \times n$ real matrices.

$$\text{(a) } A = \begin{pmatrix} 1 & 1 & 1 & 1 & \dots & 1 \\ 1 & 2 & 1 & 1 & \dots & 1 \\ 1 & 1 & 2 & 1 & \dots & 1 \\ 1 & 1 & 1 & 2 & \dots & 1 \\ \vdots & \vdots & \vdots & \vdots & \ddots & \vdots \\ 1 & 1 & 1 & 1 & \dots & 2 \end{pmatrix}, \text{ i.e., } \alpha_{ij} = \begin{cases} 2, & \text{if } i = j \geq 2; \\ 1, & \text{otherwise.} \end{cases}$$

$$\text{(b) } B = \begin{pmatrix} 1 & 2 & 3 & 4 & \dots & n \\ 0 & 1 & 2 & 3 & \dots & n-1 \\ 0 & 0 & 1 & 2 & \dots & n-2 \\ 0 & 0 & 0 & 1 & \dots & n-3 \\ \vdots & \vdots & \vdots & \vdots & \ddots & \vdots \\ 0 & 0 & 0 & 0 & \dots & 1 \end{pmatrix}, \text{ i.e., } \beta_{ij} = \begin{cases} j - i + 1, & \text{if } i \leq j; \\ 0, & \text{otherwise.} \end{cases}$$

$$\text{(c) } C = \begin{pmatrix} 1 & 1 & 1 & 1 & \dots & 1 \\ 1 & 2 & 2 & 2 & \dots & 2 \\ 1 & 2 & 3 & 3 & \dots & 3 \\ 1 & 2 & 3 & 4 & \dots & 4 \\ \vdots & \vdots & \vdots & \vdots & \ddots & \vdots \\ 1 & 2 & 3 & 4 & \dots & n \end{pmatrix}, \text{ i.e., } \gamma_{ij} = \min(i, j).$$

3.5.3. Find all right and left zero divisor pairs of the real matrices A and B, i.e., all 4×4 non-zero matrices X, Y, Z, and W satisfying $AX = YA = BZ = WB = 0$.

$$\text{(a) } A = \begin{pmatrix} 1 & 0 & 1 & 0 \\ 0 & 1 & 0 & 1 \\ 1 & 0 & 0 & 1 \\ 0 & 1 & 1 & 0 \end{pmatrix}; \qquad \text{(b) } B = \begin{pmatrix} 1 & 1 & 1 & 1 \\ 1 & 2 & 3 & 4 \\ 1 & 3 & 5 & 7 \\ 2 & 3 & 4 & 5 \end{pmatrix}.$$

3.5.4. The sum of elements in each row is 0 in an $n \times n$ matrix $A \neq 0$. Verify that A is a zero divisor and find a matrix $B \neq 0$ satisfying $AB = 0$.

3.5.5. Find all integers n such that the $n \times n$ real matrix A_n is invertible or is a zero divisor, respectively. Compute its inverse or find all its right and left zero divisor pairs, i.e., all $B, C \in \mathbf{R}^n$ satisfying $A_n B = C A_n = 0$.

$$A_n = \begin{pmatrix} 1 & 1 & 0 & 0 & \dots & 0 \\ 0 & 1 & 1 & 0 & \dots & 0 \\ 0 & 0 & 1 & 1 & \dots & 0 \\ 0 & 0 & 0 & 1 & \dots & 0 \\ \vdots & \vdots & \vdots & \vdots & \ddots & \vdots \\ 1 & 0 & 0 & 0 & \dots & 1 \end{pmatrix}, \text{ i.e., } \alpha_{ij} = \begin{cases} 1, & \text{if } i \leq j \leq i+1 \\ & \text{or } i = n, j = 1; \\ 0, & \text{otherwise.} \end{cases}$$

3.5.6. (a) Let $A \in F^{n \times n}$ and let $\mathbf{u}_j$ be the jth unit vector, i.e., the jth column of the identity matrix I_n. Prove that if exactly m of the n systems of equations $A\mathbf{x} = \mathbf{u}_j$ are solvable, then $\mathrm{rk}(A) \geq m$.

(b) Verify that (a) does not hold with equality in general: show a matrix A of rank $n - 1$ where no system of equations $A\mathbf{x} = \mathbf{u}_j$ is solvable.

3.5.7. Let $A \in F^{n \times n}$, $\det A = 0$, and assume that the systems of equations $A\mathbf{x} = \mathbf{0}$ and $A^T\mathbf{x} = \mathbf{0}$ have exactly the same solutions. Does this imply that A is a *symmetric* matrix, i.e., $A^T = A$?

3.5.8. Let $A \in F^{n \times n}$, $A \neq 0$, $\det A = 0$.

(a) Show that the right and left zero divisor pairs belonging to A are not necessarily the same, i.e., $AB = 0$ and $BA = 0$ do not imply each other (where $B \in F^{n \times n}$).

*(b) Prove the existence of a non-zero $B \in F^{n \times n}$ satisfying $AB = BA = 0$.

***(c)** Find all matrices A satisfying $AB = 0 \iff BA = 0$.

Chapter 4

Vector Spaces

Linear algebra developed from the theory of systems of linear equations. As discussed in the previous chapter, certain properties of vectors in F^k play an important role in this topic. In the present chapter, we introduce the vector space as a new algebraic structure that provides a general and abstract approach to handle such questions. The results go far beyond systems of equations and have many applications in various parts of mathematics. We discuss some of them in Chapters 9 and 10.

4.1. Vector Space

Let F be a commutative field (see Definition A.5.1). The most important examples are $\mathbf{R}$, $\mathbf{C}$, $\mathbf{Q}$ (the fields of the real, complex, and rational numbers), and the modulo p field F_p where p is a prime number.

The notion of vector space generalizes the properties of the operations of addition and multiplication by scalars to the abstract setting, extending this structure beyond vectors in the plane, space, or in F^k.

Definition 4.1.1. A non-empty set V is a *vector space* over the field F if the requirements (called *axioms*) below hold.

(A) An operation called *addition* is defined on the set V: to any pair of $\mathbf{v}, \mathbf{w} \in V$ we assign a unique element in V denoted by $\mathbf{v} + \mathbf{w}$.

(A1) Addition is *associative*: any three elements $\mathbf{v}, \mathbf{w}, \mathbf{z} \in V$ satisfy

$$(\mathbf{v} + \mathbf{w}) + \mathbf{z} = \mathbf{v} + (\mathbf{w} + \mathbf{z}).$$

(A2) Addition is *commutative*: any two elements $\mathbf{v}, \mathbf{w} \in V$ satisfy

$$\mathbf{v} + \mathbf{w} = \mathbf{w} + \mathbf{v}.$$

(A3) There exists a *zero element* $\mathbf{0} \in V$: every $\mathbf{v} \in V$ satisfies

$$\mathbf{0} + \mathbf{v} = \mathbf{v} + \mathbf{0} = \mathbf{v}.$$

(A4) Every element has a *negative* (or *additive inverse*): to every $\mathbf{v} \in V$ there exists an element $-\mathbf{v} \in V$ satisfying

$$\mathbf{v} + (-\mathbf{v}) = (-\mathbf{v}) + \mathbf{v} = \mathbf{0}.$$

(MS) An operation called *multiplication by scalars* is defined between the field F and the set V: to every pair $\lambda \in F$ and $\mathbf{v} \in V$ we assign a unique element in V denoted by $\lambda\mathbf{v}$.

(MS1) All $\lambda, \mu \in F$ and $\mathbf{v} \in V$ satisfy

$$(\lambda + \mu)\mathbf{v} = \lambda\mathbf{v} + \mu\mathbf{v}.$$

(MS2) All $\lambda \in F$ and $\mathbf{v}, \mathbf{w} \in V$ satisfy

$$\lambda(\mathbf{v} + \mathbf{w}) = \lambda\mathbf{v} + \lambda\mathbf{w}.$$

(MS3) All $\lambda, \mu \in F$ and $\mathbf{v} \in V$ satisfy

$$(\lambda\mu)\mathbf{v} = \lambda(\mu\mathbf{v}).$$

(MS4) Every $\mathbf{v} \in V$ satisfies

$$1\mathbf{v} = \mathbf{v}$$

where 1 is the identity element in the field F (i.e., $1\lambda = \lambda 1 = \lambda$ for every $\lambda \in F$). ♣

The elements of V are called *vectors*, so the word vector can mean polynomials, matrices, functions, etc. depending on the set V. The elements of the field F are called *scalars*. We denote vectors by lower case bold Roman letters and scalars by lower case Greek letters. The multiplication by scalars is often phrased as *scalar multiplication*, but we do not use this form to avoid confusion with the dot product (scalar product, inner product) of vectors in the plane, in the space, or in general Euclidean spaces. If the field is $\mathbf{R}$ or $\mathbf{C}$, then we speak about real or complex vector spaces.

To determine a vector space we have to specify the set V of the vectors, the field F of the scalars, and to define the two operations, addition and multiplication by scalars. Then we have to verify that these meet the requirements of the axioms (A1)–(A4) and (MS1)–(MS4).

Remarks on the axioms of vector spaces

Addition is an ordinary operation, i.e., a function $V \times V \to V$. Multiplication by scalars is a somewhat unusual operation; it is a function $F \times V \to V$.

To summarize the properties of addition, V is a commutative group under this operation (see Section A.8).

Axiom (MS1) resembles formally a distributive law, but the two $+$ signs denote two different operations: there is an addition in F on the left-hand side and an addition in V on the right-hand side. A similar problem arises in Axiom (MS3).

Multiplication by scalars only means $\lambda\mathbf{v}$; there is no need for $\mathbf{v}\lambda$. If somebody insists on using the latter, too, then it can either be considered as an alternative notation for $\lambda\mathbf{v}$, or introduced as a new operation together with the extra axiom $\lambda\mathbf{v} = \mathbf{v}\lambda$.

It is not necessary to require *in advance* that V is not the empty set as this is guaranteed by Axiom (A3). Still, we shall not omit the adjective non-empty in similar cases to emphasize that an algebraic structure is a non-empty set in general.

Axiom (MS4) does *not require* the existence of a new element. We *know* that the field F contains an identity element 1 such that multiplying it by any *scalar* $\lambda \in F$, the result will be the same scalar λ. Axiom (MS4) states that multiplying any *vector* $\mathbf{v} \in V$ by this *given* scalar 1, the result will be the same vector $\mathbf{v}$.

The above system of axioms of vector spaces follows the traditional trend. Note that Axiom (A2) could have been omitted as it can be deduced from the other axioms (see Exercise 4.1.13). The other axioms are independent (see Exercise 4.1.14).

ATTENTION, please! Multiplication of two vectors is *not defined* in a vector space in general. We shall study some *special* vector spaces later where we also introduce some type of multiplication for two vectors: these are *algebras* (see Section 5.6) and *Euclidean spaces* (see Chapter 8).

Examples of vector spaces

E1 The plane or space vectors starting from the origin form a real vector space under the ordinary vector addition and multiplication of a vector by real numbers.

E2 F^k is a vector space over the field F if the operations are performed componentwise. The previous example is the special case $F = \mathbf{R}$ and $k = 2$ or $k = 3$.

E3 The set $F^{k\times n}$ of the $k \times n$ matrices is a vector space over F under the usual addition and multiplication by scalars of matrices. The previous example is the special case $n = 1$.

E4 The polynomials with coefficients in F form the vector space $F[x]$ over F under the usual operations.

E5 The $\mathbf{R} \to \mathbf{R}$ functions form a real vector space under the usual operations $f + g : \alpha \mapsto f(\alpha) + g(\alpha)$ and $\lambda f : \alpha \mapsto \lambda f(\alpha)$.

E6 Infinite sequences of real numbers form a real vector space with the usual operations.

E7 The complex numbers $\mathbf{C}$ form a vector space over $\mathbf{R}$ using the operations in $\mathbf{C}$.

For further examples see Exercises 4.1.1–4.1.4.

Consequences of the axioms

It follows from the general properties of operations (see Section A.4) that the zero vector $\mathbf{0}$ and the negative of a vector are *unique*. Further, we can perform *subtraction*, i.e., to every pair of vectors $\mathbf{v}, \mathbf{w} \in V$ there exists a unique vector $\mathbf{z} \in V$ satisfying $\mathbf{w} + \mathbf{z} = \mathbf{v}$, namely $\mathbf{z} = \mathbf{v} + (-\mathbf{w})$. We denote $\mathbf{z}$ by $\mathbf{v} - \mathbf{w}$.

When adding more than two terms, we can omit the parentheses and order the terms arbitrarily as addition is associative and commutative.

Axioms (MS1)–(MS3) resembling formally the distributive and associative laws guarantee that multiplication by scalars obeys the usual rules, e.g., when multiplying a sum of vectors by a sum of scalars.

Some further simple but important consequences of the axioms are stated in the following theorem.

Theorem 4.1.2.

(i) $\lambda\mathbf{0} = \mathbf{0}$ *for every* $\lambda \in F$.

(ii) $0\mathbf{v} = \mathbf{0}$ *for every* $\mathbf{v} \in V$ *where the scalar* 0 *is the zero element of the field* F.

(iii) $(-1)\mathbf{v} = -\mathbf{v}$ *for every* $\mathbf{v} \in V$ *where the scalar* -1 *is the negative of the identity element in the field* F.

(iv) *If* $\lambda\mathbf{v} = \mathbf{0}$*, then* $\lambda = 0$ *or* $\mathbf{v} = \mathbf{0}$. ♣

Proof. We verify (i), the other statements can be proven similarly (see Exercise 4.1.10). Let $\mathbf{v} \in V$ be arbitrary. Then $\mathbf{v} + \mathbf{0} = \mathbf{v}$ by (A3). Multiplying this equality by the scalar λ we obtain $\lambda(\mathbf{v} + \mathbf{0}) = \lambda\mathbf{v}$. We rewrite the left-hand side using (MS2):

$$\lambda\mathbf{v} + \lambda\mathbf{0} = \lambda\mathbf{v}.$$

Adding the negative of $\lambda\mathbf{v}$ to both sides, we get $\mathbf{0}$ on the right-hand side and

$$-\lambda\mathbf{v} + (\lambda\mathbf{v} + \lambda\mathbf{0}) = (-\lambda\mathbf{v} + \lambda\mathbf{v}) + \lambda\mathbf{0} = \mathbf{0} + \lambda\mathbf{0} = \lambda\mathbf{0}$$

on the left-hand side. ∎

Exercises 4.1

4.1.1. Which sets of polynomials with real coefficients form a real vector space under the usual operations? We denote a general polynomial by f, its degree by $\deg f$, the coefficient of the term of degree i by α_i, and its leading coefficient by α_n (so $\alpha_n \neq 0$ if f is not the zero polynomial). We do not distinguish polynomials and polynomial functions in notation.

(a) $\{f \mid \deg f = 100 \text{ or } f = 0\}$;

(b) $\{f \mid \deg f \leq 100 \text{ or } f = 0\}$;

(c) $\{f \mid \deg f \geq 100 \text{ or } f = 0\}$;

(d) $\{f \mid x^3 + 1 \text{ divides } f\}$;

(e) $\{f \mid \text{the remainder of } f \text{ is a constant when divided by } x^3 + 1\}$;

(f) $\{f \mid f(5) = 0\}$;

(g) $\{f \mid f(5) = 1\}$;

(h) $\{f \mid f(3) = 2f(4)\}$;

(i) $\{f \mid \text{the sum of the coefficients of } f \text{ is } 0\}$;

(j) $\{f \mid \alpha_0 + \alpha_1 = 0\}$;

(k) $\{f \mid \alpha_0 + \alpha_n = 0\}$;

(l) $\{f \mid f \text{ has a real root}\}$;

(m) $\{f \mid \text{every coefficient of } f \text{ is a rational number}\}$.

4.1.2. Which sets of real infinite sequences form a real vector space under the usual operations? We denote a general sequence by $S = (\alpha_0, \dots, \alpha_n, \dots)$.

(a) $\{S \mid \alpha_0 = 2\alpha_3 + \alpha_5\}$;
(b) $\{S \mid \alpha_0 = 2\alpha_3\alpha_5\}$;
(c) $\{S \mid \alpha_{n+1} = \alpha_n + \alpha_{n-1},\ n = 1, 2, \dots\}$;
(d) bounded sequences;
(e) convergent sequences;
(f) $\{S \mid \lim_{n\to\infty} \alpha_n = 999\}$;
(g) non-decreasing sequences;
(h) monotone sequences;
(i) $\{S \mid \alpha_i = 0$ for infinitely many $i\}$;
(j) $\{S \mid \alpha_i = 0$ except for at most finitely many $i\}$;
(k) $\{S \mid \alpha_i = 0$ except for at most 100 values of $i\}$;
(l) $\{S \mid \alpha_i = 0$ except for at most the first 100 values of $i\}$;
(m) arithmetic sequences;
(n) geometric sequences (including the 0 sequence);
(o) periodic sequences.

4.1.3. Which sets of real functions defined on all real numbers form a real vector space under the usual operations? We denote a general function by f.

(a) Continuous functions;
(b) functions with at most finitely many discontinuity points;
(c) functions with at most five discontinuity points;
(d) $\{f \mid f$ has a root$\}$;
(e) $\{f \mid f$ has at most finitely many roots$\}$;
(f) even functions;
(g) polynomial functions;
(h) periodic functions;
(i) functions with an upper bound;
(j) $\{f \mid f(5) \geq 0\}$;
(k) $\{f \mid f(5) = f(8)\}$;
(l) $\{f \mid \exists a \neq b\ f(a) = f(b)\}$;
(m) $\{f \mid f(\pi)$ is an integer$\}$.

4.1.4. Generalize the vector spaces of Examples E5, E6, and E7.

4.1.5. Let V be the set of positive numbers, $F = \mathbf{R}$, and define the addition $\oplus$ and multiplication by scalars $\odot$ as

$$v \oplus w = vw, \quad \lambda \odot v = v^\lambda$$

where the usual multiplication and exponentiation of real numbers appear on the right-hand sides of the equalities ($v, w \in V, \lambda \in F$). Is this a vector space?

4.1.6. Let $V = \mathbf{C}$, $F = \mathbf{Q}$, and define the addition $\oplus$ and multiplication by scalars $\odot$ as

$$v \oplus w = v + w + 1, \quad \lambda \odot v = \lambda v + \lambda - 1$$

where the usual addition and multiplication of complex numbers appear on the right-hand sides of the equalities ($v, w \in V, \lambda \in F$). Is this a vector space?

4.1.7. Let $V = \mathbf{Z}$ with the usual addition and $F = \mathbf{Q}$. We define the multiplication by scalars $\odot$ as

$$\lambda \odot v = \lfloor \lambda v \rfloor$$

where the usual multiplication of rational numbers appears on the right-hand side of the equality and $\lfloor \quad \rfloor$ denotes the (lower) integer part of the number ($v, w \in V, \lambda \in F$). Is this a vector space?

*4.1.8. (a) Let $V = \mathbf{Z}$ with the usual addition and let $F = \mathbf{Q}$. Can we get a vector space with a suitably defined multiplication by scalars?

(b) Let $V = \mathbf{Z}$ with the usual addition. Can we get a vector space over some field F with a suitably defined multiplication by scalars?

(c) Let $V = \mathbf{Z}$ and $F = \mathbf{Q}$. Can we get a vector space with a suitably defined addition and multiplication by scalars?

(d) Let $V = \mathbf{Z}$ and $F = \mathbf{C}$. Can we get a vector space with a suitably defined addition and multiplication by scalars?

(e) Let V be the set of infinite real sequences with the usual addition and $F = \mathbf{C}$. Can we get a vector space with a suitably defined multiplication by scalars?

**(f) Let $V = \mathbf{R}$ with the usual addition and let $F = \mathbf{C}$. Can we get a vector space with a suitably defined multiplication by scalars?

4.1.9. Let V be the set of infinite complex sequences with the usual addition and let $F = \mathbf{C}$. Which axioms of vector spaces hold for the multiplication by scalars $\odot$ defined below? We denote a general sequence by $S = (\alpha_0, \alpha_1, \dots, \alpha_n, \dots)$, the usual multiplication of complex numbers appear on the right-hand sides of the equalities, and $\mathrm{Re}(\lambda)$ is the real part of λ ($S \in V, \lambda \in F$).

(a) $\lambda \odot S = S$;

(b) $\lambda \odot S = (\lambda\alpha_0, \alpha_1, \alpha_2, \dots)$;

(c) $\lambda \odot S = (\lambda\alpha_0, 0, 0, \dots)$;

(d) $\lambda \odot S = (\mathrm{Re}(\lambda)\alpha_0, \mathrm{Re}(\lambda)\alpha_1, \mathrm{Re}(\lambda)\alpha_2, \dots)$.

4.1.10. Prove the last three statements in Theorem 4.1.2.

4.1.11. True or false? (V is a vector space over the field F, $\mathbf{v}, \mathbf{w} \in V$, $\lambda, \mu \in F$.)

(a) If $\mathbf{v} \neq \mathbf{0}$ and $\lambda\mathbf{v} = \mu\mathbf{v}$, then $\lambda = \mu$.

(b) If $\lambda \neq 0$ and $\lambda\mathbf{v} = \lambda\mathbf{w}$, then $\mathbf{v} = \mathbf{w}$.

(c) If $\mathbf{v} \neq \mathbf{0}$, $\mathbf{w} \neq \mathbf{0}$, $\lambda \neq 0$, $\mu \neq 0$ and $\lambda\mathbf{v} = \mu\mathbf{w}$, then $\mathbf{v} = \mathbf{w}$ and $\lambda = \mu$.

4.1.12. Prove that replacing (MS4) by any of the two conditions below in the system of axioms (A1)–(A4) and (MS1)–(MS4), we get exactly the same notion of vector spaces.

(a) $\forall \mathbf{v} \in V \; \exists \lambda \in F \quad \lambda\mathbf{v} = \mathbf{v}$.

*(b) If $\lambda\mathbf{v} = \mathbf{0}$, then $\lambda = 0$ or $\mathbf{v} = \mathbf{0}$.

4.1.13. Prove that the commutative law for addition follows from the other axioms of vector spaces.

*4.1.14. Show that apart from the commutative law for addition, all other axioms are independent, i.e., none of them can be deduced from the others. (To verify this, show examples where exactly one of the axioms is false.)

4.2. Subspace

Definition 4.2.1. Let V be a vector space over a field F. A non-empty subset $W \subseteq V$ is a *subspace* in V if W itself is a vector space over the *same* field F under the operations in V (restricted to W). ♣

If W is a subspace in V, we denote this by $W \leq V$.

Observe that the subspace is much more than just a subset that is a vector space: W is a *substructure* of V since the field F and the operations are a priori given for W. So, e.g., the vector space in Exercise 4.1.5 is *not* a subspace of the usual vector space of $\mathbf{R}$ over itself.

A subset $W \subseteq V$ has to satisfy all axioms of vector spaces to be a subspace. First of all, the operations have to be well-defined, i.e., W has to be *closed* for the operations in V, they cannot lead out from W. It turns out that if this closure is okay, then the other axioms hold, so W is a subspace in V.

Theorem 4.2.2. *Let V be a vector space over F. A non-empty $W \subseteq V$ is a subspace in V if and only if*

(i) $\mathbf{v}, \mathbf{w} \in W \Rightarrow \mathbf{v} + \mathbf{w} \in W$;

(ii) $\mathbf{v} \in W, \lambda \in F \Rightarrow \lambda\mathbf{v} \in W$. ♣

Proof. If W is a subspace, then (i) and (ii) hold as — repeating the argument above — they express that (the restrictions of) the operations of the vector space V are operations on the set W, too.

For the converse, we have to verify that (i) and (ii) imply the validity of all axioms. The operational identities (A1)–(A2) and (MS1)–(MS4) hold for all vectors in V (and scalars in F), so they are automatically true for all elements of W (and scalars in F).

We have to show that W has a zero element and every element in W has a negative (in W). The zero vector $\mathbf{0}$ of V clearly acts as a zero element for W, too, provided it is contained in W. To prove this, let $\mathbf{v} \in W$ be arbitrary (there exists such a $\mathbf{v}$ as $W \neq \emptyset$). Then (ii) implies $\mathbf{0} = 0\mathbf{v} \in W$, and we are done. Similarly, if $\mathbf{v} \in W$, then $-\mathbf{v} = (-1)\mathbf{v} \in W$ serves as the negative in W of $\mathbf{v}$. ∎

We note that in Definition 4.2.1 we do not have to prescribe $W \neq \emptyset$ *extra* since a vector space is automatically not empty by definition. In Theorem 4.2.2, however, we cannot omit this condition as properties (i) and (ii) are satisfied by the empty set, too.

By the theorem, we do not have to go over all axioms to decide whether or not a given subset of a vector space is a subspace, it is sufficient to check the closure for the operations. Another practical criterion is stated in Exercise 4.2.5(a).

The proof of the theorem yielded that the zero element in the subspace W is the same as the one in the original vector space V, and the same applies to the negative. These do not follow from the notion or uniqueness of the zero element (see Exercises 4.2.14 and 4.2.15).

Examples of subspaces

E1 The entire vector space and the subset consisting of the vector $\mathbf{0}$ alone are subspaces in any vector space. These are called *trivial subspaces*. (We shall denote also the subspace containing only the zero vector simply by $\mathbf{0}$ instead of $\{\mathbf{0}\}$, so we shall not distinguish this subspace and the zero vector in notation.)

E2 In geometry, it is convenient to characterize the vectors starting from the origin with their endpoints. This makes possible to identify the usual vector space of the plane or space vectors with the set of points in the plane or space. In this interpretation, the non-trivial subspaces in the plane and the space are the lines containing the origin and the lines and planes containing the origin, respectively.

E3 In any vector space, all scalar multiples of a given vector form a subspace.

E4 Let $A \in F^{k\times n}$ be a fixed $k \times n$ matrix. Then

$$\operatorname{Ker} A = \{\mathbf{x} \in F^n \mid A\mathbf{x} = \mathbf{0}\}$$

is a subspace in F^n and

$$\operatorname{Im} A = \{A\mathbf{x} \mid \mathbf{x} \in F^n\} = \{\mathbf{y} \in F^k \mid \exists \mathbf{x} \in F^n\ A\mathbf{x} = \mathbf{y}\}$$

is a subspace in F^k. These subspaces are called the *kernel* (or *null space*) and the *image* (or *column space*) of the matrix A.

Exercises 4.2

In the exercises V denotes a vector space over the field F.

4.2.1. How are Exercises 4.1.1–4.1.3 related to the notion of subspaces?

4.2.2. Which subsets are subspaces in $V = F^{100\times 100}$?

(a) $\{A \in V \mid AB = BA\}$ where $B \in F^{100\times 100}$ is a fixed matrix;
(b) $\{A \in V \mid AB = 0\}$ where $B \in F^{100\times 100}$ is a fixed matrix;
(c) $\{A \in V \mid A^2 = 0\}$;
(d) *nilpotent* matrices, i.e., that have a power equal to the 0 matrix;
(e) singular matrices (including the 0 matrix);
(f) matrices of rank 3 and the 0 matrix;
(g) matrices of rank not greater than 3;
(h) *diagonal* matrices (every entry outside the main diagonal is 0);
(i) upper triangular matrices (every entry below the main diagonal is 0);
(j) *symmetric* matrices ($\alpha_{ij} = \alpha_{ji}$ for every i, j).

4.2.3. How can we find the kernel and the image of a given matrix? Determine $\operatorname{Ker} A$ and $\operatorname{Im} A$ for

$$A = \begin{pmatrix} 1 & 2 & 3 & 4 \\ 2 & 3 & 4 & 5 \\ 3 & 4 & 5 & 6 \end{pmatrix}.$$

4.2.4. Give an example for a vector space and its subset
(a) closed for addition but not for multiplication by scalars;
(b) closed for multiplication by scalars but not for addition;
(c) closed for neither addition, nor for multiplication by scalars.
Are all three cases possible for every field?

4.2.5. Let W be a non-empty subset of V. Which conditions imply that W is a subspace in V?
(a) $\mathbf{v}, \mathbf{w} \in W, \lambda, \mu \in F \Rightarrow \lambda\mathbf{v} + \mu\mathbf{w} \in W$;
(b) $\mathbf{v}, \mathbf{w} \in W, \lambda \in F \Rightarrow \lambda\mathbf{v} + \mathbf{w} \in W$;
(c) $\mathbf{v}, \mathbf{w} \in W, \lambda \in F \Rightarrow \lambda\mathbf{v} \in W$ and $\mathbf{v} - \mathbf{w} \in W$;
(d) $\mathbf{v}, \mathbf{w} \in W, \lambda \in F \Rightarrow \lambda\mathbf{v} \in W$ and $\mathbf{v} + \mathbf{w}$ or $\mathbf{v} - \mathbf{w} \in W$ (or both);
(e) $\mathbf{v} + \mathbf{w} \in W \Rightarrow \mathbf{v} \in W$ and $\mathbf{w} \in W$;
(f) $\mathbf{v} + \mathbf{w} \in W \Rightarrow \mathbf{v} \in W$ or $\mathbf{w} \in W$ (or both).

4.2.6. Let W be a non-trivial subspace in V. True or false ($\mathbf{v}, \mathbf{w} \in V, \lambda \in F$)?
(a) $\mathbf{v} + \mathbf{w} \in W \Rightarrow \mathbf{v}, \mathbf{w} \in W$;
(b) $\lambda \neq 0, \lambda\mathbf{v} \in W \Rightarrow \mathbf{v} \in W$;
(c) $\mathbf{v} \in W, \mathbf{w} \notin W \Rightarrow \mathbf{v} + \mathbf{w} \notin W$;
(d) $\mathbf{v} \notin W, \mathbf{w} \notin W \Rightarrow \mathbf{v} + \mathbf{w} \notin W$;
(e) $\mathbf{v} \notin W, \mathbf{w} \notin W \Rightarrow \mathbf{v} + \mathbf{w} \in W$.

4.2.7. Let $F = \mathbf{R}$, W a non-trivial subspace in V, and $\mathbf{v}, \mathbf{w} \in V$. What can we assert about the containment relation of W and the vectors $\mathbf{v}$ and $\mathbf{w}$? If the situation is not unique, show examples for all possible cases.
(a) $\mathbf{v} + \mathbf{w} \in W$;
(b) $\mathbf{v} + \mathbf{w} \notin W$;
(c) $2\mathbf{v} + 3\mathbf{w} \in W, \mathbf{v} + 7\mathbf{w} \in W$;
(d) $2\mathbf{v} + 3\mathbf{w} \in W, \mathbf{v} + 7\mathbf{w} \notin W$;
(e) $2\mathbf{v} + 3\mathbf{w} \notin W, \mathbf{v} + 7\mathbf{w} \notin W$.
What are the changes for other fields instead of $\mathbf{R}$?

4.2.8. Let $F = \mathbf{R}$, W a non-trivial subspace in V, and $\mathbf{v}, \mathbf{w}, \mathbf{z} \in V$. Assume

$$\mathbf{v} + \mathbf{w} \in W, \quad \mathbf{w} + 2\mathbf{z} \notin W, \quad \mathbf{z} + 3\mathbf{v} \in W.$$

What can we assert about the relation of W and the vectors $5\mathbf{v} + 3\mathbf{w} + \mathbf{z}$ and $6\mathbf{v} + 3\mathbf{w} + \mathbf{z}$? (Possible answers: the subspace necessarily contains the given vector; never contains it; both cases can occur.)

4.2.9. Characterize the vector spaces having only trivial subspaces.

4.2.10. Prove that if a vector space over an infinite field has a non-trivial subspace, then it has infinitely many subspaces.

4.2.11. Find the number of subspaces in the vector space
(a) F_2^2 (over F_2);
(b) F_p^2 (over F_p) where p is a prime.
For further generalizations see Exercise 4.6.14.

4.2.12. (a) Prove that the intersection of arbitrarily many subspaces is a subspace.
(b) Establish a (simple) necessary and sufficient condition for the union of two subspaces to be a subspace.
(c) Can the set-theoretic difference or symmetric difference of two subspaces be a subspace?
*(d) Can the union of three subspaces be a subspace if none of them contains another one?
****(e)** Let F be an infinite field. Prove that no V is the union of finitely many non-trivial subspaces.
****(f)** Let F be a finite field. Prove that no V is the union of less than $|F| + 1$ non-trivial subspaces.
****(g)** Let F be a finite field. Prove that if V has a non-trivial subspace, then V is the union of $|F| + 1$ non-trivial subspaces.

4.2.13. Let W be a subspace in V and $Z \subseteq W$. Verify that Z is a subspace in V if and only if it is a subspace in W. (So, being a subspace does not depend on how big the original vector space is.)

4.2.14. Let $V = \{c \in \mathbf{R} \mid c \geq 3\}$, $F = \mathbf{Q}$, and define addition $\oplus$ and multiplication by scalars $\odot$ as

$$u \oplus v = \max(u, v), \quad \lambda \odot v = v \quad (u, v \in V, \lambda \in F).$$

Let $W = \{c \in \mathbf{R} \mid c \geq 5\}$. Then W is closed for the operations $\oplus$ and $\odot$. The zero element in V is 3, but it is 5 in W. How is this compatible with the fact that the zero element in a subspace must be the same as the zero element of the original vector space (see the proof of Theorem 4.2.2)?

4.2.15. We present four arguments that the zero element in a subspace must be the same as the zero element of the original vector space, but only one of them is correct. Which is correct and what are the errors in the other three? (The zero elements of the vector space V and the subspace W are denoted by $\mathbf{0}_V$ and $\mathbf{0}_W$, respectively.)
(a) Since $\mathbf{0}_V + \mathbf{v} = \mathbf{v}$ holds for every vector $\mathbf{v} \in V$, it must hold also for the vectors in W. So, by definition, $\mathbf{0}_V$ is a zero element of W, too. The zero element in W is unique, hence $\mathbf{0}_W = \mathbf{0}_V$.
(b) If $\mathbf{0}_W \neq \mathbf{0}_V$, then there are two zero elements in V contradicting the uniqueness of the zero element in V.
(c) Let $\mathbf{v} \in W$ be arbitrary. Then both $\mathbf{0}_V + \mathbf{v} = \mathbf{v}$ and $\mathbf{0}_W + \mathbf{v} = \mathbf{v}$ hold, so $\mathbf{0}_V + \mathbf{v} = \mathbf{0}_W + \mathbf{v}$. Adding the negative in W of $\mathbf{v}$ to both sides, we infer $\mathbf{0}_V = \mathbf{0}_W$.
(d) Let $\mathbf{v} \in W$ be arbitrary. Then both $\mathbf{0}_V + \mathbf{v} = \mathbf{v}$ and $\mathbf{0}_W + \mathbf{v} = \mathbf{v}$ hold, so $\mathbf{0}_V + \mathbf{v} = \mathbf{0}_W + \mathbf{v}$. Adding the negative in V of $\mathbf{v}$ to both sides, we infer $\mathbf{0}_V = \mathbf{0}_W$.

4.2.16. Let W be a subspace in V and fix an arbitrary vector $\mathbf{v} \in V$. The set $\mathbf{v} + W = \{\mathbf{v} + \mathbf{w} \mid \mathbf{w} \in W\}$ is called a *translate* or a *coset* of W.

(a) Describe all translates in the usual vector space of plane or space vectors.

(b) Verify that two translates of the same subspace W either coincide or are disjoint.

*(c) Prove that if two translates of two distinct subspaces are not disjoint, then their intersection is a translate (of some subspace), too.

*(d) Show that a non-empty subset $T \subseteq V$ is a translate (of some subspace) if and only if

$$\mathbf{a}, \mathbf{b}, \mathbf{c} \in T, \lambda \in F \Rightarrow \mathbf{a} + \lambda(\mathbf{b} - \mathbf{c}) \in T.$$

*4.2.17. Let W be a subspace in V and consider the set X of all (distinct) translates of W. We define an addition $\oplus$ and a multiplication by scalars $\odot$ on X as

$$(\mathbf{v} + W) \oplus (\mathbf{w} + W) = (\mathbf{v} + \mathbf{w}) + W, \quad \lambda \odot (\mathbf{v} + W) = \lambda\mathbf{v} + W.$$

Prove that X is a vector space over F under these operations. It is called the *factor space* of V with respect to W and is denoted by V/W.

4.3. Span

Definition 4.3.1. Let V be a vector space over F, $\mathbf{a}_1, \dots, \mathbf{a}_n \in V$, $\lambda_1, \dots, \lambda_n \in F$. The vector $\lambda_1\mathbf{a}_1 + \dots + \lambda_n\mathbf{a}_n$ is called a *linear combination* of the vectors $\mathbf{a}_i$ (with scalars λ_i). ♣

It is well-known that in the (ordinary three-dimensional) space every vector is a linear combination of three (or more) fixed non-coplanar vectors. We can express this fact by saying that the given vectors *span* or *generate* the space. We generalize this notion to an arbitrary vector space.

Definition 4.3.2. The vectors $\mathbf{a}_1, \dots, \mathbf{a}_n \in V$ form a *spanning set* for the vector space V if every vector in V is a linear combination of the vectors $\mathbf{a}_i$. ♣

We may allow repetition among the vectors $\mathbf{a}_i$ but this has no significance as the set of linear combinations is not affected by including more copies of a vector. (This is not the case for linear independence; see Sections 3.3 and 4.4.)

In general, the elements of a vector space can be represented several ways as a linear combination of a given spanning set. Spanning sets yielding a *unique* representation play a special role, which are called *bases*; see Section 4.5.

Most vector spaces have many spanning sets, but it often occurs that a vector space has no (finite) spanning set at all (see Exercise 4.3.2). We shall introduce the notion of an infinite spanning set at the end of this section. However, we shall only use finite spanning sets except if indicated otherwise.

The set of all linear combinations of the vectors $\mathbf{a}_1, \dots, \mathbf{a}_n \in V$ plays an important role even if the vectors $\mathbf{a}_i$ do not form a spanning set. This gives rise to the following definition.

Definition 4.3.3. The *span* of $\mathbf{a}_1, \dots, \mathbf{a}_n \in V$ is the set of all linear combinations of the vectors $\mathbf{a}_i$ and is denoted by $\langle \mathbf{a}_1, \dots, \mathbf{a}_n \rangle$:

$$\langle \mathbf{a}_1, \dots, \mathbf{a}_n \rangle = \{\lambda_1 \mathbf{a}_1 + \dots + \lambda_n \mathbf{a}_n \mid \lambda_1, \dots, \lambda_n \in F\}. \clubsuit$$

Instead of the term span we can also say that $\langle \mathbf{a}_1, \dots, \mathbf{a}_n \rangle$ is the subspace spanned or generated by the vectors $\mathbf{a}_i$. (Theorem 4.3.4 justifies that the span is a subspace.) The vectors $\mathbf{a}_i$ are sometimes called the *generators* or *generating elements* of $\langle \mathbf{a}_1, \dots, \mathbf{a}_n \rangle$.

The span of a single vector consists of all scalar multiples of the given vector. In the (ordinary three-dimensional) space the span of two non-collinear vectors is the plane determined by them. Clearly, the vectors $\mathbf{a}_1, \dots, \mathbf{a}_n \in V$ form a spanning set for V if and only if $\langle \mathbf{a}_1, \dots, \mathbf{a}_n \rangle = V$.

Theorem 4.3.4. *$U = \langle \mathbf{a}_1, \dots, \mathbf{a}_n \rangle$ is the tightest subspace containing the vectors $\mathbf{a}_i$, i.e.,*

(i) *U is a subspace;*

(ii) *$\mathbf{a}_i \in U$, $i = 1, \dots, n$;*

(iii) *if W is a subspace and $\mathbf{a}_i \in W$, $i = 1, \dots, n$, then $U \subseteq W$.* ♣

Proof.

(i) A simple calculation shows that the sum of two linear combinations and the scalar multiple of a linear combination are again linear combinations of the given vectors.

(ii) We obtain $\mathbf{a}_i$ as a linear combination with scalars $\lambda_i = 1$ and $\lambda_j = 0$ if $j \neq i$.

(iii) A subspace is closed under linear combinations. So if W contains every $\mathbf{a}_i$, then it must contain all their linear combinations, too. ∎

By Theorem 4.3.4, properties (i)–(iii) can serve as an alternative *definition* for the subspace spanned by given vectors. A third type of characterization is described in Exercise 4.3.9.

As it is clear from (ii), a spanning set for a subspace must consist purely of vectors in the subspace itself; no external element can be taken.

Now we introduce the span or sum of two subspaces.

Definition 4.3.5. Let W and Z be subspaces in a vector space V. The *span* or *sum* of W and Z is $\{\mathbf{w} + \mathbf{z} \mid \mathbf{w} \in W, \mathbf{z} \in Z\}$. We denote it by $\langle W, Z \rangle$ or $W + Z$. ♣

Similar to Theorem 4.3.4, $\langle W, Z \rangle$ is the tightest subspace containing the subspaces W and Z (see Exercise 4.3.11).

It is an important special case if the elements in $\langle W, Z \rangle$ have a *unique* representation as $\mathbf{w} + \mathbf{z}$ ($\mathbf{w} \in W, \mathbf{z} \in Z$).

Theorem 4.3.6. *Let W and Z be subspaces in V. The representation $\mathbf{w}+\mathbf{z}$ (with $\mathbf{w} \in W$, $\mathbf{z} \in Z$) of the elements in $\langle W, Z \rangle$ is unique if and only if $W \cap Z = \mathbf{0}$.* ♣

Proof. We assume first that $W \cap Z = \mathbf{0}$ and consider an $\mathbf{x} \in \langle W, Z \rangle$ represented as

$$\mathbf{x} = \mathbf{w}_1 + \mathbf{z}_1 = \mathbf{w}_2 + \mathbf{z}_2 \quad \text{where} \quad \mathbf{w}_i \in W, \mathbf{z}_i \in Z.$$

Ordering the second equality, we obtain $\mathbf{w}_1 - \mathbf{w}_2 = \mathbf{z}_2 - \mathbf{z}_1$. The vector on the left-hand side is in W and the one on the right-hand side is in Z. So their common value is $\mathbf{0}$. Thus $\mathbf{w}_1 = \mathbf{w}_2$ and $\mathbf{z}_1 = \mathbf{z}_2$ proving the uniqueness of the representation.

For the converse, we assume the uniqueness of the representation. For any $\mathbf{v} \in W \cap Z$ we have $\mathbf{v} = \mathbf{v} + \mathbf{0} = \mathbf{0} + \mathbf{v}$. This gives two distinct representations if $\mathbf{v} \neq \mathbf{0}$. Therefore only $\mathbf{v} = \mathbf{0}$ is possible, so $W \cap Z = \mathbf{0}$. ∎

In the case $W \cap Z = \mathbf{0}$ we say that the subspaces W and Z are *disjoint* as they have no common elements except the zero vector that must be contained in every subspace.

Definition 4.3.7. If $W \cap Z = \mathbf{0}$, then the subspace $\langle W, Z \rangle$ is called the *direct sum* of W and Z and is denoted by $W \oplus Z$. ♣

Note that only disjoint subspaces have a direct sum.

The span of infinitely many vectors

The difficulty is that we cannot define the sum of infinitely many vectors in general. We can take, however, all linear combinations of all possible *finite* subsets of the given vectors.

Definition 4.3.8. Let H be any non-empty subset in the vector space V. The span $\langle H \rangle$ of H is the set of all (finite but arbitrarily long) linear combinations formed from the elements of H. ♣

Similar to the case of a finite H, $\langle H \rangle$ is the tightest subspace containing H. Also, if W and Z are subspaces, then $\langle W, Z \rangle = \langle W \cup Z \rangle$.

In this broader sense, a subset $H \subseteq V$ is a *spanning set* for V if $\langle H \rangle = V$. In other words, every $\mathbf{v} \in V$ can be represented as a linear combination of finitely many suitable vectors from H. To represent different vectors we generally need different elements of H and the number of terms used in the representations is unbounded in most cases.

In this extended interpretation, every vector space V has a spanning set as, e.g., $V = \langle V \rangle$.

In the usual vector space $F[x]$ of polynomials, $1, x, x^2, \ldots$ is a spanning set. But our intuition acquired in the finite case may be deceptive: in the usual vector space of infinite sequences, the sequences with a 1 at one place and 0 at all other places do *not* form a spanning set as no *finite* linear combination of them can produce, e.g., the sequence consisting purely of 1s.

Exercises 4.3

In the exercises V denotes a vector space.

4.3.1. Select the spanning sets for the vector space $\mathbf{C}^4$.

(a) $\begin{pmatrix}1\\0\\1\\0\end{pmatrix}, \begin{pmatrix}0\\1\\0\\1\end{pmatrix}, \begin{pmatrix}1\\0\\0\\1\end{pmatrix}, \begin{pmatrix}0\\1\\1\\0\end{pmatrix}, \begin{pmatrix}1\\1\\1\\1\end{pmatrix};$

(b) $\begin{pmatrix} 1 \\ -1 \\ 0 \\ 0 \end{pmatrix}, \begin{pmatrix} 1 \\ 0 \\ -1 \\ 0 \end{pmatrix}, \begin{pmatrix} 1 \\ 0 \\ 0 \\ -1 \end{pmatrix}, \begin{pmatrix} 0 \\ 1 \\ -1 \\ 0 \end{pmatrix}, \begin{pmatrix} 0 \\ 1 \\ 0 \\ -1 \end{pmatrix}, \begin{pmatrix} 0 \\ 0 \\ 1 \\ -1 \end{pmatrix};$

(c) $\begin{pmatrix} 1 \\ 1 \\ 1 \\ 1 \end{pmatrix}, \begin{pmatrix} 1 \\ 2 \\ 4 \\ 8 \end{pmatrix}, \begin{pmatrix} 1 \\ 3 \\ 9 \\ 27 \end{pmatrix}, \begin{pmatrix} 1 \\ 4 \\ 16 \\ 64 \end{pmatrix}.$

4.3.2. Which vector spaces in the examples in Section 4.1 and in Exercises 4.1.1–4.1.3 have a finite spanning set?

4.3.3. True or false?

(a) Extending a spanning set with a new vector we get a spanning set.

(b) Deleting any vector from a spanning set of at least two elements the remaining elements form a spanning set.

(c) In every spanning set of at least two elements there is a vector that can be deleted so that the remaining elements form a spanning set.

(d) If there are two equal vectors in a spanning set, then deleting one of them the remaining vectors form a spanning set.

(e) In a spanning set of at least two elements, there exists a vector that can be deleted so that the remaining vectors form a spanning set if and only if some element in the (original) spanning set is a linear combination of the others.

4.3.4. Let $V = \mathbf{R}[x]$ be the usual vector space of the polynomials with real coefficients. True or false?

(a) $x^3 + 7x^2 + 5x \in \langle x^3 + 2x, 3x^3 + 4x, 5x^2 + 6x \rangle$;

(b) $x^3 + 7x^2 + 5 \in \langle x^3 + 2x, 3x^3 + 4x, 5x^2 + 6x \rangle$;

(c) $x - 1 \in \langle x^3 - x, x^3 - x^2, x^3 - 1, 2x^2 - 3x + 1 \rangle$;

(d) $x + 1 \in \langle x^3 - x, x^3 - x^2, x^3 - 1, 2x^2 - 3x + 1 \rangle$;

(e) $x + 1 \in \langle x^3 - x, x^3 - x^2, x^3 - 1, 2x^2 + 3x + 1 \rangle$.

4.3.5. Assume $\mathbf{a}+\mathbf{b}+\mathbf{c} = \mathbf{0}$ for some elements $\mathbf{a}$, $\mathbf{b}$, and $\mathbf{c}$ in V. Prove $\langle \mathbf{a}, \mathbf{b} \rangle = \langle \mathbf{a}, \mathbf{c} \rangle$.

4.3.6. Assume $\mathbf{a}+\mathbf{b}+\mathbf{c}+\mathbf{d} = \mathbf{0}$ for some elements $\mathbf{a}$, $\mathbf{b}$, $\mathbf{c}$, and $\mathbf{d}$ in V. True or false?

(a) $\langle \mathbf{a}, \mathbf{b} \rangle = \langle \mathbf{a}, \mathbf{c} \rangle$;

(b) $\langle \mathbf{a}, \mathbf{b} \rangle = \langle \mathbf{c}, \mathbf{d} \rangle$;

(c) $\langle \mathbf{a}, \mathbf{b}, \mathbf{c} \rangle = \langle \mathbf{a}, \mathbf{c}, \mathbf{d} \rangle$;

(d) $\langle \mathbf{a}, \mathbf{b}, \mathbf{c} \rangle = \langle \mathbf{a}, \mathbf{d} \rangle$;

(e) $\langle \mathbf{a}, \mathbf{b}, \mathbf{c} \rangle \supseteq \langle \mathbf{a}, \mathbf{d} \rangle$.

4.3.7. Assume $\mathbf{a} \notin \langle \mathbf{b}, \mathbf{c} \rangle$, $\mathbf{b} \notin \langle \mathbf{a}, \mathbf{c} \rangle$, and $\mathbf{c} \in \langle \mathbf{a}, \mathbf{b} \rangle$ for some elements $\mathbf{a}$, $\mathbf{b}$, and $\mathbf{c}$ in V. Determine $\mathbf{c}$.

4.3.8. Prove that to given vectors $\mathbf{a}_1, \dots, \mathbf{a}_n \in V$ there exists only one U satisfying requirements (i)–(iii) in Theorem 4.3.4.

4.3.9. Verify that the span of $\mathbf{a}_1, \dots, \mathbf{a}_n \in V$ is the intersection of all subspaces containing every $\mathbf{a}_i$.

4.3.10. Let V be the usual vector space of all $\mathbf{R} \to \mathbf{R}$ functions and let f denote a general function. Characterize the span $\langle W, Z\rangle$ of subspaces W and Z.
(a) $W = \{\text{even functions}\}, \quad Z = \{\text{odd functions}\}$;
(b) $W = \{f \mid f(5) = 0\}, \quad Z = \{f \mid f(6) = 0\}$;
(c) $W = \{f \mid f(x) = 0, \text{ if } x \neq 5\}, \quad Z = \{f \mid f(x) = 0, \text{ if } x \neq 6\}$;
(d) $W = \{f \mid \forall x \in \mathbf{Q}\ f(x) = 0\}, \quad Z = \{f \mid \forall x \notin \mathbf{Q}\ f(x) = 0\}$;
(e) $W = \{f \mid \forall x, y \in \mathbf{Q}\ f(x) = f(y)\}, \quad Z = \{f \mid \forall x, y \notin \mathbf{Q}\ f(x) = f(y)\}$.
In which cases does $\langle W, Z\rangle = W \oplus Z$ hold?

4.3.11. Let W and Z be subspaces in V. Prove that $\langle W, Z\rangle$ is the tightest subspace containing W and Z. (Phrase precisely what tightest means.)

4.3.12. Let W_1, W_2, and W_3 be subspaces in V. Find the connection between
(a) $\langle W_1, W_2\rangle \cap W_3$ and $\langle W_1 \cap W_3, W_2 \cap W_3\rangle$;
(b) $\langle W_1 \cap W_2, W_3\rangle$ and $\langle W_1, W_3\rangle \cap \langle W_2, W_3\rangle$;
(c) $\langle W_1, W_2\rangle \cap W_3$ and $\langle W_1, W_2 \cap W_3\rangle$, if $W_1 \subseteq W_3$.

4.3.13. Let V be the usual vector space of real infinite sequences. We denote a general sequence by $S = (\alpha_0, \alpha_1, \dots, \alpha_n, \dots)$. Determine whether V is the direct sum of W and Z.
(a) $W = \{S \mid \alpha_5 = 0\}, \quad Z = \{S \mid \alpha_6 = 0\}$;
(b) $W = \{S \mid \alpha_5 = 0\}, \quad Z = \{S \mid \alpha_i = 0, \text{ if } i \neq 5\}$;
(c) $W = \{S \mid \alpha_5 = 0\}, \quad Z = \{S \mid \alpha_1 = \cdots = \alpha_5, \alpha_6 = \alpha_7 = \cdots = 0\}$;
(d) $W = \{S \mid \alpha_5 = 0\}, \quad Z = \{S \mid \alpha_5 \neq 0\}$;
(e) $W = \{S \mid \alpha_i = 0, \text{ if } i \text{ is even}\}, \quad Z = \{S \mid \alpha_i = 0, \text{ if } i \text{ is odd}\}$;
(f) $W = \{S \mid \alpha_i = 0, \text{ if } i \neq 5\}, \quad Z = \{S \mid \alpha_i = 0, \text{ if } i \neq 6\}$.

*4.3.14. Generalize Definition 4.3.5 and Theorem 4.3.6 for more than two (but finitely many) subspaces and use these to extend the notion of direct sums (Definition 4.3.7) for more than two subspaces.

*4.3.15. Determine the span of the subsets in Exercises 4.1.1–4.1.3 except for part (n) in Exercise 4.1.2 and part (h) in Exercise 4.1.3.

*4.3.16. Consider the vector space of all $\mathbf{R} \to \mathbf{R}$ functions over the *rational* field $\mathbf{Q}$. Let H be the set of functions assuming only integer values. Which functions are in the span $\langle H\rangle$ of H?

(a) $f(x) = \begin{cases} 5/7, & \text{if } x \in \mathbf{Q}; \\ 3/8, & \text{if } x \notin \mathbf{Q}. \end{cases}$ (b) $g(x) = \begin{cases} 1/x, & \text{if } x = 1, 2, 3, \dots; \\ 0, & \text{otherwise.} \end{cases}$

**(c) Solve the problem if the field $\mathbf{Q}$ is replaced by $\mathbf{R}$.

4.3.17. Select the spanning sets for the usual vector space of real sequences.
(a) Every term in the sequence is 0 or 1;
**(b) every term is a rational number;
(c) every term is an irrational number.

4.4. Linear Independence

We have introduced special cases of linear dependence and independence in Chapter 3 related to matrices and systems of linear equations. The Definitions 3.3.2 and 3.3.3 can be transferred literally to general vector spaces and also all basic properties remain valid. We first summarize all this.

Let V be a vector space over the field F, $\mathbf{v}_1, \dots, \mathbf{v}_n \in V$, $\lambda_1, \dots, \lambda_n \in F$ and consider the linear combination $\lambda_1\mathbf{v}_1 + \cdots + \lambda_n\mathbf{v}_n$. If every $\lambda_i = 0$, then this *trivial* linear combination is $\mathbf{0}$. It can happen, however, that we can also obtain $\mathbf{0}$ as a *non-trivial* linear combination with other coefficients. The vectors $\mathbf{v}_i$ are called *linearly dependent* in this case and *linearly independent* otherwise.

Definition 4.4.1. The vectors $\mathbf{v}_1, \dots, \mathbf{v}_m \in V$ are *linearly dependent* if there exist scalars $\lambda_1, \dots, \lambda_m \in F$ not all 0 such that $\lambda_1\mathbf{v}_1 + \cdots + \lambda_m\mathbf{v}_m = \mathbf{0}$. ♣

Definition 4.4.2. The vectors $\mathbf{v}_1, \dots, \mathbf{v}_m \in V$ are *linearly independent* if $\lambda_1\mathbf{v}_1 + \cdots + \lambda_m\mathbf{v}_m = \mathbf{0}$ holds **only** when every $\lambda_i = 0$:

$$\lambda_1\mathbf{u}_1 + \cdots + \lambda_m\mathbf{u}_m = \mathbf{0} \Rightarrow \lambda_i = 0,\ i = 1, \dots, m. \clubsuit$$

So, a system of vectors $\mathbf{u}_1, \dots, \mathbf{u}_m \in V$ is linearly independent if and only if it is not linearly dependent. For brevity we sometimes omit the adjective "linear" and say that a system of vectors is dependent or independent.

In a system of vectors the same vector may occur repeatedly (in contrast to a set of vectors). This can influence the question of independence: if a vector occurs more than once, say $\mathbf{v}_1 = \mathbf{v}_2$, then $1\mathbf{v}_1 + (-1)\mathbf{v}_2 + 0\mathbf{v}_3 + \cdots + 0\mathbf{v}_m = \mathbf{0}$, so the system $\mathbf{v}_1, \dots, \mathbf{v}_m$ is dependent.

A few simple observations. A single vector is independent if and only if it is not the zero vector. Two vectors are linearly independent if and only if none is a scalar multiple of the other. This is *false* for more than two vectors: e.g., any three vectors in the plane are dependent.

ATTENTION, please! Linear independence is a very delicate notion with lots of hidden traps. Therefore, it is essential to digest every detail thoroughly, especially at the beginning, to avoid a false conception based on some erroneous vision.

Theorem 3.3.5 remains valid in an arbitrary vector space.

Theorem 4.4.3.

I. *Deleting any vector from a linearly independent system of at least two vectors, the remaining system is linearly independent.*

II. *Extending a linearly dependent system by any vector, the new system is linearly dependent.*

III. *A system of at least two vectors is linearly dependent if and only if at least one of the vectors is a linear combination of the others.*

IV. *If extending a linearly independent system by a new vector we get a dependent system, then the new vector is a linear combination of the old ones.*

V. *Assume that a vector is a linear combination of a system of vectors. Then this representation is unique if and only if the system is linearly independent.* ♣

Proof. See Theorem 3.3.5. ∎

Definition 4.4.4. A vector $\mathbf{w}$ *depends linearly* on the vectors $\mathbf{v}_1, \dots, \mathbf{v}_m$ if $\mathbf{v}$ is a linear combination of $\mathbf{v}_1, \dots, \mathbf{v}_m$. ♣

If $\mathbf{w}$ depends linearly on $\mathbf{v}_1, \dots, \mathbf{v}_m$, then $\mathbf{w}, \mathbf{v}_1, \dots, \mathbf{v}_m$ are linearly dependent, but the converse is *false*! By statement III in Theorem 4.4.3, linear dependence means that there *exists* a vector among them depending on the others. (There are more such vectors in a dependent system in general, and it *can* happen that all vectors have this property; see Exercises 4.4.4–4.4.6.)

Using the notion of span, $\mathbf{w}$ depends linearly on $\mathbf{v}_1, \dots, \mathbf{v}_m$ if and only if $\mathbf{w} \in \langle \mathbf{v}_1, \dots, \mathbf{v}_m \rangle$.

Finally, we mention that the linear independence of *infinitely many* vectors means that *an arbitrary finite* subsystem of them is linearly independent. Note that we used a similar approach to the span of infinitely many vectors as a linear combination of infinitely many vectors makes no sense.

Exercises 4.4

(See also Exercises 3.3.)

4.4.1. True or false?

(a) If both $\mathbf{u}_1, \dots, \mathbf{u}_{100}$ and $\mathbf{v}_1, \dots, \mathbf{v}_{100}$ are linearly independent, then $\mathbf{u}_1, \dots, \mathbf{u}_{100}, \mathbf{v}_1, \dots, \mathbf{v}_{100}$ are linearly independent.

(b) If $\mathbf{u}_1, \dots, \mathbf{u}_{100}, \mathbf{v}_1, \dots, \mathbf{v}_{100}$ are linearly independent, then both $\mathbf{u}_1, \dots, \mathbf{u}_{100}$ and $\mathbf{v}_1, \dots, \mathbf{v}_{100}$ are linearly independent.

(c) If both $\mathbf{u}_1, \dots, \mathbf{u}_{100}$ and $\mathbf{v}_1, \dots, \mathbf{v}_{100}$ are linearly independent, then $\mathbf{u}_1 + \mathbf{v}_1, \dots, \mathbf{u}_{100} + \mathbf{v}_{100}$ are linearly independent.

(d) If $\mathbf{u}_1 + \mathbf{v}_1, \dots, \mathbf{u}_{100} + \mathbf{v}_{100}$ are linearly independent, then both $\mathbf{u}_1, \dots, \mathbf{u}_{100}$ and $\mathbf{v}_1, \dots, \mathbf{v}_{100}$ are linearly independent.

(e) If some $\mathbf{u}_i$ among $\mathbf{u}_1, \dots, \mathbf{u}_{100}$ is a scalar multiple of some other $\mathbf{u}_j$, then $\mathbf{u}_1, \dots, \mathbf{u}_{100}$ are linearly dependent.

(f) If any 99 vectors from $\mathbf{u}_1, \dots, \mathbf{u}_{100}$ are linearly independent, then $\mathbf{u}_1, \dots, \mathbf{u}_{100}$ are linearly independent.

(g) If $\mathbf{u}_1, \dots, \mathbf{u}_{100}$ are linearly independent, then $\mathbf{u}_1 + \mathbf{u}_2, \mathbf{u}_3, \dots, \mathbf{u}_{100}$ are linearly independent.

(h) If $\mathbf{u}_1, \dots, \mathbf{u}_{100}$ are linearly dependent, then $\mathbf{u}_1 + \mathbf{u}_2, \mathbf{u}_3, \dots, \mathbf{u}_{100}$ are linearly dependent.

(i) If $\mathbf{u}_1, \dots, \mathbf{u}_{100}$ are linearly independent, then $\mathbf{u}_1, \mathbf{u}_1 + \mathbf{u}_2, \dots, \mathbf{u}_1 + \mathbf{u}_2 + \dots + \mathbf{u}_{100}$ are linearly independent.

(j) If $\mathbf{u}_1, \mathbf{u}_2, \dots, \mathbf{u}_{100}$ are linearly dependent, then $\mathbf{u}_1, \mathbf{u}_1 + \mathbf{u}_2, \dots, \mathbf{u}_1 + \mathbf{u}_2 + \dots + \mathbf{u}_{100}$ are linearly dependent.

4.4.2. Let **a**, **b**, and **c** be non-zero vectors. What can we deduce about linear dependence and independence of **a** and **c** from the conditions below. (Possible answers: they are always independent; they are always dependent; they can be independent and can be dependent.)
(a) Both **a**, **b** and **b**, **c** are linearly dependent.
(b) **a**, **b** are linearly independent, **b**, **c** are linearly dependent.
(c) Both **a**, **b** and **b**, **c** are linearly independent.

4.4.3. Assume that in a vector space over an infinite field there are only finitely many linear combinations of $\mathbf{u}_1, \dots, \mathbf{u}_m$ resulting in $\mathbf{0}$. Does this imply that $\mathbf{u}_1, \dots, \mathbf{u}_m$ are linearly independent?

4.4.4. Assume that *exactly* one of $\mathbf{u}_1, \dots, \mathbf{u}_m$ depends linearly on the other $m - 1$ vectors. Prove that it is necessarily the zero vector.

4.4.5. Let $m \geq 2$ and $0 \leq s \leq m$. Find m distinct vectors in a suitable vector space so that *exactly* s of them have the property of depending linearly on the other $m - 1$ vectors.

4.4.6. Assume that $\mathbf{u}_1, \dots, \mathbf{u}_m$ are linearly dependent. True or false?
(a) If any $m - 1$ of the m vectors are linearly independent, then every $\mathbf{u}_i$ depends linearly on the other $m - 1$ vectors.
(b) If every $\mathbf{u}_i$ depends linearly on the other $m - 1$ vectors, then any $m - 1$ of the m vectors are linearly independent.
(c) If $\lambda_t \neq 0$ in a linear combination $\lambda_1\mathbf{u}_1 + \dots + \lambda_m\mathbf{u}_m = \mathbf{0}$, then $\mathbf{u}_t$ depends linearly on the other $m - 1$ vectors.
(d) If $\lambda_t = 0$ in a non-trivial linear combination $\lambda_1\mathbf{u}_1 + \dots + \lambda_m\mathbf{u}_m = \mathbf{0}$, then $\mathbf{u}_t$ does *not* depend linearly on the other $m - 1$ vectors.

4.4.7. Assume that **a**, **b**, **c**, **d** are linearly independent, **a**, **b**, **c**, **e** are linearly dependent, **c**, **d**, **e** are linearly dependent, and $\mathbf{e} \neq \mathbf{0}$. What can we deduce about linear dependence and independence of **a**, **b**, **d**, **e**?

4.4.8. Assume that **a**, **b**, **c** are linearly independent but the systems **a**, **b**, **d**; **a**, **c**, **d**; and **b**, **c**, **d** are linearly dependent. Determine **d**.

4.4.9. Investigate the linear dependence and independence of the vectors in the usual vector space $\mathbf{C}[x]$ over $\mathbf{C}$.
(a) $(x + 1)(x + 2), (x + 2)(x + 3), (x + 3)(x + 1)$;
(b) $(x + 1)(x + 2), (x + 2)(x + 3), (x + 3)(x + 4), (x + 4)(x + 1)$;
(c) $x^3 + ix^2 - x - i, -ix^3 - x^2 + x + i, -x^3 - ix^2 + ix + 1, ix^3 + x^2 - ix - 1$;
(d) 1000 polynomials of distinct degrees;
(e) 1000 polynomials of the same degree;
(f) 1000 irreducible polynomials over $\mathbf{R}$ in $\mathbf{R}[x]$;
(g) 1000 irreducible polynomials over $\mathbf{Q}$ in $\mathbf{Q}[x]$.

4.4.10. Let V be a real vector space, $m \geq 2$, $1 \leq k < m$, and assume that $\mathbf{u}_1, \dots, \mathbf{u}_m$ are linearly independent. Find a necessary and sufficient condition for the linear independence of
(a) $\mathbf{u}_1 + \mathbf{u}_2, \mathbf{u}_2 + \mathbf{u}_3, \dots, \mathbf{u}_m + \mathbf{u}_1$;
(b) $\mathbf{u}_1 + \mathbf{u}_2 + \dots + \mathbf{u}_k, \mathbf{u}_2 + \mathbf{u}_3 + \dots + \mathbf{u}_{k+1}, \dots, \mathbf{u}_m + \mathbf{u}_1 + \dots + \mathbf{u}_{k-1}$.

4.4.11. If every element in a k-tuple is 0 or 1, then the k-tuple can be considered as a vector in $\mathbf{Q}^k$ or $\mathbf{R}^k$ or F_p^k for some prime p. So the linear independence of such vectors may depend on the field chosen. True or false?

(a) Linear independence over $\mathbf{R}$ implies linear independence over $\mathbf{Q}$.
(b) Linear independence over $\mathbf{Q}$ implies linear independence over $\mathbf{R}$.
(c) Linear independence over $\mathbf{Q}$ implies linear independence over F_2.
(d) Linear independence over F_2 implies linear independence over $\mathbf{Q}$.
(e) Linear independence over $\mathbf{Q}$ implies linear independence over F_p except finitely many primes p.

4.4.12. Consider the real numbers as a vector space over the *rational* field under the usual operations. Prove:

(a) The logarithms (with a fixed base) of distinct primes are always linearly independent.
(b) All powers with positive integer exponents of a real number are linearly independent if and only if the number is *transcendental.* (See the definition of a transcendental number in Section A.10.)

4.4.13. Verify that the direct sum $\langle \mathbf{u} \rangle \oplus \langle \mathbf{v} \rangle$ makes sense if and only if either $\mathbf{u}$ and $\mathbf{v}$ are linearly independent, or at least one of them is $\mathbf{0}$.

4.5. Basis

Definition 4.5.1. A *basis* is a linearly independent spanning set. ♣

The definition of a spanning set and part V in Theorem 4.4.3 immediately yield the following theorem.

Theorem 4.5.2. *A system of vectors $\mathbf{u}_1, \dots, \mathbf{u}_m$ is a basis if and only if every element in the vector space is a unique linear combination of $\mathbf{u}_1, \dots, \mathbf{u}_m$.* ♣

Examples. A natural basis in F^n or $F^{k\times n}$ consists of the vectors or matrices with an element 1 and all other elements 0. A vector space has many bases in general. In the ordinary three-dimensional space any three non-coplanar vectors form a basis.

The $\mathbf{0}$ space has no basis since its only element, the $\mathbf{0}$, is linearly dependent. The usual vector space of the infinite real sequences has no (finite) basis since it has no finite spanning set. By basis we shall mean a *finite* system of vectors. The introduction of an infinite basis will be sketched at the end of this section.

The next proposition is of fundamental importance.

Theorem 4.5.3. *In a vector space, any two bases have the same number of elements.* ♣

We shall prove a stronger statement.

Theorem 4.5.4. *Let $\mathbf{f}_1, \dots, \mathbf{f}_n$ be a linearly independent system and let $\mathbf{g}_1, \dots, \mathbf{g}_k$ be a spanning set in a vector space V. Then $n \le k$.* ♣

Theorem 4.5.4 implies Theorem 4.5.3: The first basis is linearly independent, the second one is a spanning set, so the first basis cannot have more elements than the second one. Repeating the argument in reverse roles, we arrive at Theorem 4.5.3.

We give two proofs to Theorem 4.5.4.

First proof. We assume $n > k$ and derive that

$$\lambda_1 \mathbf{f}_1 + \lambda_2 \mathbf{f}_2 + \cdots + \lambda_n \mathbf{f}_n = \mathbf{0} \tag{4.5.1}$$

holds not just for $\lambda_1 = \cdots = \lambda_n = 0$, thus contradicting linear independence.

As the vectors $\mathbf{g}_j$ form a spanning set, every $\mathbf{f}_i$ is their linear combination:

$$\begin{aligned}
\mathbf{f}_1 &= \alpha_{11}\mathbf{g}_1 + \alpha_{21}\mathbf{g}_2 + \ldots + \alpha_{k1}\mathbf{g}_k;\\
\mathbf{f}_2 &= \alpha_{12}\mathbf{g}_1 + \alpha_{22}\mathbf{g}_2 + \ldots + \alpha_{k2}\mathbf{g}_k;\\
&\ \vdots\\
\mathbf{f}_n &= \alpha_{1n}\mathbf{g}_1 + \alpha_{2n}\mathbf{g}_2 + \ldots + \alpha_{kn}\mathbf{g}_k.
\end{aligned}$$

Substituting these representations into (4.5.1) and ordering the left-hand side by the vectors $\mathbf{g}_j$, we obtain:

$$\begin{aligned}
(\lambda_1\alpha_{11} + \lambda_2\alpha_{12} + \cdots + \lambda_n\alpha_{1n})\mathbf{g}_1 + (\lambda_1\alpha_{21} + \lambda_2\alpha_{22} + \cdots + \lambda_n\alpha_{2n})\mathbf{g}_2&\\
+ \cdots + (\lambda_1\alpha_{k1} + \lambda_2\alpha_{k2} + \cdots + \lambda_n\alpha_{kn})\mathbf{g}_k = \mathbf{0}.&
\end{aligned}$$

This certainly holds if the coefficient of every $\mathbf{g}_j$ is 0 (it can hold in other cases, too, as the vectors $\mathbf{g}_j$ are not necessarily independent). Writing in detail that the coefficient of every $\mathbf{g}_j$ is 0, we get

$$\begin{aligned}
\alpha_{11}\lambda_1 + \alpha_{12}\lambda_2 + \ldots + \alpha_{1n}\lambda_n &= 0\\
\alpha_{21}\lambda_1 + \alpha_{22}\lambda_2 + \ldots + \alpha_{2n}\lambda_n &= 0\\
&\ \vdots\\
\alpha_{k1}\lambda_1 + \alpha_{k2}\lambda_2 + \ldots + \alpha_{kn}\lambda_n &= 0.
\end{aligned}$$

This is a homogeneous system of linear equations where $\lambda_1, \ldots, \lambda_n$ are the n unknowns and there are only k equations, so the assumption $n > k$ guarantees a non-trivial solution. Hence (4.5.1) can be satisfied non-trivially contradicting the independence of $\mathbf{f}_1, \ldots, \mathbf{f}_n$. ∎

Second proof.

Lemma 4.5.5 (Replacement Theorem). *Let $\mathbf{f}_1, \ldots, \mathbf{f}_n$ be a linearly independent system and let $\mathbf{g}_1, \ldots, \mathbf{g}_k$ be a spanning set in a vector space V. Then every $\mathbf{f}_i$ can be replaced by a suitable $\mathbf{g}_j$ so that the system*

$$\mathbf{f}_1, \ldots, \mathbf{f}_{i-1}, \mathbf{g}_j, \mathbf{f}_{i+1}, \ldots, \mathbf{f}_n$$

is linearly independent. ♣

Proof of the Replacement Theorem. For a proof by contradiction we assume that, e.g., $\mathbf{f}_1$ cannot be replaced, i.e., any $\mathbf{g}_j$ together with $\mathbf{f}_2, \ldots, \mathbf{f}_n$ form a dependent system. Then every $\mathbf{g}_j$ is a linear combination of $\mathbf{f}_2, \ldots, \mathbf{f}_n$ as the latter are independent (Theorem 4.4.3/I,IV). This implies that every linear combination of $\mathbf{g}_1, \ldots, \mathbf{g}_k$ can be expressed with $\mathbf{f}_2, \ldots, \mathbf{f}_n$. As $\mathbf{g}_1, \ldots, \mathbf{g}_k$ is a spanning set, this means that every element in V, including $\mathbf{f}_1$ is a linear combination of $\mathbf{f}_2, \ldots, \mathbf{f}_n$, contradicting the independence of $\mathbf{f}_1, \ldots, \mathbf{f}_n$. ∎

Now we deduce Theorem 4.5.4 from the Replacement Theorem 4.5.5. We replace first $\mathbf{f}_1$ by some $\mathbf{g}_j$, then we replace $\mathbf{f}_2$ in this *new* independent system by a suitable $\mathbf{g}_m$, etc., until all vectors $\mathbf{f}_i$ are replaced. Each vector in this final independent system is some $\mathbf{g}_t$ and no two can be equal due to independence. Therefore the number of the vectors $\mathbf{g}_j$ cannot be less than the number of the vectors $\mathbf{f}_i$. ∎

The next two theorems guarantee that we can obtain a basis in many ways, namely essentially any spanning set contains a basis and every independent system can be extended to a basis.

Theorem 4.5.6. *Any (finite) spanning set in a vector space $V \neq \mathbf{0}$ contains a basis.* ♣

Proof. If the spanning set is linearly independent, then it is a basis itself. If it is dependent, then some element in it is a linear combination of the others, say $\mathbf{g}_k = \mu_1\mathbf{g}_1 + \cdots + \mu_{k-1}\mathbf{g}_{k-1}$. Deleting this $\mathbf{g}_k$, the remaining elements still span the vector space as any $\mathbf{v} \in V$ satisfies

$$\begin{aligned}\mathbf{v} &= \alpha_1\mathbf{g}_1 + \cdots + \alpha_{k-1}\mathbf{g}_{k-1} + \alpha_k\mathbf{g}_k \\ &= \alpha_1\mathbf{g}_1 + \cdots + \alpha_{k-1}\mathbf{g}_{k-1} + \alpha_k(\mu_1\mathbf{g}_1 + \cdots + \mu_{k-1}\mathbf{g}_{k-1}) \\ &= (\alpha_1 + \alpha_k\mu_1)\mathbf{g}_1 + \cdots + (\alpha_{k-1} + \alpha_k\mu_{k-1})\mathbf{g}_{k-1}.\end{aligned}$$

If the new spanning set $\mathbf{g}_1, \ldots, \mathbf{g}_{k-1}$ is independent, then we are done, otherwise we repeat the previous step. The procedure terminates at the latest when the spanning set consists of a single vector; this cannot be $\mathbf{0}$ since $V \neq \mathbf{0}$, so it is independent. ∎

Theorem 4.5.7. *If a vector space has a (finite) spanning set, then every linearly independent system can be extended to a basis.* ♣

Proof. If the independent system is a spanning set, then it is a basis itself. Otherwise, there exists some $\mathbf{v} \in V$ that is not a linear combination of these linearly independent vectors. Then, extending the linearly independent system by $\mathbf{v}$ we still have an independent system (Theorem 4.4.3/IV). If this extended system is not a basis, we repeat the previous argument. The vector space possesses a finite spanning set of, say, s elements, so there cannot be more than s independent vectors in V by Theorem 4.5.4. Hence the procedure terminates in at most s steps. ∎

The proofs of Theorems 4.5.6 and 4.5.7 also yield methods on how to select a basis from a spanning set and how to extend a linearly independent system to a basis.

Similar to spanning and linear independence, we can also extend the notion of a basis for infinitely many vectors as a linearly independent spanning set. The analogue of Theorem 4.5.2 states that a set H of vectors is a basis if and only if every element in the vector space is essentially a unique linear combination of finitely many vectors from H; the adjective "essentially" means that two such representations may differ only in terms with coefficients 0. It can be shown by transfinite methods that every vector space $V \neq \mathbf{0}$ has a basis; it is often called a *Hamel basis*. The analogue of Theorem 4.5.3 asserts that any two (Hamel) bases in a vector space are of the same *cardinality*.

Exercises 4.5

4.5.1. Find a basis in the subspaces of the usual vector space of polynomials of degree not greater than 20 with real coefficients over $\mathbf{R}$. A general polynomial or polynomial function is denoted by $f = \alpha_0 + \alpha_1 x + \cdots + \alpha_{20}x^{20}$.

(a) $\{f \mid \deg f \le 10 \text{ or } f = 0\}$;
(b) $\{f \mid x^3 + 1 \text{ divides } f\}$;
(c) $\{f \mid$ the remainder of f is a constant under long division by $x^3 + 1\}$;
(d) $\{f \mid f(5) = 0\}$;
(e) $\{f \mid$ the sum of coefficients of f is $0\}$;
(f) $\{f \mid f(3) = 2f(4)\}$;
(g) $\{f \mid \alpha_0 = \alpha_1 = \alpha_{13}\}$.

4.5.2. Which sets form a basis in the usual vector space $\mathbf{Q}^{2\times 3}$ of the 2×3 matrices with rational elements over $\mathbf{Q}$? Extend the linearly independent systems to a basis and select a basis from the spanning sets.

(a) One element is 0 and all other elements are 5;
(b) two elements are 0 and all other elements are 5;
(c) all elements in a row or column are 5 and all other elements are 0;
(d) the elements in a column are 5 and 6 (in any order) and all other elements are 0;
(e)
$$\begin{pmatrix} 1 & 2 & 3 \\ 4 & 5 & 6 \end{pmatrix}$$
and its reflections across the horizontal and vertical bisectors and the center;
(f)
$$\begin{pmatrix} 1 & 2 & 3 \\ 4 & 6 & 8 \end{pmatrix}$$
and its reflections across the horizontal and vertical bisectors and the center.

4.5.3. A linearly independent system is *maximal* if no further vector can be added to it preserving independence, and it is *of maximal size* if no independent system can have more elements. Prove that these two notions are equivalent in every vector space.

4.5.4. Define a spanning set to be *minimal* and to be *of minimal size* and verify that these are equivalent to each other and to the notions in the previous exercise.

4.5.5. Show that if a vector space has no spanning set of 19 elements, then it has a linearly independent system of 20 elements.

4.5.6. Assume $\langle \mathbf{a}_1, \mathbf{a}_2, \mathbf{a}_3 \rangle = \langle \mathbf{b}_1, \mathbf{b}_2, \mathbf{b}_3, \mathbf{b}_4 \rangle$. What can be asserted concerning linear independence or dependence of $\mathbf{a}_1, \mathbf{a}_2, \mathbf{b}_1, \mathbf{b}_2$?

4.5.7. Let W be a non-trivial subspace in a vector space V. True or false?

(a) The elements in W of any basis of V form a basis of W.
(b) Any basis of W can be extended to a basis of V assuming that V has a basis at all.

(c) If V has a basis of k elements, then W has a spanning set of k elements.
(d) If V has a basis of k elements, then W has a linearly independent system of k elements.
(e) If W has a basis of k elements, then V has a spanning set of k elements.
(f) If W has a basis of k elements, then V has a linearly independent system of k elements.

4.5.8. Let $n \geq 2$ and let $\mathbf{u}_1, \dots, \mathbf{u}_n$ be a basis in a vector space V over $\mathbf{R}$. Which vectors form a linearly independent system, a spanning set, and a basis, respectively?
(a) $\mathbf{u}_1 - \mathbf{u}_2, \mathbf{u}_2 - \mathbf{u}_3, \dots, \mathbf{u}_n - \mathbf{u}_1$;
(b) $\mathbf{u}_1 - \mathbf{u}_2, \mathbf{u}_2 - \mathbf{u}_3, \dots, \mathbf{u}_{n-1} - \mathbf{u}_n$;
(c) $\mathbf{u}_1 + \mathbf{u}_2, \mathbf{u}_2 + \mathbf{u}_3, \dots, \mathbf{u}_n + \mathbf{u}_1$;
(d) $\mathbf{u}_1 + \mathbf{u}_2, \mathbf{u}_2 + \mathbf{u}_3, \dots, \mathbf{u}_{n-1} + \mathbf{u}_n, \mathbf{u}_n$;
(e) $\mathbf{u}_1 - \mathbf{u}_2, \mathbf{u}_2 - \mathbf{u}_3, \dots, \mathbf{u}_n - \mathbf{u}_1, \mathbf{u}_1 + \mathbf{u}_2 + \dots + \mathbf{u}_n$.

4.5.9. Let $\mathbf{u}_1, \dots, \mathbf{u}_n$ be a basis in a vector space V and let $\mathbf{v} = \alpha_1\mathbf{u}_1 + \dots + \alpha_n\mathbf{u}_n$. Prove that $\mathbf{u}_1 + \mathbf{v}, \dots, \mathbf{u}_n + \mathbf{v}$ is a basis in V if and only if $\alpha_1 + \dots + \alpha_n \neq -1$.

4.5.10. Let $\mathbf{u}_1, \dots, \mathbf{u}_n$ be a basis in a vector space V and let

$$\mathbf{v}_i = \beta_{1i}\mathbf{u}_1 + \dots + \beta_{ni}\mathbf{u}_n, \quad i = 1, 2, \dots, n.$$

Verify that $\mathbf{v}_1, \dots, \mathbf{v}_n$ is a basis in V if and only if the $n \times n$ determinant of the elements β_{ij} is not zero.

4.5.11. Let $\mathbf{u}_1, \dots, \mathbf{u}_n$ and $\mathbf{v}_1, \dots, \mathbf{v}_n$ be two arbitrary bases in a vector space V. Prove that to *every* $\mathbf{u}_i$ there exists some $\mathbf{v}_j$ so that swapping $\mathbf{u}_i$ and $\mathbf{v}_j$ we get again two bases.

4.5.12. (a) Does there exist a vector space of 243 elements over the field F_3 of the modulo 3 residue classes?
(b) Does there exist a vector space of 300 elements over the field F_3 of the modulo 3 residue classes?
**(c) Prove that the number of elements in a finite field has to be a prime power.
(d) Prove that the number of elements in a finite vector space has to be a prime power.

4.5.13. Assume that a vector space has only finitely many bases (but at least one). Show that the vector space contains finitely many elements.

*4.5.14. (a) How many bases are there in the vector space F_p^2 over the field F_p? And in F_p^n?
(b) How many $n \times n$ matrices with entries from the field F_p have an inverse?
(c) Prove $n! \mid (p^n - 1)(p^n - p)\dots(p^n - p^{n-1})$ for any prime p and integer $n > 0$.

4.6. Dimension

Definition 4.6.1. The *dimension* of a vector space V is the number of elements in a basis. If V has no finite spanning set, then its dimension is infinite. Finally, the $\mathbf{0}$ space has dimension 0. ♣

The dimension of V is denoted by $\dim V$.

Definition 4.6.1 uses that every basis consists of the same number of vectors (Theorem 4.5.3), so the dimension is independent of which basis we consider. The notion of infinite dimension can be refined to the cardinalities of Hamel bases mentioned at the end of the previous section.

Examples. The ordinary three-dimensional space has dimension 3; indeed, F^n and $F^{k\times n}$ have dimensions n and kn. The usual vector space $F[x]$ of the polynomials over F is infinite-dimensional, the polynomials of degree not exceeding r form an $(r+1)$-dimensional subspace.

The finite positive dimension has several equivalent characterizations: the most important ones are the maximal number of linearly independent vectors and the minimal number of elements in a spanning set.

Theorem 4.6.2. *The conditions below are equivalent for any vector space $V \neq \mathbf{0}$ and integer $n > 0$:*

(i) $\dim V = n$;

(ii) *V contains n independent vectors, but any $n+1$ vectors are dependent;*

(iii) *V contains a spanning set of n elements, but no $n-1$ elements form a spanning set.* ♣

It is easy to see that $n+1$ can be replaced in (ii) by more than n, and $n-1$ can be replaced in (iii) by less than n.

Proof.

(i)⇒(ii): The condition means that V has a basis of n elements. This consists of n independent vectors, by definition. There cannot be more independent vectors by Theorem 4.5.4.

(ii)⇒(i): We show that the n independent vectors form a basis. By the assumption, extending them by any vector we get a dependent system. Then the new vector is a linear combination of the original n vectors by Theorem 4.4.3/IV. This holds for an arbitrary vector, so the original n vectors form a spanning set. They are independent, so it is a basis.

The equivalence of (i) and (iii) can be proved similarly. ■

The simple observation below is often useful.

Theorem 4.6.3. *Let $n > 0$ be an integer and let* $\dim V = n$. *Then any independent system of n vectors is a basis in V. The same holds also for any spanning set of n elements.* ♣

Proof. If the independent system of n elements were not a basis, then it could be extended to a basis by Theorem 4.5.7. This new basis would have more than n elements contradicting $\dim V = n$. The assertion for spanning sets can be verified similarly. ∎

By Theorem 4.6.3, n vectors form a basis in an n-dimensional vector space if and only if they are linearly independent. This makes it easier to determine whether given n vectors form a basis or not, since we have to check only one of the two requirements in the definition of a basis. (And if the number of vectors differs from the dimension, then they cannot be a basis for sure.) This observation can yield a simpler solution also for several exercises in the previous section.

The next theorem examines the relation of the dimensions of a vector space and its subspace. The results are consistent with our intuition.

Theorem 4.6.4.

I. *Let W be a subspace in V. Then* $\dim W \leq \dim V$.

II. *If V is finite-dimensional, W is a subspace in V, and* $\dim W = \dim V$, *then* $W = V$. ♣

Proof.

I. The statement is obvious if $W = \mathbf{0}$ or $\dim V = \infty$. In the other cases V has a basis. There cannot be more independent vectors in W than the number of elements in a basis of V, since such vectors would be independent in V, too, contradicting Theorem 4.5.4. Thus, Theorem 4.6.2 implies $\dim W \leq \dim V$.

II. The statement is obvious for $V = \mathbf{0}$. Otherwise let $\dim V = \dim W = n (\neq 0)$ and consider a basis in W (of n elements). This is an independent system in V, so it is a basis in V, too, by Theorem 4.6.3. Thus, this spanning set of W is a spanning set also for V, hence V cannot be larger than W. ∎

Part II in Theorem 4.6.4 is false if the dimension of V is infinite: e.g., the polynomials divisible by x form a proper subspace in the usual vector space of polynomials while the dimensions are equal.

Now we generalize the notion of the rank of a matrix.

Definition 4.6.5. The *rank* of a system of vectors $\mathbf{a}_1, \dots, \mathbf{a}_n$ is r if there exist r linearly independent vectors $\mathbf{a}_i$, but any $r + 1$ among them are linearly dependent. ♣

Thus the rank is the maximal number of linearly independent elements in a system of vectors. In general, there are more than one such independent subsystems of maximal size among the given vectors.

Now we apply the rank to describe the dimension of a subspace spanned by some given vectors.

Theorem 4.6.6. *The dimension of the subspace $\langle \mathbf{a}_1, \dots, \mathbf{a}_n \rangle$ is the rank of the system of vectors $\mathbf{a}_1, \dots, \mathbf{a}_n$.* ♣

Proof. Let, e.g., $\mathbf{a}_1, \dots, \mathbf{a}_r$ be an independent system of maximal size. We show that this is a basis in $W = \langle \mathbf{a}_1, \dots, \mathbf{a}_n \rangle$. Independence holds, so we have to show that every vector in W is a linear combination of $\mathbf{a}_1, \dots, \mathbf{a}_r$. Clearly, it is sufficient to verify this for the generators of W. Extending the independent vectors $\mathbf{a}_1, \dots, \mathbf{a}_r$ by any $\mathbf{a}_i$, these $r+1$ vectors are dependent, so $\mathbf{a}_i$ is a linear combination of $\mathbf{a}_1, \dots, \mathbf{a}_r$. ∎

We can apply Theorem 4.6.6 to obtain a new proof for the solvability of a system of linear equations (see the first statement in Theorem 3.4.3).

Theorem 4.6.7. *A system of linear equations is solvable if and only if the rank of the coefficient matrix equals the rank of the augmented matrix.* ♣

Proof. We write the system of equations as

$$x_1\mathbf{a}_1 + \cdots + x_n\mathbf{a}_n = \mathbf{b},$$

where $\mathbf{a}_i$ is the ith column of the coefficient matrix. The system is solvable if and only if

$$\mathbf{b} \in \langle \mathbf{a}_1, \dots, \mathbf{a}_n \rangle.$$

This is equivalent to

$$\langle \mathbf{a}_1, \dots, \mathbf{a}_n \rangle = \langle \mathbf{a}_1, \dots, \mathbf{a}_n, \mathbf{b} \rangle$$

since two subspaces are equal if and only if they contain mutually the generators of each other. Finally, the subspace on the left-hand side is part of the one on the right-hand side and both have finite dimensions, so by Theorem 4.6.4/II, they are equal exactly if their dimensions are equal, i.e.,

$$\dim\langle \mathbf{a}_1, \dots, \mathbf{a}_n \rangle = \dim\langle \mathbf{a}_1, \dots, \mathbf{a}_n, \mathbf{b} \rangle.$$

By Theorem 4.6.6 this means that the ranks of the two systems of vectors are the same. These are just the ranks of the coefficient and augmented matrices yielding the theorem. ∎

Exercises 4.6

4.6.1. Compute the dimensions of the vector spaces (with the usual operations).

(a) The complex numbers over $\mathbf{R}$;

(b) the complex numbers over $\mathbf{Q}$;

(c) the symmetric $n \times n$ matrices;

(d) the polynomials over the field F_p;

(e) the polynomial *functions* over the field F_p;

(f) the homogeneous polynomials of degree 6 in 4 variables over $\mathbf{R}$ (i.e., where the total degree of every term is exactly 6) and the 0;

(g) the polynomials of degree at most 6 in 4 variables over $\mathbf{R}$ (i.e., where the total degree of every term is at most 6) and the 0;

(h) the polygonal line functions defined in the interval $[0, 1]$ changing directions at most at the rational numbers with denominator 19;

(i) the solutions of a homogeneous system of k linear equations in n variables where the rank of the coefficient matrix is r;

(j) the kernel of a matrix (see Example E2 in Section 4.2);

(k) the image of a matrix (see Example E2 in Section 4.2).

4.6.2. Which are bases in the usual vector space of polynomials of degree not greater than 12 with real coefficients?
(a) $(x-1)(x-2)(x-3)\dots(x-12)$, $(x-2)(x-3)(x-4)\dots(x-13)$, $\dots$, $(x-13)(x-14)(x-15)\dots(x-24)$;
(b) $(x-1)^{12}, (x-2)^{12}, \dots, (x-13)^{12}$;
(c) $(x-1)^{12}, (x-1)^{11}(x-2), \dots, (x-2)^{12}$;
(d) $(x^2-1)^6, (x^2-2)^6, \dots, (x^2-8)^6, (x-1)^{12}, \dots, (x-5)^{12}$.

4.6.3. Let $0 \le k \le n$ be arbitrary integers. Demonstrate that there exists a k-dimensional subspace in every n-dimensional vector space.

4.6.4. Let V be the usual vector space $F^{n\times n}$ of the $n\times n$ matrices over the field F and let $B \in F^{n\times n}$ be a fixed matrix. Consider the subspace W of the matrices A satisfying $BA = 0$. Verify that n divides $\dim W$.

4.6.5. Let W_1 and W_2 be subspaces in V, $\dim V = 40$, $\dim W_1 = 23$ and $\dim W_2 = 18$. Show $W_1 \cap W_2 \neq \mathbf{0}$.

4.6.6. Let W_1 and W_2 be subspaces in V. Prove
(a) $\dim\langle W_1, W_2\rangle \le \dim W_1 + \dim W_2$;
(b) $\dim(W_1 \oplus W_2) = \dim W_1 + \dim W_2$;
(c) $\dim\langle W_1, W_2\rangle = \dim W_1 + \dim W_2 - \dim(W_1 \cap W_2)$.

***4.6.7.** (a) Let V be a 100-dimensional vector space over $\mathbf{R}$. How many vectors are there in V such that *any* 100 of them form a basis?
(b) Solve the same problem for vector spaces over the fields F_2, F_{97}, and F_{101}.

4.6.8. The sequence of *Fibonacci numbers* is defined by the recursion

$$\varphi_0 = 0, \quad \varphi_1 = 1, \quad \varphi_{i+1} = \varphi_i + \varphi_{i-1}, \quad i = 1, 2, \dots.$$

Find an explicit formula for φ_n. (Hint: Consider all real sequences $\alpha_0, \alpha_1, \dots$ satisfying $\alpha_{i+1} = \alpha_i + \alpha_{i-1}$, $i = 1, 2, \dots$. (i) Compute the dimension of this vector space. (ii) Then find a basis consisting of "nice" sequences and express the Fibonacci sequence with this basis.)

4.6.9. Give a parametric characterization of all 3×3 *magic squares* of real numbers where the sums in all rows, columns, and diagonals are equal.

4.6.10. Compute the dimensions of the spans of the system of vectors in Exercise 4.3.1.

4.6.11. We know that for some elements $\mathbf{a}_1, \dots, \mathbf{a}_k$ in a vector space V, the vectors $\mathbf{a}_i + \mathbf{a}_j$, $1 \le i < j \le k$ form a basis in V. Prove that $\dim V = 1$ or 3.

4.6.12. Verify that the rank of a system of vectors remains the same if
(a) one of the vectors is multiplied by a scalar $\lambda \neq 0$;
(b) a scalar multiple of another vector is added to one of the vectors.

4.6.13. Prove that the rank of the sum of two matrices is not greater than the sum of the ranks, i.e., $\mathrm{rk}(A+B) \le \mathrm{rk}(A) + \mathrm{rk}(B)$ if $A, B \in F^{k\times n}$.

*4.6.14. How many r-dimensional subspaces are there in the vector space F_p^n over the field F_p?

**4.6.15. How many $k \times n$ matrices with elements from the field F_p have rank r?

4.6.16. (a) Prove that if every element of a matrix is 0 or 1, then its rank over F_2 cannot exceed its rank over $\mathbf{R}$.

(b) Show an example of a 0–1 matrix such that its rank over F_2 is by 1000 less than its rank over $\mathbf{R}$.

**(c) Determine the smallest integer n for which there exists an $n \times n$ 0–1 matrix with the property in part (b).

4.7. Coordinates

Definition 4.7.1. Let $\mathbf{b}_1, \dots, \mathbf{b}_n$ be a fixed basis in a vector space V. Then every vector $\mathbf{v} \in V$ has a unique representation $\mathbf{v} = \alpha_1\mathbf{b}_1 + \cdots + \alpha_n\mathbf{b}_n$. The scalars α_i are called the *coordinates* of the vector $\mathbf{v}$ in the basis $\mathbf{b}_1, \dots, \mathbf{b}_n$. ♣

In the ordinary three-dimensional space we get the usual coordinates if the basis consists of the standard orthogonal unit vectors. In F^n the coordinates of a vector are just its components if the basis consists of the vectors having one component 1 and all other components 0. The coordinates vary in general if we modify the basis.

After fixing a basis, it is convenient to work with coordinates instead of vectors. The coordinates of the sum of two vectors are the sums of the corresponding coordinates, and a similar statement is valid for the scalar multiples, too. Thus we reduced any n-dimensional vector space V over a field F to the single vector space F^n. In other words, V and F^n are identical from an algebraic point of view, just the elements and operations are denoted differently. Such vector spaces are called *isomorphic*(=having the same form). We shall clarify this notion in Section 5.2 in more detail.

Exercises 4.7

4.7.1. How do the coordinates of a vector change if

(a) two elements of the basis are swapped;

(b) an element of the basis is multiplied by $\lambda(\neq 0)$;

(c) λ times another element is added to an element of the basis?

4.7.2. Determine all vectors that have the same coordinates in every basis.

4.7.3. Consider the basis

$$\begin{pmatrix}1\\2\\5\end{pmatrix}, \begin{pmatrix}3\\7\\8\end{pmatrix}, \begin{pmatrix}2\\5\\2\end{pmatrix}$$

in the vector space $\mathbf{C}^3$. Compute the coordinates of the vectors

$$\begin{pmatrix}1\\0\\0\end{pmatrix}, \begin{pmatrix}0\\1\\0\end{pmatrix}, \text{ and } \begin{pmatrix}0\\0\\1\end{pmatrix}.$$

4.7.4. Let $\mathbf{v} \neq \mathbf{0}$ be a given vector in an n-dimensional vector space and consider all bases $\mathbf{b}_1, \dots, \mathbf{b}_n$ where the first two coordinates of $\mathbf{v}$ are 1. What are the possible values of $\mathbf{b}_1$?

4.7.5. (a) The Sultan thought of a basis in $\mathbf{R}^{1001}$ and Scheherazade has to figure it out in 1001 nights or she must die. Every night she can choose a vector and ask its coordinates. Will she remain alive?

(b) What happens if she can always ask only the first coordinate and she can survive if she figures out the first basis vector?

Chapter 5

Linear Maps

A linear map is an operation-preserving function between two vector spaces. An important special case is a bijective linear map, called isomorphism, implying that the two vector spaces are essentially the same, just the elements and operations are denoted differently. The nice structure of vector spaces is shown by the fact that the dimension determines a vector space completely, i.e., any two vector spaces (over the same field) having equal dimensions are isomorphic.

The linear maps are closely related to matrices including the operations defined for them. The characterization of linear maps by matrices is of fundamental importance both from theoretical and practical points of view.

5.1. Linear Map

Definition 5.1.1. Let V_1 and V_2 be vector spaces over the *same* commutative field F. A function $\mathcal{A}$ from V_1 to V_2 is called a *linear map* (or a vector space *homomorphism*) if it *preserves the operations*, i.e.,

(i) $\mathcal{A}(\mathbf{u}+\mathbf{v}) = \mathcal{A}\mathbf{u} + \mathcal{A}\mathbf{v}$ for every $\mathbf{u}, \mathbf{v} \in V_1$; and

(ii) $\mathcal{A}(\lambda\mathbf{u}) = \lambda(\mathcal{A}\mathbf{u})$ for every $\mathbf{u} \in V_1, \lambda \in F$. ♣

A linear map thus assigns a *unique* vector in V_2 to *every* vector in V_1. It may assign the same vector of V_2 to *several* vectors of V_1, i.e., a vector in V_2 may have more *inverse images* in V_1. On the other hand, not every vector in V_2 occurs necessarily as an image of a vector in V_1, it is possible that some elements in V_2 have no inverse image in V_1.

The two signs + in (i) mean two different operations, namely additions in V_1 and V_2. The same applies to the multiplication by a scalar in (ii).

We shall denote the linear maps by calligraphic capital letters. The notion is a special case of *homomorphisms* between algebraic structures.

A linear map also preserves zero, negatives, and linear combinations.

Theorem 5.1.2.

I. $\mathcal{A}\mathbf{0}_1 = \mathbf{0}_2$ *where* $\mathbf{0}_i$ *is the zero vector of* V_i.

II. $\mathcal{A}(-\mathbf{u}) = -(\mathcal{A}\mathbf{u})$.

III. $\mathcal{A}(\lambda_1\mathbf{u}_1 + \cdots + \lambda_k\mathbf{u}_k) = \lambda_1\mathcal{A}\mathbf{u}_1 + \cdots + \lambda_k\mathcal{A}\mathbf{u}_k$. ♣

Proof. The addition-preserving property implies $\mathcal{A}\mathbf{u} = \mathcal{A}(\mathbf{u} + \mathbf{0}_1) = \mathcal{A}\mathbf{u} + \mathcal{A}\mathbf{0}_1$. Adding the negative of $\mathcal{A}\mathbf{u}$ to both sides we have I. Applying the linear map $\mathcal{A}$ to $\mathbf{u} + (-\mathbf{u}) = \mathbf{0}$ and using I yields II. (Or we could have applied $\mathcal{A}$ to $-\mathbf{u} = (-1)\mathbf{u}$.) Finally, III follows immediately from repeated applications of (i) and (ii). ∎

In the sequel we shall denote both zero vectors of V_1 and V_2 by $\mathbf{0}$.

A linear map $\mathcal{A}$ induces two important sets.

Definition 5.1.3. Let $\mathcal{A}$ be a linear map from V_1 to V_2. The *image* (or *range space*) $\operatorname{Im}\mathcal{A}$ is the set of all image vectors in V_2:

$$\operatorname{Im}\mathcal{A} = \{\mathbf{y} \in V_2 \mid \exists \mathbf{x} \in V_1\ \mathcal{A}\mathbf{x} = \mathbf{y}\} = \{\mathcal{A}\mathbf{x} \mid \mathbf{x} \in V_1\}. ♣$$

Definition 5.1.4. Let $\mathcal{A}$ be a linear map from V_1 to V_2. The *kernel* (or *null space*) $\operatorname{Ker}\mathcal{A}$ is the set of all vectors in V_1 mapped to the zero vector of V_2:

$$\operatorname{Ker}\mathcal{A} = \{\mathbf{x} \in V_1 \mid \mathcal{A}\mathbf{x} = \mathbf{0}\}. ♣$$

Thus $\operatorname{Im}\mathcal{A}$ is a subset of V_2 and $\operatorname{Ker}\mathcal{A}$ is a subset of V_1. As expected, they are subspaces.

Theorem 5.1.5. $\operatorname{Im}\mathcal{A}$ *is a subspace in* V_2 *and* $\operatorname{Ker}\mathcal{A}$ *is a subspace in* V_1. ♣

Proof. The kernel is not empty as $\mathcal{A}\mathbf{0} = \mathbf{0}$. Further, it is closed under the operations: If $\mathcal{A}\mathbf{u} = \mathcal{A}\mathbf{v} = \mathbf{0}$, then

$$\mathcal{A}(\mathbf{u} + \mathbf{v}) = \mathcal{A}\mathbf{u} + \mathcal{A}\mathbf{v} = \mathbf{0} + \mathbf{0} = \mathbf{0} \quad \text{and} \quad \mathcal{A}(\lambda\mathbf{u}) = \lambda(\mathcal{A}\mathbf{u}) = \lambda\mathbf{0} = \mathbf{0}.$$

The statement for the image can be verified similarly. ∎

Examples of linear maps

E1 Let $V_1 = V_2$ be the usual vector space of plane vectors ($F = \mathbf{R}$). Some linear maps are:

(a) rotations around the origin by arbitrary angles;
(b) magnifications from the origin by arbitrary factors;
(c) reflections across lines passing through the origin;
(d) projections in arbitrary directions onto lines passing through the origin.

In examples (a), (b), and (c) we have $\operatorname{Ker}\mathcal{A} = \mathbf{0}$ and $\operatorname{Im}\mathcal{A} = V_2$, whereas in (d) the image is the target line of the projection and the kernel is the line parallel to the direction of the projection passing through the origin.

Note that translations are *not* linear maps since the image of the zero vector is not zero. Also, there are many other linear maps of the plane besides congruences and similarities.

E2 For any V_1 and V_2 we map every element of V_1 to the zero vector of V_2. This linear map is called the *zero map* and is denoted by $\mathcal{O}$. Its kernel is the entire V_1 and its image consists only of the zero vector $\mathbf{0}$ of V_2.

E3 Let $V = V_1 = V_2$ and we assign to every element itself. This linear map is the *identity map* and is denoted by $\mathcal{I}$. Its kernel is $\mathbf{0}$ and its image is the entire V.

E4 Let $A \in F^{k\times n}$, $V_1 = F^n$, $V_2 = F^k$, and the linear map is the multiplication by the matrix A, i.e., $\mathcal{A}\mathbf{x} = A\mathbf{x}$. The kernel and the image are just the kernel and the image of the matrix A (see example E4 in Section 4.2).

E5 Let V_1 be an n-dimensional vector space and let $V_2 = F^n$. We fix a basis in V_1 and express every vector in V_1 as a linear combination of the basis vectors. Now, we assign to every vector its *coordinate vector* in F^n formed of its coordinates in the given basis. The kernel of this linear map is $\mathbf{0}$ and its image is the entire F^n.

E6 The various fields of mathematics are abundant in important linear maps. We just indicate a few in analysis and leave the precise definition of the maps, the verification of their linearity, and the determination of the kernels and images to the reader.

We assign to (suitable real) functions their values assumed at a given place; their derivatives; their (definite or indefinite) integrals; their restrictions to a given subset of their domain; their product by a given function; we assign to (suitable real infinite) sequences their limits; the (infinite) sum of their terms; suitable subsequences; etc.

For further examples see Exercises 5.1.1–5.1.4.

Definition 5.1.6. A $V \to V$ linear map (i.e., when $V_1 = V_2 = V$) is called a *linear transformation* of V. ♣

Examples E1 and E3 are linear transformations. The projections in E1 show that the image of a linear transformation is not necessarily the entire V and several vectors can have the same image.

We note that some books use the expression linear transformation for any linear map and do not apply a special terminology for the case $V_1 = V_2$.

Exercises 5.1

5.1.1. Let $V = V_1 = V_2$ be the usual vector space of polynomials with real coefficients of degree not greater than 100 (including the zero polynomial). Which maps are linear transformations of V? For each linear transformation, find its kernel and image and determine their dimensions. We denote a general polynomial by f or $f(x)$; α_i is the coefficient of the term of degree i; and the leading coefficient is α_n (so $\alpha_n \neq 0$ except if f is the zero polynomial).

(a) $f \mapsto f'$;

(b) $f(x) \mapsto xf(x)$;

(c) $f(x) \mapsto f(x) - xf'(x)$;

(d) $f(x) \mapsto f(x+1) - f(x)$;

(e) $f \mapsto \alpha_0 x$;
(f) $f \mapsto \alpha_n x^2$;
(g) $f \mapsto (\alpha_0 + \alpha_1 + \cdots + \alpha_n)(x + x^2)$;
(h) $f \mapsto (\deg f)x^3$;
(i) $f \mapsto$ the remainder of f under division by $x^7 + 4x + 1$;
(j) $f \mapsto \alpha_n + \alpha_{n-1}x + \cdots + \alpha_0 x^n$.

5.1.2. Let $F = \mathbf{R}$ and $V = V_1 = V_2 = \mathbf{C}$ with the usual operations. Which maps are linear transformations of V? For each linear transformation, find its kernel and image and determine their dimensions. To each complex number we assign
(a) its real part;
(b) the maximum of its real and imaginary parts;
(c) its absolute value;
(d) its angle;
(e) its conjugate;
(f) its product by a fixed complex number;
(g) its square;
(h) the difference of $(1 + i)$ times its real part and $(1111i - \sqrt{2})$ times its imaginary part.

5.1.3. Let F be the field $\mathbf{Z}_2$ of the modulo 2 residue classes, $V_1 = F^{3\times3}$, and $V_2 = F^3$ with the usual operations. Which are linear maps from V_1 to V_2? For each linear map find its kernel and image and determine their dimensions. To each matrix we assign
(a) its middle column;
(b) the vector with all coordinates equal to the determinant of the matrix;
(c) the vector with all coordinates equal to the trace of the matrix(=the sum of elements in the main diagonal);
(d) the vector with all coordinates equal to the modulo 2 residue of the rank of the matrix;
(e) its product with the vector with every coordinate 1;
(f) the vector with every coordinate 1 if the matrix is non-singular and the zero vector if the matrix is singular.

5.1.4. Let $V = V_1 = V_2$ be the usual vector space of infinite rational sequences over $\mathbf{Q}$. We denote a general sequence by $S = (\alpha_0, \alpha_1, \ldots, \alpha_n, \ldots)$. Find the kernel and image of each linear transformation and determine their dimensions.
(a) Right shift: $(\alpha_0, \alpha_1, \ldots, \alpha_n, \ldots) \mapsto (0, \alpha_0, \alpha_1, \ldots, \alpha_{n-1}, \ldots)$;
(b) Left shift: $(\alpha_0, \alpha_1, \ldots, \alpha_n, \ldots) \mapsto (\alpha_1, \alpha_2, \alpha_3, \ldots, \alpha_{n+1}, \ldots)$;
(c) Double: $(\alpha_0, \alpha_1, \ldots, \alpha_n, \ldots) \mapsto (\alpha_0, \alpha_0, \alpha_1, \alpha_1, \ldots, \alpha_n, \alpha_n, \ldots)$;
(d) Decimation: $(\alpha_0, \alpha_1, \ldots, \alpha_n, \ldots) \mapsto (\alpha_0, \alpha_{10}, \alpha_{20}, \ldots, \alpha_{10n}, \ldots)$;
(e) $(\alpha_0, \alpha_1, \ldots, \alpha_n, \ldots) \mapsto (\alpha_0 - \alpha_1, \alpha_1 - \alpha_2, \ldots, \alpha_n - \alpha_{n+1}, \ldots)$;
(f) $(\alpha_0, \alpha_1, \ldots, \alpha_n, \ldots) \mapsto (\alpha_0 + \alpha_1, \alpha_0 - \alpha_1, \alpha_2 + \alpha_3, \alpha_2 - \alpha_3, \ldots)$;
(g) $(\alpha_0, \alpha_1, \ldots, \alpha_n, \ldots) \mapsto (\alpha_0 + \alpha_1, \alpha_2 + \alpha_3, \alpha_0 + \alpha_2, \alpha_1 + \alpha_3, \alpha_4 + \alpha_5, \alpha_6 + \alpha_7, \alpha_4 + \alpha_6, \alpha_5 + \alpha_7, \ldots)$.

5.1.5. Let W be a non-trivial subspace in the vector space V. Are the functions linear transformations of V?

(a) $\mathcal{A}\mathbf{x} = \begin{cases} \mathbf{x}, & \text{if } \mathbf{x} \in W; \\ \mathbf{0}, & \text{if } \mathbf{x} \notin W. \end{cases}$

(b) $\mathcal{B}\mathbf{x} = \begin{cases} \mathbf{x}, & \text{if } \mathbf{x} \notin W; \\ \mathbf{0}, & \text{if } \mathbf{x} \in W. \end{cases}$

5.1.6. Show examples for functions between vector spaces V_1 and V_2 (over the same field F)

(a) preserving multiplication by scalars but not preserving addition;
(b) preserving addition but not preserving multiplication by scalars;
(c) preserving neither addition, nor multiplication by scalars.

5.1.7. Find all vector spaces V with the property: If V' is an arbitrary vector space over the same field and a map from V to V' preserves multiplication by scalars, then this is necessarily a linear map.

*5.1.8. Prove that if a map between vector spaces over (a) $\mathbf{Z}_p$; (b) $\mathbf{Q}$ preserves addition, then it is necessarily linear. Does the same hold for $\mathbf{R}$?

5.1.9. Let $\mathcal{A}$ be a linear map from V_1 to V_2, $\mathbf{c}_i \in V_1$. True or false?

(a) If $\mathbf{c}_1, \dots, \mathbf{c}_k$ are linearly independent, then $\mathcal{A}\mathbf{c}_1, \dots, \mathcal{A}\mathbf{c}_k$ are linearly independent.
(b) If $\mathcal{A}\mathbf{c}_1, \dots, \mathcal{A}\mathbf{c}_k$ are linearly independent, then $\mathbf{c}_1, \dots, \mathbf{c}_k$ are linearly independent.
(c) If $\mathbf{c}_1, \dots, \mathbf{c}_k$ span V_1, then $\mathcal{A}\mathbf{c}_1, \dots, \mathcal{A}\mathbf{c}_k$ span V_2.
(d) If $\mathbf{c}_1, \dots, \mathbf{c}_k$ span V_1, then $\mathcal{A}\mathbf{c}_1, \dots, \mathcal{A}\mathbf{c}_k$ span $\operatorname{Im} \mathcal{A}$.
(e) If $\mathcal{A}\mathbf{c}_1, \dots, \mathcal{A}\mathbf{c}_k$ span $\operatorname{Im} \mathcal{A}$, then $\mathbf{c}_1, \dots, \mathbf{c}_k$ span V_1.

5.1.10. Let $\mathcal{A}$ be a linear map from V_1 to V_2 and $\mathbf{c}_1, \dots, \mathbf{c}_k$ a spanning set in V_1. Prove that the rank of the system $\mathcal{A}\mathbf{c}_1, \dots, \mathcal{A}\mathbf{c}_k$ is equal to $\dim \operatorname{Im} \mathcal{A}$.

5.1.11. Let $\mathcal{A}$ be a linear map from V_1 to V_2. Verify

$$\mathcal{A}\mathbf{u} = \mathcal{A}\mathbf{v} \iff \mathbf{u} - \mathbf{v} \in \operatorname{Ker} \mathcal{A}.$$

5.1.12. Prove that every vector in $\operatorname{Im} \mathcal{A}$ has the same number of inverse images. What are the possible values of this number in vector spaces over $\mathbf{Z}_{101}$?

5.1.13. Assume that $\mathcal{A}$ is a linear map from V_1 to V_2 and $\mathbf{u}_1, \dots, \mathbf{u}_k$ are linearly independent vectors in V_1 having the same image $\mathcal{A}\mathbf{u}_1 = \dots = \mathcal{A}\mathbf{u}_k$. Prove $\dim \operatorname{Ker} \mathcal{A} \geq k - 1$.

5.1.14. Let $0 < \dim V_1 < \infty$ and let $\mathcal{A}$ be an arbitrary non-zero linear map from V_1 to some V_2. Show that V_1 has a basis $\mathbf{b}_1, \dots, \mathbf{b}_n$ such that no $\mathcal{A}\mathbf{b}_i$ is $\mathbf{0}$.

5.1.15. Let $0 < \dim V < \infty$ and let $\mathcal{A} : V \to Z$ be a linear map. Prove that all elements of some basis in V have the same image if and only if $\dim \operatorname{Im} \mathcal{A} \leq 1$.

5.1.16. Let $\mathcal{A}$ be a linear map from V_1 to V_2. The image of a subset $H \subseteq V_1$ is the subset $\mathcal{A}H = \{\mathcal{A}\mathbf{x} \mid \mathbf{x} \in H\} \subseteq V_2$. Let U and Z be two subspaces in V_1. What is the relation between

(a) $\mathcal{A}U \cap \mathcal{A}Z$ and $\mathcal{A}(U \cap Z)$;
(b) $\mathcal{A}\langle U, Z \rangle$ and $\langle \mathcal{A}U, \mathcal{A}Z \rangle$?

5.2. Isomorphism

Definition 5.2.1. A bijective linear map is called an *isomorphism*. A vector space V_1 is *isomorphic* to a vector space V_2 if there exists a $V_1 \to V_2$ isomorphism. This relation is denoted by $V_1 \cong V_2$. ♣

Thus an isomorphism is a $V_1 \to V_2$ linear map where every element of V_2 has *exactly one* inverse image. In other words, distinct elements of V_1 are mapped to distinct elements of V_2, and every element of V_2 occurs among the images.

Isomorphic vector spaces are indistinguishable from an algebraic point of view: they are like identical twins, just the elements and operations have different names.

Among the examples in the previous section, E1(a)–(c), E3, and E5 are isomorphisms. A simple observation shows that an isomorphism can be recognized from its kernel and image.

Theorem 5.2.2. *A linear map* $\mathcal{A} : V_1 \to V_2$ *is an isomorphism if and only if* $\operatorname{Ker} \mathcal{A} = \mathbf{0}$ *and* $\operatorname{Im} \mathcal{A} = V_2$. ♣

Proof. $\operatorname{Im} \mathcal{A} = V_2$ means, by definition, that every element of V_2 is an image. We verify that the condition on the kernel is equivalent to distinct vectors in V_1 having distinct images in V_2. First assume that distinct vectors have distinct images. As $\mathbf{0}$ is mapped to $\mathbf{0}$, no other element can be mapped to $\mathbf{0}$, so $\operatorname{Ker} \mathcal{A} = \mathbf{0}$. Conversely, assume $\operatorname{Ker} \mathcal{A} = \mathbf{0}$ and let $\mathcal{A}\mathbf{u} = \mathcal{A}\mathbf{v}$. By Exercise 5.1.11 this implies $\mathbf{u} - \mathbf{v} \in \operatorname{Ker} \mathcal{A} = \mathbf{0}$, so $\mathbf{u} = \mathbf{v}$. ■

We can replace the asymmetric form "*U* is isomorphic to *V*" by a symmetric version "*U* and *V* are isomorphic", due to the following theorem.

Theorem 5.2.3. *For vector spaces, U isomorphic to V is an equivalence relation.* ♣

Proof. We show that the relation is reflexive, symmetric, and transitive.

Reflexivity: The identical linear transformation $\mathcal{I} : U \to U$ is clearly an isomorphism.

Symmetry: If $\mathcal{A} : U \to V$ is an isomorphism, then we show that its inverse $\mathcal{A}^{-1} : V \to U$ is an isomorphism, too. It is clearly bijective, so we need that it preserves the operations. For addition, consider $\mathbf{v}_1, \mathbf{v}_2 \in V$. Then $\mathcal{A}^{-1}\mathbf{v}_i$ is the (uniquely determined) $\mathbf{u}_i \in U$ satisfying $\mathcal{A}\mathbf{u}_i = \mathbf{v}_i$. Since $\mathcal{A}$ preserves addition, we have

$$\mathcal{A}(\mathbf{u}_1 + \mathbf{u}_2) = \mathcal{A}\mathbf{u}_1 + \mathcal{A}\mathbf{u}_2 = \mathbf{v}_1 + \mathbf{v}_2,$$

hence

$$\mathcal{A}^{-1}(\mathbf{v}_1 + \mathbf{v}_2) = \mathbf{u}_1 + \mathbf{u}_2 = \mathcal{A}^{-1}\mathbf{v}_1 + \mathcal{A}^{-1}\mathbf{v}_2.$$

We can verify preserving multiplication by scalars similarly.

Transitivity: We can show by similar arguments that the composition of two isomorphisms is an isomorphism. ■

As indicated in Section 4.7, every n-dimensional vector space ($0 < n < \infty$) over F is isomorphic to F^n.

Theorem 5.2.4. *If* $0 < n < \infty$ *and* V *is an* n*-dimensional vector space over* F*, then* $V \cong F^n$. ♣

Proof. The map in Example E5 in Section 5.1 is an isomorphism. ∎

This implies that over a given field there exists only one n-dimensional vector space up to isomorphism.

Theorem 5.2.5. *Two finite-dimensional vector spaces over a field* F *are isomorphic if and only if their dimensions are equal. So, if* U *and* V *have finite dimensions, then*

$$U \cong V \iff \dim U = \dim V. \text{ ♣}$$

Proof. This is obvious if one of the vector spaces is $\mathbf{0}$. Otherwise, if both dimensions are n, then both vector spaces are isomorphic to F^n, so they are isomorphic to each other, too. For the converse, assume $U \cong V$, i.e., there exists an isomorphism $\mathcal{A} : U \to V$. Let $\mathbf{u}_1, \dots, \mathbf{u}_n$ be a basis in U. We show that $\mathcal{A}\mathbf{u}_1, \dots, \mathcal{A}\mathbf{u}_n$ is a basis in V thus proving the equality of the dimensions. We have to verify that every $\mathbf{v} \in V$ has a unique decomposition

$$\mathbf{v} = \lambda_1(\mathcal{A}\mathbf{u}_1) + \cdots + \lambda_n(\mathcal{A}\mathbf{u}_n).$$

This is equivalent to

$$\mathbf{v} = \mathcal{A}(\lambda_1\mathbf{u}_1 + \cdots + \lambda_n\mathbf{u}_n)$$

since $\mathcal{A}$ is linear. Further, $\mathcal{A}$ is bijective, so $\mathbf{v}$ has exactly one inverse image $\mathbf{u} \in U$ satisfying $\mathcal{A}\mathbf{u} = \mathbf{v}$. Therefore the previous condition is equivalent to

$$\mathbf{u} = \lambda_1\mathbf{u}_1 + \cdots + \lambda_n\mathbf{u}_n.$$

As $\mathbf{u}_1, \dots, \mathbf{u}_n$ is a basis in U, thus there exists exactly one such system of coefficients $\lambda_1, \dots, \lambda_n$. ∎

Theorem 5.2.5 remains valid for infinite dimensions, too, if dimension means the *cardinality* of the (Hamel) basis (cf. the end of Section 4.5).

Exercises 5.2

5.2.1. Which maps in Exercises 5.1.1–5.1.4 are isomorphisms?

5.2.2. Which well-known vector spaces are isomorphic to the ones occurring in Exercises 4.1.5 and 4.1.6?

5.2.3. Which conditions imply that an $\mathcal{A} : U \to V$ linear map is an isomorphism?

(a) The image of every linearly independent system in U is linearly independent in V.

(b) The image of every spanning set in U is a spanning set in V.

(c) The image of every basis in U is a basis in V.

(d) There exists a basis in U such that its image is a basis in V.

(e) There exists a linearly independent system in U such that its image is linearly independent in V, and there exists a spanning set in U such that its image is a spanning set in V.

5.2.4. Verify that for any two finite-dimensional vector spaces over the same field one of them is isomorphic to a subspace of the other one.

5.2.5. How many pairwise non-isomorphic subspaces are there in an n-dimensional vector space?

5.2.6. Which vector spaces are isomorphic? They are over $\mathbf{R}$ and operations are as usual. The polynomials in (a)–(e) have real coefficients and contain the zero polynomial, too.

(a) Polynomials of degree at most 20 containing terms only with prime exponents.
(b) Polynomials of degree at most 15 that are even $\mathbf{R} \to \mathbf{R}$ functions.
(c) Polynomials of degree at most 9 having π as a root.
(d) Polynomials of degree at most 9 having the complex number i as a root.
(e) Polynomials of degree at most 100 divisible both by $x^{29}+1$ and $2x^{29}+1$.
(f) $\mathbf{C}^4$ (over $\mathbf{R}$!).
(g) 3×4 real matrices with equal first and last rows.
(h) 7×7 real matrices with equal elements in the main diagonal.
(i) 7×7 real matrices with equal elements outside the main diagonal.
(j) Infinite real sequences where the sum of arbitrary nine consecutive terms is 0.
(k) $\mathbf{R} \to \mathbf{R}$ functions that assume 0 everywhere except perhaps for $x = 1, 2, \ldots, 8$.

5.3. Characterization of Linear Maps by the Image of a Basis

The linearity of a map does not seem to be a very severe constraint at first sight. This impression, however, turns out to be false immediately if we try to define a linear map by elements. Even if we do this by a simple natural rule, it is nearly certain that the resulting map will not be linear (see, e.g., Exercise 5.1.5). We get easily entangled into the strong invisible net of the structure of the map created by the laws of preserving the operations. We face the same problem when we want to handle a linear map effectively. We can get completely lost in the labyrinth of the generally infinitely many elements obeying the seemingly chaotic jungle induced by the rules of preserving the operations.

These difficulties are eliminated perfectly by characterizing linear maps with the image of a (fixed) basis. This means that, on the one hand, the image of a basis can be chosen arbitrarily, without any restrictions, and on the other hand, the image of a basis determines the image of every other element uniquely.

Theorem 5.3.1. *Let $\mathbf{b}_1, \ldots, \mathbf{b}_n$ be a basis in a vector space V_1 and let $\mathbf{c}_1, \ldots, \mathbf{c}_n$ be arbitrary elements in a vector space V_2 over the same field. Then there exists exactly one linear map $\mathcal{A}: V_1 \to V_2$ satisfying*

$$\mathcal{A}\mathbf{b}_i = \mathbf{c}_i, \quad i = 1, 2, \ldots, n,$$

i.e., that assigns the prescribed vectors $\mathbf{c}_i$ to the basis vectors $\mathbf{b}_i$. ♣

Proof. Every vector $\mathbf{u} \in V_1$ has a unique decomposition $\mathbf{u} = \beta_1\mathbf{b}_1 + \cdots + \beta_n\mathbf{b}_n$. *If there exists* a linear map $\mathcal{A}$ with the given property, then the condition and the law of preserving the operations imply

$$\mathcal{A}\mathbf{u} = \mathcal{A}(\beta_1\mathbf{b}_1 + \cdots + \beta_n\mathbf{b}_n) = \beta_1(\mathcal{A}\mathbf{b}_1) + \cdots + \beta_n(\mathcal{A}\mathbf{b}_n) = \beta_1\mathbf{c}_1 + \cdots + \beta_n\mathbf{c}_n.$$

This means that $\mathcal{A}\mathbf{u}$ is uniquely determined, so there exists at most one such $\mathcal{A}$. Moreover, *only* the map defined by

$$\mathcal{A}\mathbf{u} = \beta_1\mathbf{c}_1 + \cdots + \beta_n\mathbf{c}_n$$

can work. Thus we have to show that this is a linear map, indeed. First note that the coefficients β_i exist and are unique due to $\mathbf{b}_i$ forming a basis, so $\mathcal{A}\mathbf{u}$ is a uniquely defined vector in V_2. To verify preserving the addition, let $\mathbf{v} = \gamma_1\mathbf{b}_1 + \cdots + \gamma_n\mathbf{b}_n$, then $\mathbf{u} + \mathbf{v} = (\beta_1 + \gamma_1)\mathbf{b}_1 + \cdots + (\beta_n + \gamma_n)\mathbf{b}_n$. By the definition of map $\mathcal{A}$, we have

$$\mathcal{A}(\mathbf{u} + \mathbf{v}) = (\beta_1 + \gamma_1)\mathbf{c}_1 + \cdots + (\beta_n + \gamma_n)\mathbf{c}_n = \mathcal{A}\mathbf{u} + \mathcal{A}\mathbf{v}.$$

Preserving the multiplication by scalars can be shown similarly. ∎

Based on this theorem, linear maps will be generally defined by selecting the image of a basis. This will make possible the characterization of linear maps by matrices (see Section 5.7).

Exercises 5.3

5.3.1. Show that every subspace is the (a) image; (b) kernel of a suitable linear transformation: If W is a subspace of a finite-dimensional vector space V, then there exists a linear transformation of V with W as its (a) image; (b) kernel (cf. Exercise 5.1.5).

5.3.2. Find all linear maps $V_1 \to V_2$ completely determined by their (a) image; (b) kernel (i.e., no other $V_1 \to V_2$ linear map can have the same kernel or image).

5.3.3. (a) Let $\mathbf{u}_1, \dots, \mathbf{u}_n$ be a spanning set in V_1 and let $\mathbf{c}_1, \dots, \mathbf{c}_n$ be *arbitrary* elements in V_2 (over the same field). How many $\mathcal{A} : V_1 \to V_2$ linear maps satisfy $\mathcal{A}\mathbf{u}_i = \mathbf{c}_i$, $i = 1, 2, \dots, n$?

(b) Investigate the same question if the vectors $\mathbf{u}_i$ are linearly independent (instead of spanning V_1).

5.3.4. How many linear transformations $\mathcal{A}$ of the vector space $\mathbf{R}^2$ satisfy

(a) $\mathcal{A}\begin{pmatrix}1\\3\end{pmatrix} = \begin{pmatrix}0\\0\end{pmatrix}$ and $\mathcal{A}\begin{pmatrix}2\\6\end{pmatrix} = \begin{pmatrix}1\\0\end{pmatrix}$;

(b) $\mathcal{A}\begin{pmatrix}1\\3\end{pmatrix} = \begin{pmatrix}0\\0\end{pmatrix}$ and $\mathcal{A}\begin{pmatrix}2\\3\end{pmatrix} = \begin{pmatrix}1\\0\end{pmatrix}$;

(c) $\mathcal{A}\begin{pmatrix}1\\3\end{pmatrix} = \begin{pmatrix}6\\5\end{pmatrix}$ and $\mathcal{A}\begin{pmatrix}2\\6\end{pmatrix} = \begin{pmatrix}12\\10\end{pmatrix}$?

5.3.5. Let k and n be positive integers and let F be the field $\mathbf{Z}_p$ of the modulo p residue classes. How many $\mathcal{A} : F^n \to F^k$ linear maps exist?

5.3.6. Let V_1 and V_2 be arbitrary finite-dimensional vector spaces over a field F. Prove that there exists a linear map $\mathcal{A} : V_1 \to V_2$ satisfying at least one of the conditions $\operatorname{Ker}\mathcal{A} = \mathbf{0}$ and $\operatorname{Im}\mathcal{A} = V_2$.

5.4. Dimension Theorem

If the kernel of a linear map is large, then many vectors have the same image by Exercise 5.1.11, so our intuition suggests that the image space is small. A precise quantitative relation is expressed in the following theorem.

Theorem 5.4.1 (Dimension Theorem). *Let V_1 and V_2 be vector spaces over F, where* $\dim V_1 < \infty$, *and let* $\mathcal{A}: V_1 \to V_2$ *be a linear map. Then*

$$\dim \operatorname{Ker} \mathcal{A} + \dim \operatorname{Im} \mathcal{A} = \dim V_1. \clubsuit$$

The dimension of $\operatorname{Ker} \mathcal{A}$ is sometimes called the *nullity* of $\mathcal{A}$, and the dimension of $\operatorname{Im} \mathcal{A}$ is the *rank* of $\mathcal{A}$ (cf. Exercises 5.1.10 and 5.7.11). The Dimension Theorem states that the sum of the nullity and the rank is the dimension of V_1. It is clear that the dimension of V_2 plays no part in the relation as V_2 can have an arbitrarily large part outside the image of $\mathcal{A}$.

The proof will also yield that if $\dim V_1 = \infty$, then at least one of the dimensions of the kernel and image has to be infinite, so the theorem remains valid for this case, too. Moreover, it holds even in the stronger form when we make a distinction between infinite dimensions by refining the definition to the *cardinalities* of (Hamel) bases (see the remark after Definition 4.6.1).

Proof. The assertion is obvious for $\mathcal{A} = \mathcal{O}$. Otherwise, let $\mathcal{A}\mathbf{c}_1, \dots, \mathcal{A}\mathbf{c}_r$ be a basis in $\operatorname{Im} \mathcal{A}$ and let $\mathbf{b}_1, \dots, \mathbf{b}_s$ be a basis in $\operatorname{Ker} \mathcal{A}$. (If $\operatorname{Ker} \mathcal{A} = \mathbf{0}$, then we consider it as $s = 0$, and the further arguments remain equally valid. We leave the details to the Reader.) The theorem follows if we show that the vectors $\mathbf{b}_1, \dots, \mathbf{b}_s, \mathbf{c}_1, \dots, \mathbf{c}_r$ form a basis in V_1, i.e., every $\mathbf{u} \in V_1$ is a unique linear combination of them.

Using the basis in the image, we have

$$\mathcal{A}\mathbf{u} = \sum_{i=1}^{r} \lambda_i (\mathcal{A}_i \mathbf{c}_i) = \mathcal{A}\left(\sum_{i=1}^{r} \lambda_i \mathbf{c}_i\right)$$

with unique coefficients λ_i. By Exercise 5.1.11, this is equivalent to

$$\mathbf{u} - \sum_{i=1}^{r} \lambda_i \mathbf{c}_i \in \operatorname{Ker} \mathcal{A}, \qquad \text{so} \qquad \mathbf{u} - \sum_{i=1}^{r} \lambda_i \mathbf{c}_i = \sum_{j=1}^{s} \mu_j \mathbf{b}_j$$

with unique coefficients μ_j. Thus $\mathbf{u}$ is a unique linear combination

$$\mathbf{u} = \sum_{i=1}^{r} \lambda_i \mathbf{c}_i + \sum_{j=1}^{s} \mu_j \mathbf{b}_j$$

of the given vectors as claimed. ∎

An important corollary is the following theorem.

Theorem 5.4.2. *Let V be a finite-dimensional vector space and let $\mathcal{A}$ be a linear transformation of V. Then*

$$\operatorname{Im} \mathcal{A} = V \iff \operatorname{Ker} \mathcal{A} = \mathbf{0}. \clubsuit$$

Proof. If $\operatorname{Im} \mathcal{A} = V$, then $\dim \operatorname{Im} \mathcal{A} = \dim V$, so $\dim \operatorname{Ker} \mathcal{A} = 0$, i.e., $\operatorname{Ker} \mathcal{A} = \mathbf{0}$. The converse can be proved similarly (Theorem 4.6.4/II is needed in the last step). ∎

Theorem 5.4.2 shows that for a linear transformation of a finite-dimensional vector space, any of the two conditions for isomorphism in Theorem 5.2.2 implies the other. This is false for an infinite dimension, see, e.g., Exercise 5.1.4.

Exercises 5.4

The vector spaces in these exercises are assumed to have *finite dimensions.*

5.4.1. Which vector spaces have a linear transformation where the kernel and image coincide?

5.4.2. Let $\mathcal{A} : V_1 \to V_2$ and $\mathcal{B} : V_2 \to V_1$ be linear maps. Do the conditions imply that V_1 and V_2 are isomorphic?
(a) $\operatorname{Im} \mathcal{A} = V_2$ and $\operatorname{Im} \mathcal{B} = V_1$;
(b) $\operatorname{Ker} \mathcal{A} = \mathbf{0}$ and $\operatorname{Ker} \mathcal{B} = \mathbf{0}$;
(c) $\operatorname{Im} \mathcal{A} = V_2$ and $\operatorname{Ker} \mathcal{B} = \mathbf{0}$.

5.4.3. Let $\mathcal{A}$ be a linear transformation of V and let $\mathcal{A}\mathbf{u}_1, \dots, \mathcal{A}\mathbf{u}_k$ be a spanning set of V. Does this imply that $\mathbf{u}_1, \dots, \mathbf{u}_k$ is a spanning set of V?

5.4.4. Solve Exercise 5.2.3 for the case when the two vector spaces are the same.

5.4.5. A linear map $\mathcal{A} : V_1 \to V_2$ satisfies that
(i) the images of any four vectors are linearly dependent; and
(ii) no six linearly independent vectors in V_1 can all be mapped to zero.
Prove $\dim V_1 \le 8$.

5.4.6. Let $\mathcal{A}, \mathcal{B} : V_1 \to V_2$ be linear maps with $\operatorname{Ker} \mathcal{A} \subseteq \operatorname{Ker} \mathcal{B}$ and $\operatorname{Im} \mathcal{A} \subseteq \operatorname{Im} \mathcal{B}$. Verify that $\operatorname{Ker} \mathcal{A} = \operatorname{Ker} \mathcal{B}$ and $\operatorname{Im} \mathcal{A} = \operatorname{Im} \mathcal{B}$.

5.4.7. Assume that the linear maps $\mathcal{A}, \mathcal{B} : V_1 \to V_2$ satisfy
$$V_1 = \operatorname{Ker} \mathcal{A} \oplus \operatorname{Ker} \mathcal{B} \quad \text{and} \quad V_2 = \operatorname{Im} \mathcal{A} \oplus \operatorname{Im} \mathcal{B}.$$
Prove $V_1 \cong V_2$.

5.5. Addition and Multiplication by Scalars of Linear Maps

These operations for linear maps conform to the usual definition of operations for functions.

Definition 5.5.1. The *sum* of linear maps $\mathcal{A}$ and $\mathcal{B} : V_1 \to V_2$ denoted by $\mathcal{A} + \mathcal{B}$ assigns $\mathcal{A}\mathbf{u} + \mathcal{B}\mathbf{u} \in V_2$ to every vector $\mathbf{u} \in V_1$:
$$(\mathcal{A} + \mathcal{B})\mathbf{u} = \mathcal{A}\mathbf{u} + \mathcal{B}\mathbf{u}. \clubsuit$$

The sum of two linear maps is defined only for maps acting from the *same* vector space V_1 to the *same* vector space V_2, and their sum maps from V_1 to V_2, too.

The two plus signs + have different meanings: the one on the left-hand side is the addition just defined, whereas the one on the right-hand side refers to the addition of vectors in V_2. (So the formula cannot be considered as some type of distributive law.)

Definition 5.5.2. *Multiplication* of a linear map $\mathcal{A} : V_1 \to V_2$ *by a scalar* $\lambda \in F$ is the map denoted by $\lambda\mathcal{A}$ assigning $\lambda(\mathcal{A}\mathbf{u}) \in V_2$ to every vector $\mathbf{u} \in V_1$:

$$(\lambda\mathcal{A})\mathbf{u} = \lambda(\mathcal{A}\mathbf{u}). \clubsuit$$

The remarks on the sum apply mutatis mutandis to the multiplication by scalars, too.

Theorem 5.5.3. *Let V_1 and V_2 be two arbitrary vector spaces over the same field F. Then all $V_1 \to V_2$ linear maps form a vector space over F under the operations defined above. We denote this vector space by* $\operatorname{Hom}(V_1, V_2)$, *and by* $\operatorname{Hom} V$ *if $V_1 = V_2 = V$.* $\clubsuit$

Proof. We have to show first that the sum of two linear maps and the scalar multiple of a linear map are linear maps. We verify that $\mathcal{A} + \mathcal{B}$ preserves addition, the other three requirements can be proved similarly.

$$(\mathcal{A}+\mathcal{B})(\mathbf{u}+\mathbf{v}) = \mathcal{A}(\mathbf{u}+\mathbf{v})+\mathcal{B}(\mathbf{u}+\mathbf{v}) = \mathcal{A}\mathbf{u}+\mathcal{A}\mathbf{v}+\mathcal{B}\mathbf{u}+\mathcal{B}\mathbf{v} = (\mathcal{A}+\mathcal{B})\mathbf{u}+(\mathcal{A}+\mathcal{B})\mathbf{v}.$$

We used the definition of addition of linear maps, the addition-preserving property of $\mathcal{A}$ and $\mathcal{B}$, and reordering the terms of sums in V_2.

The zero element of $\operatorname{Hom}(V_1, V_2)$ is the $\mathcal{O}$ map that assigns $\mathbf{0} \in V_2$ to every element of V_1. This linear map satisfies the requirement for a zero element.

The negative of a linear map $\mathcal{A}$ is the linear map $-\mathcal{A}$ defined by $(-\mathcal{A})\mathbf{u} = -(\mathcal{A}\mathbf{u})$ for every $\mathbf{u} \in V_1$. We leave to the Reader to check that this map is linear and satisfies the requirement for the negative.

All other vector space axioms are operational identities. We verify $(\lambda + \mu)\mathcal{A} = \lambda\mathcal{A} + \mu\mathcal{A}$ in detail, the others can be proved similarly. On the one hand,

$$[(\lambda + \mu)\mathcal{A}]\mathbf{u} = (\lambda + \mu)(\mathcal{A}\mathbf{u}),$$

and on the other hand,

$$(\lambda\mathcal{A} + \mu\mathcal{A})\mathbf{u} = (\lambda\mathcal{A})\mathbf{u} + (\mu\mathcal{A})\mathbf{u} = \lambda(\mathcal{A}\mathbf{u}) + \mu(\mathcal{A}\mathbf{u}).$$

(We used only the definitions of the operations for linear maps.) The vectors on the right-hand side are equal just because the corresponding axiom holds in the vector space V_2. ∎

Exercises 5.5

5.5.1. Prove for any $\mathcal{A}, \mathcal{B} \in \operatorname{Hom}(V_1, V_2)$:
- (a) $\operatorname{Ker}(\mathcal{A} + \mathcal{B}) \supseteq \operatorname{Ker}\mathcal{A} \cap \operatorname{Ker}\mathcal{B}$;
- (b) $\operatorname{Im}(\mathcal{A} + \mathcal{B}) \subseteq \langle \operatorname{Im}\mathcal{A}, \operatorname{Im}\mathcal{B}\rangle$;
- (c) $\operatorname{Ker}(\lambda\mathcal{A}) = \operatorname{Ker}\mathcal{A}$, if $\lambda \neq 0$;
- (d) $\operatorname{Im}(\lambda\mathcal{A}) = \operatorname{Im}\mathcal{A}$, if $\lambda \neq 0$.

Find examples in (a) and (b) both for equality and strict containment.

5.5.2. Let $V_1 = V_2$ be the usual vector space of plane vectors. Describe $\mathcal{A} + \mathcal{B}$ if $\mathcal{A}$ and $\mathcal{B}$ are
- (a) reflections across the x and y axes;
- (b) perpendicular projections onto the x and y axes;

(c) rotations around the origin by +60 and −60 degrees;
(d) rotations around the origin by angles Φ and $-\Phi$;
(e) the identity and the rotation around the origin by +90 degrees.

5.5.3. Which linear maps form a subspace in $\mathrm{Hom}(V_1, V_2)$?
(a) The kernel is a fixed subspace U_1 in V_1.
(b) The dimension of the image is at most one.
(c) The image of every element is a scalar multiple of a fixed vector.
(d) Every element in a given subspace of V_1 is mapped into a given subspace U_2 of V_2.
(e) A given vector in V_1 is mapped to a given vector in V_2.

5.5.4. Let $V_1 = V_2 = V$. Which linear transformations form a subspace in $\mathrm{Hom}\, V$?
(a) All isomorphisms.
(b) All non-isomorphisms.
(c) All transformations mapping a given vector to itself.
(d) All transformations where the image is part of the kernel.

5.5.5. Let $\mathcal{A}, \mathcal{B} \in \mathrm{Hom}(V_1, V_2)$. True or false?
(a) $\mathrm{Im}\,\mathcal{A} \cap \mathrm{Im}\,\mathcal{B} = \mathbf{0} \Rightarrow \mathrm{Ker}(\mathcal{A} + \mathcal{B}) = \mathrm{Ker}\,\mathcal{A} \cap \mathrm{Ker}\,\mathcal{B}$.
(b) $\mathrm{Ker}(\mathcal{A} + \mathcal{B}) = \mathrm{Ker}\,\mathcal{A} \cap \mathrm{Ker}\,\mathcal{B} \Rightarrow \mathrm{Im}\,\mathcal{A} \cap \mathrm{Im}\,\mathcal{B} = \mathbf{0}$.

5.5.6. Let $\mathbf{a}_1, \dots, \mathbf{a}_n$ be a basis in V_1 and let $\mathbf{b}_1, \dots, \mathbf{b}_k$ be a basis in V_2. We define $\mathcal{C}_{ij} \in \mathrm{Hom}(V_1, V_2)$ by

$$\mathcal{C}_{ij}\mathbf{a}_r = \begin{cases} \mathbf{b}_i, & \text{if } r = j; \\ \mathbf{0}, & \text{if } r \neq j. \end{cases} \qquad 1 \le j \le n, \quad 1 \le i \le k.$$

Verify that the linear maps $\mathcal{C}_{ij}$ form a basis in $\mathrm{Hom}(V_1, V_2)$.

5.5.7. Prove $\dim \mathrm{Hom}(V_1, V_2) = \dim V_1 \cdot \dim V_2$ for finite-dimensional vector spaces V_i.

5.5.8. Let $\mathcal{A}_1, \dots, \mathcal{A}_r \in \mathrm{Hom}(V_1, V_2)$. True or false?
(a) If $\mathcal{A}_1, \dots, \mathcal{A}_r$ are linearly dependent in $\mathrm{Hom}(V_1, V_2)$, then the vectors $\mathcal{A}_1\mathbf{x}, \dots, \mathcal{A}_r\mathbf{x}$ are linearly dependent in V_2 for every $\mathbf{x} \in V_1$.
(b) If the vectors $\mathcal{A}_1\mathbf{x}, \dots, \mathcal{A}_r\mathbf{x}$ are linearly dependent in V_2 for every $\mathbf{x} \in V_1$, then $\mathcal{A}_1, \dots, \mathcal{A}_r$ are linearly dependent in $\mathrm{Hom}(V_1, V_2)$.

5.5.9. Let $V_1 = \mathbf{C}^4$, $V_2 = \mathbf{C}^2$, and $\mathcal{A}_{ij} \in \mathrm{Hom}(V_1, V_2)$ defined by

$$\mathcal{A}_{ij}\begin{pmatrix} \alpha_1 \\ \alpha_2 \\ \alpha_3 \\ \alpha_4 \end{pmatrix} = \begin{pmatrix} \alpha_i \\ \alpha_j \end{pmatrix} \qquad 1 \le i, j \le 4.$$

(a) Prove that any three (distinct) maps $\mathcal{A}_{ij}$ are linearly independent in $\mathrm{Hom}(V_1, V_2)$.
(b) Show four (distinct) linearly dependent maps $\mathcal{A}_{ij} \in \mathrm{Hom}(V_1, V_2)$.
*(c) What is the maximal number of linearly independent maps $\mathcal{A}_{ij}$ in $\mathrm{Hom}(V_1, V_2)$?

5.6. Multiplication of Linear Maps

The multiplication of linear maps is defined as their composition and plays a particularly important role in studying $V \to V$ linear transformations. These latter form a special algebraic structure called *algebra* under this multiplication plus the addition and multiplication by scalars of linear transformations. Multiplication of linear maps shows very similar properties to the features of matrix multiplication. The deeper reason for this similarity will be revealed in the next section.

Definition 5.6.1. Let V_1, V_2, and V_3 be vector spaces over a field F, $\mathcal{A} \in \operatorname{Hom}(V_2, V_3)$ and $\mathcal{B} \in \operatorname{Hom}(V_1, V_2)$. The *product* $\mathcal{AB}$ of the linear maps $\mathcal{A}$ and $\mathcal{B}$ is their composition, i.e., a $V_1 \to V_3$ map assigning $\mathcal{A}(\mathcal{B}\mathbf{u}) \in V_3$ to every $\mathbf{u} \in V_1$:

$$(\mathcal{AB})\mathbf{u} = \mathcal{A}(\mathcal{B}\mathbf{u}). \clubsuit$$

Note that to perform $\mathcal{AB}$ we apply the *second* factor $\mathcal{B}$ first and apply $\mathcal{A}$ afterwards. This seemingly artificial order is immediately explained by the formula $(\mathcal{AB})\mathbf{u} = \mathcal{A}(\mathcal{B}\mathbf{u})$. The natural order $\mathcal{A}$ first and $\mathcal{B}$ next would occur in operator notation where the map stands *behind* the vector and not in front of it: in this case the definition would be $\mathbf{u}(\mathcal{AB}) = (\mathbf{u}\mathcal{A})\mathcal{B}$. We stay, however, at the traditional function notation $\mathcal{A}\mathbf{u}$ of analysis (and as a consequence, a few definitions might seem to be somewhat artificial).

We emphasize that the product of two linear maps *only* makes sense if the conditions of Definition 5.6.1 hold, i.e., the second factor maps into the same vector space where the first factor is defined.

Theorem 5.6.2. *The product of two linear maps is linear, i.e.,* $\mathcal{A} \in \operatorname{Hom}(V_2, V_3)$ *and* $\mathcal{B} \in \operatorname{Hom}(V_1, V_2)$ *imply* $\mathcal{AB} \in \operatorname{Hom}(V_1, V_3)$. ♣

Proof. By the definition of multiplication and the linearity of $\mathcal{B}$ and $\mathcal{A}$, we have

$$(\mathcal{AB})(\lambda\mathbf{u}) = \mathcal{A}(\mathcal{B}(\lambda\mathbf{u})) = \mathcal{A}(\lambda(\mathcal{B}\mathbf{u})) = \lambda(\mathcal{A}(\mathcal{B}\mathbf{u})) = \lambda((\mathcal{AB})\mathbf{u}).$$

Preserving addition can be verified similarly. ∎

Investigating the properties of multiplication we start with the question of *commutativity*. If $\mathcal{A} \in \operatorname{Hom}(V_2, V_3)$, $\mathcal{B} \in \operatorname{Hom}(V_1, V_2)$, and $V_3 \neq V_1$, then the product $\mathcal{BA}$ is *not even defined*. If $V_3 = V_1$ but $V_1 \neq V_2$, then $\mathcal{AB} \in \operatorname{Hom}(V_1, V_1)$, whereas $\mathcal{BA} \in \operatorname{Hom}(V_2, V_2)$, so they cannot be compared. Finally, if $V_1 = V_2 = V_3 = V$, then $\mathcal{AB}$ and $\mathcal{BA}$ are still not equal in general (see, e.g., Exercises 5.6.1–5.6.4). So the multiplication of linear maps is *non-commutative* in a very strong sense.

The other usual operational identities, however, are valid.

Theorem 5.6.3. *If* $\lambda \in F$ *and* $\mathcal{A}$, $\mathcal{B}$, $\mathcal{C}$ *are any linear maps such that one side of an equality below makes sense, then the other side makes sense and equality holds.*

I. $\mathcal{A}(\mathcal{BC}) = (\mathcal{AB})\mathcal{C}$ (*associative law*);

II. $\mathcal{A}(\mathcal{B} + \mathcal{C}) = \mathcal{AB} + \mathcal{AC}$,

$(\mathcal{A} + \mathcal{B})\mathcal{C} = \mathcal{AC} + \mathcal{BC}$ (*distributive laws*);

III. $\lambda(\mathcal{AB}) = (\lambda\mathcal{A})\mathcal{B} = \mathcal{A}(\lambda\mathcal{B})$. ♣

As multiplication is not commutative, we have to verify the two distributive laws separately. Also, in III, we can only change the position of the scalar $\lambda \in F$ but must not modify the order of $\mathcal{A}$ and $\mathcal{B}$.

Proof. Any of the two sides in I makes sense if and only if $\mathcal{A} \in \operatorname{Hom}(V_3, V_4)$, $\mathcal{B} \in \operatorname{Hom}(V_2, V_3)$, and $\mathcal{C} \in \operatorname{Hom}(V_1, V_2)$. In this case

$$(\mathcal{A}(\mathcal{B}\mathcal{C}))\mathbf{x} = \mathcal{A}((\mathcal{B}\mathcal{C})\mathbf{x}) = \mathcal{A}(\mathcal{B}(\mathcal{C}\mathbf{x}))$$

and

$$((\mathcal{A}\mathcal{B})\mathcal{C})\mathbf{x} = (\mathcal{A}\mathcal{B})(\mathcal{C}\mathbf{x}) = \mathcal{A}(\mathcal{B}(\mathcal{C}\mathbf{x}))$$

for every $\mathbf{x} \in V_1$. So the two sides of I mean the same map. This shows that the associative law holds for any composition, since we have to apply the functions one after the other in the given order in both cases.

We can verify II and III similarly, but we also need the linearity of the maps besides the definition of multiplication (see Exercise 5.6.6). ∎

Now we summarize the properties of operations for linear transformations of a vector space.

Theorem 5.6.4. Hom V *is a vector space under the addition and multiplication by scalars of linear maps, it is a ring under the addition and multiplication of linear maps, further*

$$\lambda(\mathcal{A}\mathcal{B}) = (\lambda\mathcal{A})\mathcal{B} = \mathcal{A}(\lambda\mathcal{B})$$

for every $\mathcal{A}$, $\mathcal{B} \in \operatorname{Hom} V$, $\lambda \in F$. ♣

Proof. Vector space property is a special case of Theorem 5.5.3. Multiplication is an operation on Hom V by Theorem 5.6.2. The operational identities involving multiplication and the one connecting multiplication and multiplication by scalars follow from Theorem 5.6.3. ∎

The features stated in Theorem 5.6.4 represent a new algebraic structure.

Definition 5.6.5. A non-empty set A is an *algebra* over the commutative field F if

(i) an addition, a multiplication, and a multiplication by the elements of F on A are defined on A;

(ii) A is a ring under the addition and multiplication;

(iii) A is a vector space under the addition and multiplication by the elements of F; and

(iv) $\lambda(ab) = (\lambda a)b = a(\lambda b)$ for every $\lambda \in F$, $a, b \in A$. ♣

Examples of algebras

E1 Hom V: the linear transformations of a vector space over F under the corresponding operations.

E2 $F^{n \times n}$: the $n \times n$ matrices over F under the usual operations.

E3 $F[x]$: the polynomials over F under the usual operations.

E4 The complex numbers over **R** under the usual operations. More generally: If F_1 is a *subfield* of F_2 (i.e., the field F_1 is not just a subset of F_2 but also the operations in F_1 are the same as the operations in F_2), then F_2 is an algebra over F_1 under the operations induced by F_2 (see in more detail at field extensions, Section A.10). We get an algebra even if we only assume that F_2 is a commutative ring.

E5 *Quaternions* are a generalization of complex numbers. These are expressions of the form $\alpha_0 + \alpha_1 i + \alpha_2 j + \alpha_3 k$ with $\alpha_i \in \mathbf{R}$. Addition and multiplication by real numbers are performed componentwise, so we get a 4-dimensional vector space over **R** where $1, i, j, k$ is a basis. The multiplication of basis vectors is defined by

$$i^2 = j^2 = k^2 = -1,\ ij = k, ji = -k, jk = i, kj = -i, ik = -j, ki = j$$

and 1 is an identity element. To obtain the product of two arbitrary quaternions, we multiply each term in the first quaternion by each term in the second one, commute the real coefficients with i, j, and k, and multiply the basis vectors by the above rules.

It can be shown that the quaternions form a four-dimensional algebra over **R**. Moreover, multiplication has an identity and every non-zero quaternion has an inverse (see Exercise 5.6.22), so this algebra is a *non-commutative field*.

A famous theorem of Frobenius states that, in a certain sense, quaternions can be considered as an ultimate completion of the concept of number:

> Let A be a finite-dimensional algebra over **R** that is a (not necessarily commutative) field. Then A is an isomorphic algebra to the real or complex numbers or to the quaternions. Instead of the field property it is sufficient to assume that there are no zero divisors in A (see Exercise 5.6.25).

For further examples of algebras see Exercise 5.6.20.

Further properties of multiplication in Hom V

As indicated earlier, multiplication is *not commutative* in Hom V.

Clearly, the identical transformation $\mathcal{I}$ is an *identity*.

Now we investigate the question of (multiplicative) inverses and zero divisors.

Theorem 5.6.6. *A linear transformation $\mathcal{A} \in$ Hom V has an inverse if and only if $\mathcal{A}$ is an isomorphism.* ♣

Proof. If $\mathcal{A}$ is an isomorphism, then we saw in the proof of Theorem 5.2.3 that the inverse $\mathcal{A}^{-1}$ of the bijection $\mathcal{A}$ is linear, so it is an element of Hom V. Conversely, let $\mathcal{B}$ be the inverse of $\mathcal{A}$: $\mathcal{A}\mathcal{B} = \mathcal{B}\mathcal{A} = \mathcal{E}$. Then $\mathbf{u} = \mathcal{A}(\mathcal{B}\mathbf{u})$ for every $\mathbf{u} \in V$, thus $\mathbf{u} \in \operatorname{Im} \mathcal{A}$, i.e., $\operatorname{Im} \mathcal{A} = V$. Also, if $\mathcal{A}\mathbf{u} = \mathcal{A}\mathbf{v}$, then $\mathbf{u} = \mathcal{B}(\mathcal{A}\mathbf{u}) = \mathcal{B}(\mathcal{A}\mathbf{v}) = \mathbf{v}$, i.e., the images by $\mathcal{A}$ of distinct vectors are distinct. Thus $\mathcal{A}$ is an isomorphism, indeed. ■

Theorems 5.6.6 and 5.2.2 yield that $\mathcal{A} \in \operatorname{Hom} V$ has an inverse if and only if $\operatorname{Ker} \mathcal{A} = \mathbf{0}$ and $\operatorname{Im} \mathcal{A} = V$. If the dimension of V is *finite*, then each of these last two conditions imply the other, by Theorem 5.4.2.

It follows from the proof of Theorem 5.6.6 that if $\mathcal{A}$ has a right inverse, then necessarily $\operatorname{Im} \mathcal{A} = V$, and if $\mathcal{A}$ has a left inverse, then necessarily $\operatorname{Ker} \mathcal{A} = \mathbf{0}$. Combining this with the above results, we get that if the dimension is *finite*, then the existence of a one-sided inverse implies that it is necessarily a two-sided inverse. This is false for infinite dimensions, but it can be shown that $\operatorname{Im} \mathcal{A} = V$ or $\operatorname{Ker} \mathcal{A} = \mathbf{0}$ is not only a necessary but also a sufficient condition for the existence of a left or right inverse.

We formulate the result for zero divisors only in the finite-dimensional case. Concerning infinite dimension and one-sided zero divisors, we meet an analogous situation as seen at the inverses.

Theorem 5.6.7. *Let V be a finite-dimensional vector space. If $\mathcal{A} \in$ Hom V is a left or right zero divisor, then* $\operatorname{Ker} \mathcal{A} \neq \mathbf{0}$. *Conversely, if* $\operatorname{Ker} \mathcal{A} \neq \mathbf{0}$ *and $\mathcal{A} \neq \mathcal{O}$, then $\mathcal{A}$ is a two-sided zero divisor.* ♣

Proof. We shall use the simple observation that $\mathcal{A}\mathcal{B} = \mathcal{O}$ if and only if $\operatorname{Ker} \mathcal{A} \supseteq \operatorname{Im} \mathcal{B}$. Indeed, $\mathcal{A}(\mathcal{B}\mathbf{u}) = \mathbf{0}$ holds for every $\mathbf{u} \in V$ exactly if every element $\mathcal{B}\mathbf{u}$ of $\operatorname{Im} \mathcal{B}$ falls into $\operatorname{Ker} \mathcal{A}$.

First, let $\mathcal{A} \neq \mathcal{O}$ be a left zero divisor, i.e., $\mathcal{A}\mathcal{C} = \mathcal{O}$ for some $\mathcal{C} \neq \mathcal{O}$. Then $\operatorname{Ker} \mathcal{A} \supseteq \operatorname{Im} \mathcal{C} \neq \mathbf{0}$, so $\operatorname{Ker} \mathcal{A} \neq \mathbf{0}$. Now assume that $\mathcal{A}$ is a right zero divisor, i.e., $\mathcal{D}\mathcal{A} = \mathcal{O}$ for some $\mathcal{D} \neq \mathcal{O}$. Then $V \neq \operatorname{Ker} \mathcal{D} \supseteq \operatorname{Im} \mathcal{A}$, so $\operatorname{Im} \mathcal{A} \neq V$. As $\dim V < \infty$, this is equivalent to $\operatorname{Ker} \mathcal{A} \neq \mathbf{0}$.

To prove the converse, let $\mathcal{A} \neq \mathcal{O}$ and $\operatorname{Ker} \mathcal{A} \neq \mathbf{0}$. We show first that $\mathcal{A}$ is a left zero divisor, i.e., $\mathcal{A}\mathcal{C} = \mathcal{O}$ for some $\mathcal{C} \neq \mathcal{O}$. We define $\mathcal{C}$ by the image of a basis of V. Let $\mathbf{b}_1, \dots, \mathbf{b}_s$ be a basis in $\operatorname{Ker} \mathcal{A}$ and we extend it to a basis of V by the vectors $\mathbf{b}_{s+1}, \dots, \mathbf{b}_n$. We define $\mathcal{C}$ by

$$\mathcal{C}\mathbf{b}_i = \begin{cases} \mathbf{b}_i, & \text{if } 1 \le i \le s; \\ \mathbf{0}, & \text{if } i > s. \end{cases}$$

Clearly $\mathcal{C} \neq \mathcal{O}$ and $\mathcal{A}\mathcal{C} = \mathcal{O}$.

We can show by similar arguments that $\mathcal{A}$ is a right zero divisor. We use the condition $\operatorname{Im} \mathcal{A} \neq V$, extend a basis of $\operatorname{Im} \mathcal{A}$ to a basis of V, and define a linear transformation $\mathcal{D} \neq \mathcal{O}$ satisfying $\mathcal{D}\mathcal{A} = \mathcal{O}$. We leave the details to the reader. ■

For some further questions about inverses and zero divisors, see Exercises 5.6.10–5.6.17.

Exercises 5.6

5.6.1. Let $V_1 = V_2$ be the usual vector space of plane vectors. Does $\mathcal{AB} = \mathcal{BA}$ hold if $\mathcal{A}$ and $\mathcal{B}$ are
 (a) reflections across the x axis and the line $y = x$;
 (b) perpendicular projections onto the x axis and the line $y = x$;
 (c) rotations around the origin by $+60$ and -90 degrees;
 (d) a magnification from the origin and a rotation around the origin by $+90$ degrees?

5.6.2. Let $V = \mathbf{C}^3$ and define $\mathcal{A}$ and $\mathcal{B} \in \operatorname{Hom} V$ as

$$\mathcal{A}\begin{pmatrix}\alpha_1\\ \alpha_2\\ \alpha_3\end{pmatrix} = \begin{pmatrix}\alpha_2\\ \alpha_1\\ \alpha_3\end{pmatrix} \quad \text{and} \quad \mathcal{B}\begin{pmatrix}\alpha_1\\ \alpha_2\\ \alpha_3\end{pmatrix} = \begin{pmatrix}\alpha_1\\ \alpha_3\\ \alpha_2\end{pmatrix}.$$

Determine the transformations $\mathcal{AB}$, $\mathcal{BA}$, $\mathcal{A}^{101}$, and $(\mathcal{AB})^{101}$.

5.6.3. Let V be a finite-dimensional vector space. Find all $\mathcal{A} \in \operatorname{Hom} V$ commuting with every linear transformation of V (i.e., $\mathcal{AB} = \mathcal{BA}$ for every $\mathcal{B} \in \operatorname{Hom} V$).

5.6.4. For which finite-dimensional vector spaces V is multiplication in $\operatorname{Hom} V$ commutative?

5.6.5. Let $\mathcal{A} \in \operatorname{Hom}(V_2, V_3)$ and $\mathcal{B} \in \operatorname{Hom}(V_1, V_2)$. What is the relation between
 (a) $\operatorname{Ker} \mathcal{AB}$ and $\operatorname{Ker} \mathcal{B}$; (b) $\operatorname{Im} \mathcal{AB}$ and $\operatorname{Im} \mathcal{A}$?

5.6.6. Prove statements II and III in Theorem 5.6.3.

5.6.7. Let $\mathcal{A} \in \operatorname{Hom}(V_2, V_3)$ and $\mathcal{B} \in \operatorname{Hom}(V_1, V_2)$. Prove

$$\dim \operatorname{Im} \mathcal{B} = \dim \operatorname{Im} \mathcal{AB} + \dim(\operatorname{Ker} \mathcal{A} \cap \operatorname{Im} \mathcal{B}).$$

5.6.8. Let V be a finite-dimensional vector space and $\mathcal{A} \in \operatorname{Hom} V$. Verify

$$\operatorname{Ker} \mathcal{A}^2 = \operatorname{Ker} \mathcal{A} \iff \operatorname{Im} \mathcal{A}^2 = \operatorname{Im} \mathcal{A} \iff \operatorname{Ker} \mathcal{A} \cap \operatorname{Im} \mathcal{A} = \mathbf{0}.$$

*5.6.9. Let A be a 5×5 real matrix and assume $A^{1000} = 0$. Prove $A^5 = 0$.

5.6.10. Let V be the usual vector space of polynomials $f(x)$ with real coefficients. We define $\mathcal{A}$ and $\mathcal{B} \in \operatorname{Hom} V$ by $\mathcal{A}[f(x)] = xf(x)$ and $\mathcal{B}[f(x)] = [f(x) - \alpha_0]/x$ where α_0 denotes the constant term of $f(x)$.
 (a) Find the number of left and right inverses of each transformation.
 (b) Determine for each transformation whether it is a left or right zero divisor.

5.6.11. Let $\mathbf{b}_1, \dots, \mathbf{b}_n$ be a basis in a vector space V. Which linear transformations have a left or right inverse and which are left or right zero divisors? Construct the inverse if it exists, and find a non-zero transformation to each zero divisor such that their product in the appropriate order is zero.
 (a) $\mathcal{A} : \mathbf{b}_1 \mapsto \mathbf{b}_1 - \mathbf{b}_2, \mathbf{b}_2 \mapsto \mathbf{b}_2 - \mathbf{b}_3, \dots, \mathbf{b}_n \mapsto \mathbf{b}_n - \mathbf{b}_1$;
 (b) $\mathcal{B} : \mathbf{b}_1 \mapsto \mathbf{b}_1, \mathbf{b}_2 \mapsto \mathbf{b}_1 + \mathbf{b}_2, \dots, \mathbf{b}_n \mapsto \mathbf{b}_1 + \mathbf{b}_n$;
 (c) $\mathcal{C} : \mathbf{b}_1 \mapsto \mathbf{b}_1 + \mathbf{b}_2 + \dots + \mathbf{b}_n, \dots, \mathbf{b}_n \mapsto \mathbf{b}_1 + \mathbf{b}_2 + \dots + \mathbf{b}_n$.

5.6.12. For which finite-dimensional vector spaces V does every non-zero $\mathcal{A} \in \operatorname{Hom} V$ have an inverse?

5.6.13. For which finite-dimensional vector spaces V is Hom V free of zero divisors?

5.6.14. Which finite-dimensional vector spaces V have linear transformations $\mathcal{A}$ and $\mathcal{B} \in \text{Hom } V$ satisfying

(a) $\mathcal{AB} = \mathcal{O}$ and $\mathcal{BA} \neq \mathcal{O}$;

(b) $\mathcal{AB} = \mathcal{O}$ and $\mathcal{BA} = \mathcal{E}$?

5.6.15. Let V be a finite-dimensional vector space and let $\mathcal{A} \in \text{Hom } V$ be a zero divisor. Prove that there exists a non-zero $\mathcal{B} \in \text{Hom } V$ satisfying both $\mathcal{AB} = \mathcal{O}$ and $\mathcal{BA} = \mathcal{O}$ simultaneously.

5.6.16. Let V be a finite-dimensional vector space and $\mathcal{A}, \mathcal{B} \in \text{Hom } V$. True or false?

(a) If each of $\mathcal{A}$ and $\mathcal{B}$ has an inverse, then $\mathcal{AB}$ has an inverse.

(b) If $\mathcal{AB}$ has an inverse, then each of $\mathcal{A}$ and $\mathcal{B}$ has an inverse.

(c) If both $\mathcal{A}$ and $\mathcal{B}$ are left zero divisors and $\mathcal{AB} \neq \mathcal{O}$, then $\mathcal{AB}$ is a left zero divisor.

(d) If $\mathcal{AB}$ is a left zero divisor, then both $\mathcal{A}$ and $\mathcal{B}$ are left zero divisors.

(e) If $\mathcal{A} + \mathcal{B}$ has an inverse, then at least one of $\mathcal{A}$ and $\mathcal{B}$ has an inverse.

(f) If $\mathcal{A} + \mathcal{B}$ is a left zero divisor, then at least one of $\mathcal{A}$ and $\mathcal{B}$ is a left zero divisor.

5.6.17. Let V be a finite-dimensional vector space and $\mathcal{A} \in \text{Hom } V$. We define two subsets in Hom V:

$$L = \{\mathcal{C} \in \text{Hom } V \mid \mathcal{CA} = \mathcal{O}\}; \qquad R = \{\mathcal{D} \in \text{Hom } V \mid \mathcal{AD} = \mathcal{O}\}$$

(i.e., L and R consist of all left and right zero divisor pairs of $\mathcal{A}$ including $\mathcal{O}$). Verify that L and R are subspaces in Hom V and compute their dimensions.

5.6.18. Let V be a finite-dimensional vector space. A linear transformation $\mathcal{P} \in \text{Hom } V$ is a *projection* if $\mathcal{P}^2 = \mathcal{P}$.

(a) Are there other projections besides $\mathcal{O}$ and $\mathcal{I}$?

(b) Which projections have an inverse?

(c) Verify that $\mathcal{P}$ is a projection if and only if $\mathcal{I} - \mathcal{P}$ is a projection.

(d) Let $F = \mathbf{R}$. Prove that $\mathcal{P}$ is a projection if and only if the inverse of $2\mathcal{P} - \mathcal{I}$ is itself. Demonstrate that this is not true over every field.

(e) Show that if $\mathcal{P}$ is a projection and $\lambda \neq 0, -1$, then $\mathcal{P} + \lambda\mathcal{E}$ has an inverse.

*(f) Prove that $\mathcal{P}$ is a projection if and only if V has a direct sum decomposition $V = U_1 \oplus U_2$ such that $\mathcal{P}$ maps every element of U_1 into itself and every element of U_2 into $\mathbf{0}$.

*5.6.19. Let V be a finite-dimensional vector space and $\mathcal{A} \in \text{Hom } V$. Show that there exists a linear transformation $\mathcal{B} \in \text{Hom } V$ satisfying $\mathcal{ABA} = \mathcal{A}$.

5.6.20. Which are algebras?

(a) An arbitrary vector space if the product of any two vectors is the zero vector.

(b) The ordinary three-dimensional space vectors under the usual addition, multiplication by scalars, and the cross product.

(c) Infinite sequences of real numbers if operations are defined componentwise.

(d) F^n if multiplication is defined componentwise.

(e) All $\mathbf{R} \to \mathbf{R}$ functions under the usual operations of functions.
(f) The previous example if multiplication is defined as composition.
(g) The real numbers $\alpha + \beta\sqrt{2}$, $\alpha, \beta \in \mathbf{Q}$, over $\mathbf{Q}$ under the usual operations.
(h) An arbitrary ring over the modulo 2 field if multiplication by scalars is defined by $0\mathbf{a} = \mathbf{0}$ and $1\mathbf{a} = \mathbf{a}$.
(i) The 2×2 *complex* matrices

$$\begin{pmatrix} z & w \\ -\overline{w} & \overline{z} \end{pmatrix}$$

over the field of *real* numbers under the usual operations.
(j) The previous matrices over the *complex* field.

5.6.21. Perform the quaternion operations.
(a) $(1 + i)(1 + j) - (1 + j)(1 + k)$;
(b) $(i + j + k)^{100}$;
(c) $(1 + i - j - k)(1 - i + j - k)(1 - i - j + k)$.

5.6.22. The *conjugate* of a quaternion $v = \alpha_0 + \alpha_1 i + \alpha_2 j + \alpha_3 k$ is the quaternion $\overline{v} = \alpha_0 - \alpha_1 i - \alpha_2 j - \alpha_3 k$.
(a) Compute the products $v\overline{v}$ and $\overline{v}v$.
(b) How can we use part (a) to determine the (multiplicative) inverse of a non-zero quaternion?

5.6.23. How many quaternions satisfy the equation $x^2 + 1 = 0$? How is this compatible with the theorem that a polynomial cannot have more zeroes than its degree?

***5.6.24.** Let $n > 1$ and let v be an arbitrary quaternion that is not a real number. How many nth roots does v have among the quaternions?

5.6.25. Prove that a finite-dimensional algebra of at least two elements is a (not necessarily commutative) field if and only if it is free of zero divisors.

5.7. The Matrix of a Linear Map

A linear map between finite-dimensional vector spaces can be defined by the image of a basis in V_1 and each image vector is a unique linear combination of the elements of a basis in V_2. So the linear map can be characterized by the matrix of the coefficients in these linear combinations. It turns out that this matrix representation preserves the operations, too; this explains the similar properties of maps and matrices.

This relationship between linear maps and matrices is very useful in both directions. We can verify statements about maps by matrices and vice versa. In practical applications of linear maps, we generally work with their matrices.

Definition 5.7.1. Let $\mathbf{a}_1, \ldots, \mathbf{a}_n$ be a basis in V_1 and $\mathbf{b}_1, \ldots, \mathbf{b}_k$ be a basis in V_2. The *matrix* with respect to the bases $\mathbf{a}_1, \ldots, \mathbf{a}_n$ and $\mathbf{b}_1, \ldots, \mathbf{b}_k$ of a linear map $\mathcal{A} \in \operatorname{Hom}(V_1, V_2)$ is a $k \times n$ matrix where the elements of the jth column are the coordinates of $\mathcal{A}\mathbf{a}_j$ in the basis $\mathbf{b}_1, \ldots, \mathbf{b}_k$. This matrix is denoted by $[\mathcal{A}]_{a,b}$.

In more detail, if

$$\begin{aligned}\mathcal{A}\mathbf{a}_1 &= \alpha_{11}\mathbf{b}_1 + \alpha_{21}\mathbf{b}_2 + \ldots + \alpha_{k1}\mathbf{b}_k,\\ \mathcal{A}\mathbf{a}_2 &= \alpha_{12}\mathbf{b}_1 + \alpha_{22}\mathbf{b}_2 + \ldots + \alpha_{k2}\mathbf{b}_k,\\ &\vdots\\ \mathcal{A}\mathbf{a}_n &= \alpha_{1n}\mathbf{b}_1 + \alpha_{2n}\mathbf{b}_2 + \ldots + \alpha_{kn}\mathbf{b}_k,\end{aligned}$$

then

$$[\mathcal{A}]_{a,b} = \begin{pmatrix} \alpha_{11} & \alpha_{12} & \ldots & \alpha_{1n} \\ \alpha_{21} & \alpha_{22} & \ldots & \alpha_{2n} \\ \vdots & \vdots & & \vdots \\ \alpha_{k1} & \alpha_{k2} & \ldots & \alpha_{kn} \end{pmatrix}. \clubsuit$$

Thus the columns of the matrix $[\mathcal{A}]_{a,b}$ are the images of the basis vectors $\mathbf{a}_j \in V_1$ expressed with the help of the basis vectors $\mathbf{b}_i \in V_2$. Instead of the phrase "with respect to the bases" we often use the shorter term "in the bases".

The matrix of a linear map depends on the selection of the bases in the two vector spaces, so in the case of another pair of bases the matrix will be different in general.

Please note that it is an important natural requirement for $V \to V$ linear transformations that the two occurring bases in the definition of the matrix must be the *same*. This gives a correct picture of the effect of the transformation by describing how the images are related to the original vectors, e.g., the identical transformation $\mathcal{I}$ leaving the vectors unchanged corresponds to the identity matrix I. Using the same basis in both roles is essential in proving that matrix representation preserves multiplication (see Theorems 5.7.6 and 5.7.7).

We shall also need the matrix of a vector (this notion already occurred in Example E5 in Section 5.1).

Definition 5.7.2. Let $\mathbf{c}_1, \ldots, \mathbf{c}_r$ be a basis in a vector space V. Then every $\mathbf{v} \in V$ has a unique decomposition $\mathbf{v} = \gamma_1\mathbf{c}_1 + \cdots + \gamma_r\mathbf{c}_r$. The (coordinate) *matrix* or *coordinate vector* of $\mathbf{v}$ with respect to the basis $\mathbf{c}_1, \ldots, \mathbf{c}_r$ is the (column) matrix

$$[\mathbf{v}]_c = \begin{pmatrix} \gamma_1 \\ \gamma_2 \\ \vdots \\ \gamma_r \end{pmatrix}. \clubsuit$$

Also the matrix of a vector depends on the selection of the basis.

If there is no ambiguity about the basis or pair of bases, then we can omit the index referring to the bases in the matrices of vectors or linear maps.

We verify first that, with fixed bases, the matrix of an image vector is the product of the matrices of the linear map and the original vector.

Theorem 5.7.3. *Let $\mathbf{a}_1, \ldots, \mathbf{a}_n$ and $\mathbf{b}_1, \ldots, \mathbf{b}_k$ be bases in the vector spaces V_1 and V_2, $\mathcal{A} \in \operatorname{Hom}(V_1, V_2)$, and $\mathbf{v} \in V_1$. Then*

$$[\mathcal{A}\mathbf{v}]_b = [\mathcal{A}]_{a,b} \cdot [\mathbf{v}]_a$$

where the right-hand side is the product of the two matrices. ♣

Proof. Let

$$[\mathbf{v}]_a = \begin{pmatrix} \lambda_1 \\ \lambda_2 \\ \vdots \\ \lambda_n \end{pmatrix} \quad \text{and} \quad [\mathcal{A}]_{a,b} = \begin{pmatrix} \alpha_{11} & \alpha_{12} & \dots & \alpha_{1n} \\ \alpha_{21} & \alpha_{22} & \dots & \alpha_{2n} \\ \vdots & \vdots & & \vdots \\ \alpha_{k1} & \alpha_{k2} & \dots & \alpha_{kn} \end{pmatrix}.$$

Then

$$\begin{aligned} \mathcal{A}\mathbf{v} &= \mathcal{A}(\lambda_1\mathbf{a}_1 + \dots + \lambda_n\mathbf{a}_n) \\ &= \lambda_1(\mathcal{A}\mathbf{a}_1) + \dots + \lambda_n(\mathcal{A}\mathbf{a}_n) \\ &= \lambda_1(\alpha_{11}\mathbf{b}_1 + \dots + \alpha_{k1}\mathbf{b}_k) + \dots + \lambda_n(\alpha_{1n}\mathbf{b}_1 + \dots + \alpha_{kn}\mathbf{b}_k) \\ &= (\alpha_{11}\lambda_1 + \dots + \alpha_{1n}\lambda_n)\mathbf{b}_1 + \dots + (\alpha_{k1}\lambda_1 + \dots + \alpha_{kn}\lambda_n)\mathbf{b}_k, \end{aligned}$$

so the coefficient of $\mathbf{b}_i$ is the product of the ith row of $[\mathcal{A}]$ and the (only) column of $[\mathbf{v}]$ as claimed. ∎

Next we show that, with a fixed pair of bases in V_1 and V_2, assigning to every $V_1 \to V_2$ linear map its matrix is an isomorphism between the vector spaces of linear maps and matrices (of appropriate size).

Theorem 5.7.4. *If* $\dim V_1 = n$ *and* $\dim V_2 = k$, *then* $\operatorname{Hom}(V_1, V_2) \cong F^{k\times n}$. ♣

Proof. Let $\mathbf{a}_1, \dots, \mathbf{a}_n$, and $\mathbf{b}_1, \dots, \mathbf{b}_k$ be bases in V_1 and V_2, and assign to $\mathcal{A} \in \operatorname{Hom}(V_1, V_2)$ its matrix with respect to the given pair of bases:

$$\mathcal{A} \mapsto [\mathcal{A}]_{a,b}.$$

We prove that this is a $\operatorname{Hom}(V_1, V_2) \to F^{k\times n}$ isomorphism.

1. We assigned a unique $k \times n$ matrix to every $\mathcal{A} \in \operatorname{Hom}(V_1, V_2)$ since, for every linear map $\mathcal{A}$, the images of the (basis) elements of V_1 are uniquely determined and these images can be expressed uniquely with the help of the given basis of V_2.

2. Every matrix has exactly one inverse image. By definition, the matrix determines uniquely the images of the basis vectors $\mathbf{a}_j$ and, by Theorem 5.3.1, there exists exactly one linear map $\mathcal{A}$ assigning these images to the given basis vectors.

3. Preserving addition: By the definition of addition of linear maps, $(\mathcal{A} + \mathcal{B})\mathbf{a}_j = \mathcal{A}\mathbf{a}_j + \mathcal{B}\mathbf{a}_j$. Hence, the jth column of $[\mathcal{A} + \mathcal{B}]$ is equal to the sum of the jth columns of $[\mathcal{A}]$ and $[\mathcal{B}]$. Therefore $[\mathcal{A} + \mathcal{B}] = [\mathcal{A}] + [\mathcal{B}]$, as claimed. Preserving multiplication by scalars can be verified similarly. ∎

We emphasize that all these are valid only if the pair of bases is *fixed*, i.e., we form the matrices of all linear maps with respect to the same pair of bases.

A simple consequence of the previous theorem is the following theorem.

Theorem 5.7.5. *If the vector spaces are finite-dimensional, then*

$$\dim \operatorname{Hom}(V_1, V_2) = \dim V_1 \cdot \dim V_2. ♣$$

Proof. Let $\dim V_1 = n$ and $\dim V_2 = k$. Then $\operatorname{Hom}(V_1, V_2) \cong F^{k\times n}$ and the dimension of $F^{k\times n}$ is kn (verified in Sections 4.5–4.6). ∎

We note that the statement of Theorem 5.7.5 and an alternative proof were contained in Exercises 5.5.6–5.5.7. Observe that applying the isomorphism of Theorem 5.7.4, the basis of linear maps defined in Exercise 5.5.6 corresponds exactly to the usual basis of matrices.

We turn to the property of preserving multiplication.

Theorem 5.7.6. *Let* $\mathbf{a}_1, \dots, \mathbf{a}_n$, $\mathbf{b}_1, \dots, \mathbf{b}_k$, *and* $\mathbf{c}_1, \dots, \mathbf{c}_r$ *be bases in* V_1, V_2, *and* V_3, *and let* $\mathcal{A} \in \operatorname{Hom}(V_2, V_3)$ *and* $\mathcal{B} \in \operatorname{Hom}(V_1, V_2)$. *Then*

$$[\mathcal{AB}]_{a,c} = [\mathcal{A}]_{b,c} \cdot [\mathcal{B}]_{a,b}. \clubsuit$$

Proof. Let

$$[\mathcal{A}]_{b,c} = \begin{pmatrix} \alpha_{11} & \alpha_{12} & \dots & \alpha_{1k} \\ \alpha_{21} & \alpha_{22} & \dots & \alpha_{2k} \\ \vdots & \vdots & & \vdots \\ \alpha_{r1} & \alpha_{r2} & \dots & \alpha_{rk} \end{pmatrix} \quad \text{and} \quad [\mathcal{B}]_{a,b} = \begin{pmatrix} \beta_{11} & \dots & \beta_{1n} \\ \beta_{21} & \dots & \beta_{2n} \\ \beta_{31} & \dots & \beta_{3n} \\ \vdots & \vdots & \vdots \\ \beta_{k1} & \dots & \beta_{kn} \end{pmatrix}.$$

We have to verify that the ith coordinate of $(\mathcal{AB})\mathbf{a}_j$ is equal to the product of the ith row of $[\mathcal{A}]$ and the jth column of $[\mathcal{B}]$.

$$\begin{aligned} (\mathcal{AB})\mathbf{a}_j &= \mathcal{A}(\mathcal{B}\mathbf{a}_j) \\ &= \mathcal{A}(\beta_{1j}\mathbf{b}_1 + \dots + \beta_{kj}\mathbf{b}_k) \\ &= \beta_{1j}(\mathcal{A}\mathbf{b}_1) + \dots + \beta_{kj}(\mathcal{A}\mathbf{b}_k) \\ &= \beta_{1j}(\alpha_{11}\mathbf{c}_1 + \dots + \alpha_{r1}\mathbf{c}_r) + \dots + \beta_{kj}(\alpha_{1k}\mathbf{c}_1 + \dots + \alpha_{rk}\mathbf{c}_r) \\ &= (\beta_{1j}\alpha_{11} + \dots + \beta_{kj}\alpha_{1k})\mathbf{c}_1 + \dots + (\beta_{1j}\alpha_{r1} + \dots + \beta_{kj}\alpha_{rk})\mathbf{c}_r. \end{aligned}$$

Here the coefficient of $\mathbf{c}_i$ is $\beta_{1j}\alpha_{i1} + \dots + \beta_{kj}\alpha_{ik}$ and this is the product of the ith row of $[\mathcal{A}]$ and the jth column of $[\mathcal{B}]$, indeed. ∎

Finally, we study the matrices of linear transformations $\mathcal{A} \in \operatorname{Hom} V$. As explained after Definition 5.7.1, the two bases used in the matrix representation are required to be the *same*.

Theorem 5.7.7. *Let* $\mathbf{a}_1, \dots, \mathbf{a}_n$ *be a fixed basis in the vector space* V. *Then* $\mathcal{A} \mapsto [\mathcal{A}]_a$ *is an isomorphism between the algebras* $\operatorname{Hom} V$ *and* $F^{n \times n}$. ♣

Proof. By Theorem 5.7.4, this function is bijective and preserves addition and multiplication by scalars. By Theorem 5.7.6 it preserves multiplication, too. ∎

Based on Theorem 5.7.7, we can get a new proof, e.g., for Theorem 5.6.7 (see Exercise 5.7.10). Moreover, for finite-dimensional vector spaces, arbitrary properties of multiplication of linear transformations can be reduced to the corresponding properties of matrix multiplication. Of course, we can also use the other direction of this relationship, e.g., to get a natural explanation for the associative property of matrix multiplication or the seemingly artificial definition of matrix multiplication itself.

Exercises 5.7

5.7.1. Let V be the usual vector space of polynomials of degree not greater than 6 with real coefficients and $\mathcal{A} \in \operatorname{Hom} V$ assigning to each polynomial its derivative.

(a) Write the matrix of $\mathcal{A}$ in the usual basis.

(b) Is there a basis in V such that every element of $[\mathcal{A}]$ is 0 or 1?

*(c) Is there a basis in V such that no element in $[\mathcal{A}]$ is 0?

(d) Is there a basis in V such that all elements in the last two columns of $[\mathcal{A}]$ are 0?

(e) Is there a basis in V such that all elements in the last two rows of $[\mathcal{A}]$ are 0?

5.7.2. Write the matrix of the well-known transformations of the plane with respect to several bases.

5.7.3. Let V_1 and V_2 be the usual vector spaces of polynomials with complex coefficients of degree at most 5 and 2. We denote a general polynomial by f and make no distinction between polynomials and polynomial functions. Find the matrices of the linear maps (with respect to suitably chosen pairs of bases).

(a) $f \mapsto f(0) + f(1)x + f(2)x^2$;

(b) the image of f is its remainder under division by $x^3 + 1$;

(c) the image of f is the (unique) polynomial of degree at most 2 assuming at 0, 1, and 2 the same values as f.

5.7.4. Let $\mathbf{a}_1, \dots, \mathbf{a}_n$ and $\mathbf{b}_1, \dots, \mathbf{b}_k$ be bases of V_1 and V_2. Describe the changes in the matrix of $\mathcal{A} \in \operatorname{Hom}(V_1, V_2)$ corresponding to the modification of the bases.

(a) We swap (a1) $\mathbf{a}_1$ and $\mathbf{a}_2$; (a2) $\mathbf{b}_1$ and $\mathbf{b}_2$.

(b) We replace (b1) $\mathbf{a}_3$ by $\lambda\mathbf{a}_3$; (b2) $\mathbf{b}_3$ by $\lambda\mathbf{b}_3$ (where $\lambda \neq 0$).

(c) We replace (c1) $\mathbf{a}_3$ by $\mathbf{a}_3 + \mu\mathbf{a}_2$; (c2) $\mathbf{b}_3$ by $\mathbf{b}_3 + \mu\mathbf{b}_2$.

5.7.5. Let V be a two-dimensional vector space over $\mathbf{R}$. Which pairs of matrices belong to the representation of some $\mathcal{A} \in \operatorname{Hom} V$ in two different bases?

(a) $\begin{pmatrix} 1 & 0 \\ 0 & 1 \end{pmatrix}$ and $\begin{pmatrix} 1 & 0 \\ 0 & 2 \end{pmatrix}$; (b) $\begin{pmatrix} 1 & 0 \\ 0 & -1 \end{pmatrix}$ and $\begin{pmatrix} 0 & 1 \\ 1 & 0 \end{pmatrix}$;

(c) $\begin{pmatrix} 1 & 1 \\ 1 & 1 \end{pmatrix}$ and $\begin{pmatrix} 2 & 0 \\ 0 & 0 \end{pmatrix}$; (d) $\begin{pmatrix} \frac{1}{2} & \frac{\sqrt{3}}{2} \\ -\frac{\sqrt{3}}{2} & \frac{1}{2} \end{pmatrix}$ and $\begin{pmatrix} 0 & -1 \\ 1 & 1 \end{pmatrix}$;

(e) $\begin{pmatrix} 1 & 0 \\ 0 & 2 \end{pmatrix}$ and $\begin{pmatrix} 2 & 6 \\ 1 & 3 \end{pmatrix}$; (f) $\begin{pmatrix} 1 & 0 \\ 0 & 0 \end{pmatrix}$ and $\begin{pmatrix} 0 & 0 \\ 1 & 0 \end{pmatrix}$.

5.7.6. Let $V_1 \neq V_2$ and let $\mathcal{A} \in \operatorname{Hom}(V_1, V_2)$ be an arbitrary linear map. Prove the existence of a pair of bases such that every element in the main diagonal of $[\mathcal{A}]$ is 0 or 1 and all other elements are 0.

5.7.7. Let $V_1 \neq V_2$. Prove that $\mathcal{A} \in \operatorname{Hom}(V_1, V_2)$ is an isomorphism if and only if the matrix of $\mathcal{A}$ is the identity matrix in a suitable pair of bases.

5.7.8. Let $\dim V = 2$ and $\mathcal{A} \in \operatorname{Hom} V$. Verify that if $\mathcal{A} \neq \mathcal{O}$ but $\mathcal{A}^2 = \mathcal{O}$, then $[\mathcal{A}] = \begin{pmatrix} 0 & 1 \\ 0 & 0 \end{pmatrix}$ in a suitable basis of V.

5.7.9. Which linear transformations $\mathcal{A} \in \operatorname{Hom} V$ have the same matrix in every basis?

5.7.10. Give a new proof to Theorem 5.6.7 based on Theorem 5.7.7.

5.7.11. Verify that the rank of the matrix of a linear map $\mathcal{A}$ in an arbitrary pair of bases is the dimension of $\operatorname{Im} \mathcal{A}$. (This uniquely determined integer is the *rank* of the linear map $\mathcal{A}$.)

5.7.12. Let $A \in F^{k \times n}$ and $B \in F^{n \times s}$. Prove the inequalities for the ranks.
(a) $\operatorname{rk}(AB) \leq \min(\operatorname{rk}(A), \operatorname{rk}(B)$;
(b) $AB = 0 \Rightarrow r(A) + r(B) \leq n$.

5.7.13. Does $A^{100} = A^{-1}$ imply for a real 2×2 matrix A that A is the identity matrix?

5.7.14. When does a matrix of a given linear transformation determine uniquely the basis used for the matrix representation? Prove that there exists such a matrix if and only if F is the modulo 2 field.

5.8. Switching Bases

Given the matrix of a linear map, we can switch to the matrix with respect to another pair of bases by the following theorem.

Theorem 5.8.1. *Let* $\mathbf{a}_1, \ldots, \mathbf{a}_n$ *and* $\mathbf{b}_1, \ldots, \mathbf{b}_k$ *be the old bases of* V_1 *and* V_2, *and* $\mathbf{a}'_1, \ldots, \mathbf{a}'_n$ *and* $\mathbf{b}'_1, \ldots, \mathbf{b}'_k$ *be the new bases. We need the transformations accompanying the changes in the bases:* $\mathcal{S} \in \operatorname{Hom} V_1$ *is the (unique) linear transformation of* V_1 *satisfying* $\mathcal{S}(\mathbf{a}_j) = \mathbf{a}'_j$, $j = 1, \ldots, n$, *and similarly* $\mathcal{T} \in \operatorname{Hom} V_2$ *is defined by* $\mathcal{T}(\mathbf{b}_i) = \mathbf{b}'_i$, $i = 1, \ldots, k$. *Then for any* $\mathcal{A} \in \operatorname{Hom}(V_1, V_2)$, *we have*

$$[\mathcal{A}]_{a',b'} = [\mathcal{T}]_b^{-1} \cdot [\mathcal{A}]_{a,b} \cdot [\mathcal{S}]_a. \clubsuit$$

Thus the new matrix of a linear map $\mathcal{A}$ is obtained by multiplying the old matrix by the matrix of the accompanying transformation $\mathcal{S} \in \operatorname{Hom} V_1$ from the right and by the *inverse* of the matrix of the accompanying transformation $\mathcal{T} \in \operatorname{Hom} V_2$.

Note that the matrices of the accompanying transformations are the same both in the old and new bases (see Exercise 5.8.3).

If $V_1 = V_2 = V$, i.e $\mathcal{A}$ is a linear transformation of V, then we have only one old and one new basis, and we obtain a special case of Theorem 5.8.1.

Theorem 5.8.1A. *Let* $\mathbf{a}_1, \ldots, \mathbf{a}_n$ *and* $\mathbf{a}'_1, \ldots, \mathbf{a}'_n$ *be the old and new bases of* V *and* $\mathcal{S} \in \operatorname{Hom} V$ *the (unique) accompanying linear transformation satisfying* $\mathcal{S}(\mathbf{a}_j) = \mathbf{a}'_j$, $j = 1, \ldots, n$. *Then for any* $\mathcal{A} \in \operatorname{Hom} V$, *we have*

$$[\mathcal{A}]_{a'} = [\mathcal{S}]_a^{-1} \cdot [\mathcal{A}]_a \cdot [\mathcal{S}]_a. \clubsuit$$

We turn to the proof of Theorem 5.8.1.

Proof. We denote the elements of the new matrix $A' = [\mathcal{A}]_{a',b'}$ by α'_{ij}. By definition, the jth column of A' is

$$\mathcal{A}(\mathbf{a}'_j) = \alpha'_{1j}\mathbf{b}'_1 + \cdots + \alpha'_{kj}\mathbf{b}'_k.$$

Using $(\mathcal{A}\mathcal{S})(\mathbf{a}_j) = \mathcal{A}(\mathcal{S}\mathbf{a}_j)$, $\mathcal{S}\mathbf{a}_j = \mathbf{a}'_j$, $\mathbf{b}'_i = \mathcal{T}\mathbf{b}_i$, and the linearity of $\mathcal{T}$, this is equivalent to

$$\begin{aligned}(\mathcal{A}\mathcal{S})(\mathbf{a}_j) &= \mathcal{A}(\mathcal{S}\mathbf{a}_j)\\ &= \mathcal{A}(\mathbf{a}'_j)\\ &= \alpha'_{1j}\mathbf{b}'_1 + \cdots + \alpha'_{kj}\mathbf{b}'_k\\ &= \alpha'_{1j}(\mathcal{T}\mathbf{b}_1) + \cdots + \alpha'_{kj}(\mathcal{T}\mathbf{b}_k)\\ &= \mathcal{T}(\alpha'_{1j}\mathbf{b}_1 + \cdots + \alpha'_{kj}\mathbf{b}_k).\end{aligned}$$

Multiplying $(\mathcal{A}\mathcal{S})(\mathbf{a}_j) = \mathcal{T}(\alpha'_{1j}\mathbf{b}_1 + \cdots + \alpha'_{kj}\mathbf{b}_k)$ by $\mathcal{T}^{-1}$ from the left, we obtain

$$(\mathcal{T}^{-1}\mathcal{A}\mathcal{S})(\mathbf{a}_j) = \alpha'_{1j}\mathbf{b}_1 + \cdots + \alpha'_{kj}\mathbf{b}_k.$$

By definition, this means that the old matrix of the linear map $\mathcal{T}^{-1}\mathcal{A}\mathcal{S}$ is equal to $A' = [\mathcal{A}]_{a',b'}$, i.e., $[\mathcal{A}]_{a',b'} = [\mathcal{T}^{-1}\mathcal{A}\mathcal{S}]_{a,b}$. Finally, the right-hand side can be written as $[\mathcal{T}]_b^{-1} \cdot [\mathcal{A}]_{a,b} \cdot [\mathcal{S}]_a$ by Theorem 5.7.6. ∎

For another proof, see Exercise 5.8.6.

Exercises 5.8

5.8.1. Let V be the usual vector space of polynomials with real coefficients of degree at most 2 and let $\mathcal{A} \in \operatorname{Hom} V$ be the linear transformation of differentiation. Compute the matrices of $\mathcal{A}$ in the bases
(a) $1 + x, x + x^2, x^2 + 1$;
(b) $x^2 + 1, -2x^2 + 2x, x^2 - 1$;
(c) $x^2 + x + 1, 2x + 1, -x^2 - x + 1$.

5.8.2. Using Theorem 5.8.1, find new solutions to Exercises 5.7.4, 5.7.5, and 5.7.9.

5.8.3. Show that switching bases, the accompanying transformation has the same matrix both in the old and new bases.

5.8.4. Verify that all matrices of a linear transformation have the same determinant.

5.8.5. Let V be a vector space over $\mathbf{R}$ with $2 \le \dim V < \infty$. Does every $\mathcal{A} \in \operatorname{Hom} V$ have a matrix that is
(a) symmetric;
(b) diagonal;
(c) lack of two equal elements;
(d) upper triangular (i.e., all elements below the main diagonal are 0)?

5.8.6. Give another proof for Theorem 5.8.1 based on the ideas below.
(A) Verify first the theorem for the elementary accompanying transformations: (i) we multiply a basis vector by a (non-zero) scalar; (ii) we add to a basis vector a scalar multiple of another basis vector; (iii) we swap two basis vectors (see Exercise 5.7.4).

(B) Show that if the theorem holds for accompanying transformations $\mathcal{S}_1$ and $\mathcal{S}_2$ and for any $\mathcal{A}$, then it remains valid if the accompanying transformation is the product $\mathcal{S}_1\mathcal{S}_2$ of the accompanying transformations $\mathcal{S}_1$ and $\mathcal{S}_2$. Prove the analogous statement also for the accompanying transformations $\mathcal{T}$.

(C) Similar to Gaussian elimination, verify that every accompanying transformation can be obtained as a composition of elementary transformations in (A).

5.8.7. Let $V_1 \neq V_2$ and V be finite-dimensional vector spaces over an *infinite field* F.

(a) Which linear maps $\mathcal{A} \in \operatorname{Hom}(V_1, V_2)$ have a matrix without 0 elements?

*(b) Which linear transformations $\mathcal{A} \in \operatorname{Hom} V$ have a matrix without 0 elements?

Remark. In part (a), we can omit the restriction $V_1 \neq V_2$ if, as an *exception*, we allow the use of two different bases even in the case of $V_1 = V_2$.

*5.8.8. Let V be a vector space satisfying $2 \leq \dim V < \infty$ over an *infinite field* F and let $\mathcal{A} \in \operatorname{Hom} V$ be an arbitrary linear transformation. Prove that there are infinitely many bases in V that are not scalar multiples of each other and induce the same matrix $[\mathcal{A}]$.

Chapter 6

Eigenvalues and Minimal Polynomials

Eigenvalues play a central role in describing linear transformations of finite-dimensional vector spaces and also in various applications. Eigenvalues are closely related both to the characteristic polynomial and the minimal polynomial. We shall use these to find the "nicest possible" matrix of a transformation.

6.1. Eigenvalue and Eigenvector

Throughout this chapter V denotes a finite-dimensional non-zero vector space over the commutative field F and $\mathcal{A} \in \operatorname{Hom} V$ is an arbitrary linear transformation. We shall heavily use the finiteness of dimension in general. We suggest the Reader think through which notions and statements remain valid for infinite dimension, too.

If $\mathcal{A}$ maps a non-zero vector to its scalar multiple (or informally, the vector remains in its line after the transformation), then this vector is an eigenvector of $\mathcal{A}$ and the corresponding scalar (i.e., the factor of dilation) is an eigenvalue.

Definition 6.1.1. A scalar $\lambda \in F$ is an *eigenvalue* of the linear transformation $\mathcal{A}$ if there exists a *non-zero* vector $\mathbf{v} \in V$ satisfying $\mathcal{A}\mathbf{v} = \lambda\mathbf{v}$. ♣

Definition 6.1.2. A *non-zero* vector $\mathbf{v} \in V$ is an *eigenvector* of the linear transformation $\mathcal{A}$ if there exists a scalar $\lambda \in F$ satisfying $\mathcal{A}\mathbf{v} = \lambda\mathbf{v}$. ♣

We had to exclude the zero vector in the definition of the eigenvalue since $\mathcal{A}\mathbf{0} = \lambda\mathbf{0}$ holds for every λ, i.e., every element in the field would be an eigenvalue without the restriction.

It is appropriate to exclude the zero vector also from the eigenvectors, otherwise the corresponding eigenvalue could be any element of the field.

ATTENTION, please! We do **not exclude**, however, the *scalar* 0 from the eigen*values*. The definitions imply that 0 is an eigenvalue of $\mathcal{A}$ if and only if Ker $\mathcal{A} \neq \mathbf{0}$ and the corresponding eigenvectors are the non-zero elements of the kernel.

Examples. The only eigenvalue of $\mathcal{I}$ is 1 and every non-zero vector is an eigenvector. In the plane, a rotation around the origin has no eigenvalues or eigenvectors except if the angle of rotation is a multiple of π. The eigenvalues of a reflection over a line containing the origin are 1 and -1 and those of a projection onto a line are 1 and 0. We leave the determination of eigenvectors to the reader.

If $\mathbf{v} \neq \mathbf{0}$ and $\mathcal{A}\mathbf{v} = \lambda\mathbf{v}$, then we say that λ is the eigenvalue belonging to $\mathbf{v}$ and $\mathbf{v}$ is an eigenvector belonging to λ. The difference in the use of the definite and indefinite articles is justified by the following theorem.

Theorem 6.1.3.

I. *An eigenvector determines the corresponding eigenvalue uniquely.*

II. *All eigenvectors belonging to the eigenvalue λ together with the $\mathbf{0}$ form a subspace. This is called the eigenspace of λ.* ♣

Note that, by the definition of eigenvalue, an eigenspace cannot consist solely of the zero vector.

Proof.

I. If $\mathcal{A}\mathbf{v} = \lambda\mathbf{v} = \mu\mathbf{v}$ for some $\mathbf{v} \neq \mathbf{0}$ and $\lambda, \mu \in F$, then $\mathbf{v} \neq \mathbf{0}$ implies $\lambda = \mu$ (see Exercise 4.1.11(a)).

II. The given set consists exactly of the vectors $\mathbf{v}$ satisfying $\mathcal{A}\mathbf{v} = \lambda\mathbf{v}$. We have to verify that this obviously non-empty set is closed under addition and multiplication by scalars. Let $\mathcal{A}\mathbf{v} = \lambda\mathbf{v}$ and $\mathcal{A}\mathbf{z} = \lambda\mathbf{z}$. Then

$$\mathcal{A}(\mathbf{v} + \mathbf{z}) = \mathcal{A}\mathbf{v} + \mathcal{A}\mathbf{z} = \lambda\mathbf{v} + \lambda\mathbf{z} = \lambda(\mathbf{v} + \mathbf{z}).$$

We get $\mathcal{A}(\alpha\mathbf{v}) = \lambda(\alpha\mathbf{v})$ similarly. ■

Remark. Since $\mathcal{A}\mathbf{v} = \lambda\mathbf{v} \iff (\mathcal{A} - \lambda\mathcal{I})\mathbf{v} = \mathbf{0}$, the eigenspace belonging to λ is just $\operatorname{Ker}(\mathcal{A} - \lambda\mathcal{I})$. This also yields another proof that the eigenspace is a subspace.

If the elements of a basis are eigenvectors, then the linear transformation has a nice matrix: every element outside the main diagonal is 0, i.e., the matrix is diagonal. Such a basis is called an *eigenbasis*.

Theorem 6.1.4. *A matrix of a linear transformation is diagonal if and only if the matrix was created using an eigenbasis. In this case the elements in the main diagonal are just the eigenvalues belonging to the basis vectors.* ♣

Proof.

$$[\mathcal{A}]_a = \begin{pmatrix} \lambda_1 & 0 & \dots & 0 \\ 0 & \lambda_2 & \dots & 0 \\ \vdots & \vdots & \ddots & \vdots \\ 0 & 0 & \dots & \lambda_n \end{pmatrix}$$

holds if and only if $\mathcal{A}\mathbf{a}_1 = \lambda_1\mathbf{a}_1, \dots, \mathcal{A}\mathbf{a}_n = \lambda_n\mathbf{a}_n$. ■

By the close relationship of linear transformations and matrices, we can define the notions of eigenvalues and eigenvectors for square matrices.

Definition 6.1.5. Let $A \in F^{n \times n}$. If $A\mathbf{x} = \lambda\mathbf{x}$ for some scalar $\lambda \in F$ and *non-zero* $\mathbf{x} \in F^n$, then λ is an eigenvalue and $\mathbf{x}$ is an eigenvector of the matrix A. ♣

Remark. It is customary to distinguish left and right eigenvectors of matrices. Definition 6.1.5 means a *right* eigenvector. A *left* eigenvector is a non-zero $\mathbf{z}^T$ *row* vector satisfying

$$(*) \qquad \mathbf{z}^T A = \lambda \mathbf{z}^T$$

(the letter T in the exponent means the transpose). Transposing the equality $(*)$, we get the equivalent equality $A^T\mathbf{z} = \lambda\mathbf{z}$, i.e., the left eigenvectors of A are just the right eigenvectors of the transpose of A. We shall see in the next section that the eigenvalues of A and A^T are the same (Exercise 6.2.11), so we do not have to distinguish left and right eigenvalues. In the sequel, an eigenvector of a matrix will always mean a right eigenvector conforming to Definition 6.1.5.

Definition 6.1.6. The square matrices A and B of the same size are *similar* if they are the matrices of the same linear transformation with respect to two bases. We denote this relation by $A \sim B$. ♣

Thus Exercise 5.7.5 asks whether or not the given pairs of matrices are similar.

By Theorem 5.8.1A we can formulate the similarity of two matrices without referring to linear transformations: A and B are similar if and only if $B = S^{-1}AS$ for some invertible matrix S. We can deduce either from this or from Definition 6.1.6 that similarity of matrices is an equivalence relation.

Definitions 6.1.1, 6.1.2, and 6.1.5 clearly imply that a linear transformation and any of its matrices have the same eigenvalues: If $\mathbf{c}_1, \dots, \mathbf{c}_n$ is a basis in V, then $\mathcal{A}\mathbf{v} = \lambda\mathbf{v} \iff [\mathcal{A}]_c[\mathbf{v}]_c = \lambda[\mathbf{v}]_c$. This equivalence provides the relation of the eigenvectors of a transformation and its matrix.

It follows that similar matrices have the same eigenvalues. To describe the connection between their eigenvectors, let $B = S^{-1}AS$, i.e., $SB = AS$. Then $\mathbf{x} \neq \mathbf{0}$ is an eigenvector of B if and only if $S\mathbf{x}$ is an eigenvector of A. Indeed, multiplying $B\mathbf{x} = \lambda\mathbf{x}$ by S from the left, we get $SB\mathbf{x} = S(\lambda\mathbf{x})$, i.e., $AS\mathbf{x} = \lambda S\mathbf{x}$. As S is invertible, this is equivalent to the original equality and $\mathbf{x} \neq 0 \iff S\mathbf{x} \neq \mathbf{0}$.

Exercises 6.1

6.1.1. Let V be the usual vector space of polynomials with real coefficients of degree not greater than 6 (including the 0 polynomial). We denote a general polynomial by f. Determine the eigenvalues and eigenvectors of the linear transformations and compute the dimensions of the eigenspaces. Which transformations have a diagonal matrix?

(a) $f \mapsto f'$;

(b) $f \mapsto xf'$;

(c) $f \mapsto f(6)x^6$;

(d) $f \mapsto$ the remainder of f under division by $x^2 + 3$.

6.1.2. Let $\mathcal{A}, \mathcal{B} \in \operatorname{Hom} V$ and let α be a common eigenvalue of both $\mathcal{A}$ and $\mathcal{B}$. Does this imply that (a) $\mu\mathcal{A}$; (b) $\mathcal{A} + \mathcal{B}$; (c) $\mathcal{A}^2$; (d) $\mathcal{A}\mathcal{B}$; (e) $\mathcal{A}^{-1}$ has an eigenvalue? If yes, how does this eigenvalue depend on α?

6.1.3. Let $\mathcal{A}, \mathcal{B} \in \operatorname{Hom} V$ and let $\mathbf{v}$ be a common eigenvector of both $\mathcal{A}$ and $\mathcal{B}$. Does this imply that $\mathbf{v}$ is an eigenvector of (a) $\mu\mathcal{A}$; (b) $\mathcal{A} + \mathcal{B}$; (c) $\mathcal{A}^2$; (d) $\mathcal{A}\mathcal{B}$; (e) $\mathcal{A}^{-1}$? If yes, what are the corresponding eigenvalues?

6.1.4. True or false?
(a) If $\mathbf{v}$ is an eigenvector of $\mathcal{A}^2$, then $\mathbf{v}$ is an eigenvector of $\mathcal{A}$.
(b) If 0 is an eigenvalue of $\mathcal{A}^2$, then 0 is an eigenvalue of $\mathcal{A}$.
(c) If $\mu^2 = \lambda$ and λ is an eigenvalue of $\mathcal{A}^2$, then at least one of μ and $-\mu$ is an eigenvalue of $\mathcal{A}$.

6.1.5. True or false?
(a) If $\mathcal{A} + \mathcal{B} = \mathcal{I}$, then $\mathcal{A}$ and $\mathcal{B}$ have the same eigenvectors.
(b) If $\mathcal{A}\mathcal{B} = \mathcal{O}$, then $\mathcal{A}$ and $\mathcal{B}$ have the same eigenvectors.
(c) $\mathcal{A}^2 = \mathcal{O}$ holds if and only if 0 is the only eigenvector of $\mathcal{A}$.
(d) $\mathcal{A} \neq \mathcal{O}$ is a zero divisor if and only if 0 is an eigenvalue of $\mathcal{A}$.
(e) Every eigenvector of $\mathcal{A}$ is contained in at least one of $\operatorname{Ker}\mathcal{A}$ and $\operatorname{Im}\mathcal{A}$.
(f) $\mathcal{A}^2 = \mathcal{A} \neq \mathcal{O}$ holds if and only if $\operatorname{Im}\mathcal{A}$ is an eigenspace of $\mathcal{A}$.

6.1.6. Show linear transformations in the ordinary three-dimensional space having 1, 2, and 3 (distinct) eigenvalues.

6.1.7. Let $\mathbf{u}$ and $\mathbf{v}$ be eigenvectors of $\mathcal{A}$. Find a simple necessary and sufficient condition for $\mathbf{u} + \mathbf{v}$ to be an eigenvector of $\mathcal{A}$.

6.1.8. Determine all linear transformations where every non-zero vector is an eigenvector.

6.1.9. Let $\mathbf{v}_1, \dots, \mathbf{v}_k$ be eigenvectors of $\mathcal{A}$ with k distinct eigenvalues. Show that $\mathbf{v}_1, \dots, \mathbf{v}_k$ are linearly independent.

6.1.10. Prove that if $\dim V = n$, then any $\mathcal{A} \in \operatorname{Hom} V$ can have at most n (distinct) eigenvalues.

6.1.11. A matrix of a linear transformation is

$$\begin{pmatrix} 1 & 1 & 1 \\ 1 & 1 & 1 \\ 1 & 1 & 1 \end{pmatrix}$$

in a basis. Show that a matrix of the same transformation is

$$\begin{pmatrix} 3 & 0 & 0 \\ 0 & 0 & 0 \\ 0 & 0 & 0 \end{pmatrix}$$

in some other basis.

6.1.12. Prove.
(a) The eigenvalues of $\mathcal{A}\mathcal{B}$ and $\mathcal{B}\mathcal{A}$ are the same.
(b) If $\mathcal{B}$ is invertible, then the eigenvalues of $\mathcal{A}$ and $\mathcal{B}^{-1}\mathcal{A}\mathcal{B}$ are the same.

6.2. Characteristic Polynomial

Theorem 6.2.1. *Let $\mathbf{a}_1, \dots, \mathbf{a}_n$ be a basis in V and $\mathcal{A} \in \operatorname{Hom} V$. A scalar $\lambda \in F$ is an eigenvalue of $\mathcal{A}$ if and only if the determinant of the matrix $[\mathcal{A} - \lambda\mathcal{I}]_a$ is 0: $\det([\mathcal{A} - \lambda\mathcal{I}]_a) = 0$.* ♣

Proof. λ is an eigenvalue if and only if $\mathcal{A}\mathbf{x} = \lambda\mathbf{x}$, i.e., $(\mathcal{A} - \lambda\mathcal{I})\mathbf{x} = \mathbf{0}$ for some vector $\mathbf{x} \neq \mathbf{0}$. By Theorem 5.7.3, this is equivalent to the equality of matrices $[\mathcal{A}-\lambda\mathcal{I}][\mathbf{x}] = [\mathbf{0}]$. So, λ is an eigenvalue exactly if this homogeneous system of linear equations has a non-trivial solution. This is true if and only if the determinant $\det([\mathcal{A} - \lambda\mathcal{I}]_a)$ of the coefficient matrix is 0. ∎

The theorem enables us to find the eigenvalues: considering λ as a variable, the determinant of $[\mathcal{A} - \lambda\mathcal{I}]_a$ is a polynomial in λ of degree n and the eigenvalues are the zeroes of this polynomial. The proof also provides an algorithm to determine the eigenvectors: these are the non-trivial solutions of the corresponding homogeneous systems of linear equations and can be computed by Gaussian elimination.

As the roots are the eigenvalues, these do not depend on the selection of the basis in creating the matrix of the transformation. But even a much stronger statement is true: not only the roots but the determinant polynomial itself is independent of the basis (this does not hold for the matrix, of course). When switching bases, we get a connection between the two matrices by Theorem 5.8.1A, and applying the product theorem for determinants we get the equality of the polynomials (cf. Exercise 5.8.4). We leave the details to the reader.

This uniquely determined polynomial is called the characteristic polynomial of the transformation.

Definition 6.2.2. Let

$$[\mathcal{A}] = \begin{pmatrix} \alpha_{11} & \alpha_{12} & \dots & \alpha_{1n} \\ \alpha_{21} & \alpha_{22} & \dots & \alpha_{2n} \\ \vdots & \vdots & \ddots & \vdots \\ \alpha_{n1} & \alpha_{n2} & \dots & \alpha_{nn} \end{pmatrix}$$

be a matrix of $\mathcal{A} \in \operatorname{Hom} V$ (in some basis). The *characteristic polynomial* of $\mathcal{A}$ is

$$k_{\mathcal{A}}(x) = \det[\mathcal{A} - x\mathcal{I}] = \begin{vmatrix} \alpha_{11} - x & \alpha_{12} & \dots & \alpha_{1n} \\ \alpha_{21} & \alpha_{22} - x & \dots & \alpha_{2n} \\ \vdots & \vdots & \ddots & \vdots \\ \alpha_{n1} & \alpha_{n2} & \dots & \alpha_{nn} - x \end{vmatrix}. ♣$$

By the above, this polynomial depends only on the transformation $\mathcal{A}$ and is independent of the matrix, i.e., of the selection of the basis. Its leading coefficient is $(-1)^n$, the coefficient of the term of degree $n-1$ is $(-1)^{n-1}$ times the *trace*(= the sum of elements in the main diagonal) of $[\mathcal{A}]$, and the constant term is the determinant of $[\mathcal{A}]$. This shows that neither the trace nor the determinant depend on the basis used for the matrix.

By Definition 6.2.2, Theorem 6.2.1 states that the eigenvalues of a transformation are exactly the zeroes of its characteristic polynomial.

We can define the characteristic polynomial of a matrix similarly to Definition 6.2.2 and it is also true for a matrix that its eigenvalues are just the zeroes of its characteristic polynomial. The connection between transformations and matrices yields that similar matrices have the same characteristic polynomial.

Exercises 6.2

6.2.1. Find the characteristic polynomials of the transformations in Exercise 6.1.1.

6.2.2. Let V be the usual vector space of plane vectors. Determine the characteristic polynomials of the transformations.
(a) Reflection across a line containing the origin.
(b) Projection in a given direction onto a line containing the origin.
(c) Rotation around the origin by 90 degrees.
(d) Rotation around the origin by 60 degrees.
(e) Identity.
(f) Reflection through the origin.
(g) Magnification from the origin by a factor of 5.

6.2.3. Let $f(x)$ be the characteristic polynomial of $\mathcal{A}$. Compute the characteristic polynomial of $\mu\mathcal{A}$.

6.2.4. Give new proofs to Exercises 6.1.10 and 6.1.9 (in this order).

6.2.5. Show that every linear transformation of a finite-dimensional vector space over $\mathbf{C}$ has an eigenvalue.

6.2.6. (a) Does there exist a linear transformation in the (ordinary) plane without eigenvectors?
(b) Does there exist a linear transformation in the (ordinary) space without eigenvectors?

6.2.7. Let $F = \mathbf{R}$ and let $\mathbf{b}_1, \dots, \mathbf{b}_4$ be a basis in the vector space V. Compute the characteristic polynomials, eigenvalues and eigenvectors of the linear transformations. Which transformations have a diagonal matrix?
(a) $\mathbf{b}_1 \mapsto \mathbf{b}_2, \quad \mathbf{b}_2 \mapsto \mathbf{b}_3, \quad \mathbf{b}_3 \mapsto \mathbf{b}_4, \quad \mathbf{b}_4 \mapsto \mathbf{b}_1;$
(b) $\mathbf{b}_1 \mapsto \mathbf{b}_2, \quad \mathbf{b}_2 \mapsto \mathbf{b}_1, \quad \mathbf{b}_3 \mapsto \mathbf{b}_4, \quad \mathbf{b}_4 \mapsto \mathbf{b}_4;$
(c) $\mathbf{b}_1 \mapsto \mathbf{b}_1 + \mathbf{b}_2, \quad \mathbf{b}_2 \mapsto \mathbf{b}_2 + \mathbf{b}_3, \quad \mathbf{b}_3 \mapsto \mathbf{b}_3 + \mathbf{b}_4, \quad \mathbf{b}_4 \mapsto \mathbf{b}_4 + \mathbf{b}_1.$
Solve the problem also over $F = \mathbf{C}$.

6.2.8. How can we determine the number of diagonal matrices of a linear transformation (assuming that it has at least one)?

6.2.9. Let $\dim V = n$ and assume that $\mathcal{A} \in \operatorname{Hom} V$ has n distinct eigenvalues. Prove that the sum and product of the eigenvalues are the trace and the determinant of the matrix $[\mathcal{A}]$ (in any basis).

6.2.10. Give a new solution to Exercise 5.7.5.

6.2.11. Verify that a matrix and its transpose have the same characteristic polynomial, so their eigenvalues are the same, too.

6.3. Minimal Polynomial

Let $V \neq \mathbf{0}$ be a finite-dimensional vector space over a field F, $\dim V = n$, $\mathcal{A} \in \operatorname{Hom} V$, and $f = \alpha_0 + \alpha_1 x + \cdots + \alpha_k x^k \in F[x]$.

We shall define the value $f(\mathcal{A})$ of f assumed at $\mathcal{A}$; this, too, will be a $V \to V$ linear transformation.

It is straightforward to define the 0th power of $\mathcal{A}$ by $\mathcal{A}^0 = \mathcal{I}$ and this yields the natural definition $f(\mathcal{A}) = \alpha_0 \mathcal{I} + \alpha_1 \mathcal{A} + \cdots + \alpha_k \mathcal{A}^k$. Clearly, $f(\mathcal{A}) \in \operatorname{Hom} V$.

It is easy to verify the usual rules

$$(f+g)(\mathcal{A}) = f(\mathcal{A}) + g(\mathcal{A}) \qquad \text{and} \qquad (fg)(\mathcal{A}) = f(\mathcal{A})g(\mathcal{A}).$$

We use the notion of a *root* of a polynomial in the usual sense: $\mathcal{A}$ is a root of f if $f(\mathcal{A}) = \mathcal{O}$.

Among the polynomials having a given transformation $\mathcal{A}$ as a root, the ones having the minimal degree are of special importance.

Definition 6.3.1. A *minimal polynomial* of a transformation $\mathcal{A}$ is a non-zero polynomial of smallest degree among the polynomials having $\mathcal{A}$ as a root. Notation: $m_{\mathcal{A}}$ or $m_{\mathcal{A}}(x)$. In formulas:

$$m_{\mathcal{A}}(\mathcal{A}) = \mathcal{O} \quad \text{and} \quad g(\mathcal{A}) = \mathcal{O} \Rightarrow \deg g \geq \deg m_{\mathcal{A}} \text{ or } g = 0. \clubsuit$$

Examples. In the plane, minimal polynomials of a reflection $\mathcal{T}$ (over a line containing the origin), projection $\mathcal{P}$ (onto a line containing the origin), and rotation $\mathcal{R}$ by 90 degrees (around the origin) are $m_{\mathcal{T}} = x^2 - 1$, $m_{\mathcal{P}} = x^2 - x$, and $m_{\mathcal{R}} = x^2 + 1$. In any vector space, minimal polynomials of the zero and identical transformations are $m_{\mathcal{O}} = x$ and $m_{\mathcal{I}} = x - 1$.

Theorem 6.3.2. *Every $\mathcal{A}$ has a minimal polynomial and it is unique up to a constant factor.* ♣

We shall use the notation $m_{\mathcal{A}}$ for any copy of the minimal polynomial and shall not distinguish them as they differ only in a constant factor and behave equivalently in all related respects. If preferred, absolute uniqueness could be achieved by selecting the monic one (=with leading coefficient 1) from these polynomials.

Proof. We verify uniqueness first. Let both $f = \alpha_0 + \alpha_1 x + \cdots + \alpha_k x^k$ and $g = \beta_0 + \beta_1 x + \cdots + \beta_k x^k$ be minimal polynomials of $\mathcal{A}$, $\alpha_k, \beta_k \neq 0$. To eliminate the term x^k, we construct the polynomial $h = \alpha_k g - \beta_k f$. Clearly

$$h(\mathcal{A}) = \alpha_k g(\mathcal{A}) - \beta_k f(\mathcal{A}) = \alpha_k \mathcal{O} - \beta_k \mathcal{O} = \mathcal{O}.$$

By the definition of the minimal polynomial, h cannot have degree less than k, so only $h = 0$ is possible. This yields $f = \gamma g$ where $\gamma = \alpha_k / \beta_k$.

We now prove the existence of a minimal polynomial. It is sufficient to show that there exists a non-zero polynomial having $\mathcal{A}$ as a root; among all these polynomials the ones of smallest degree are minimal polynomials.

Consider the n^2+1 transformations

$$\mathcal{I}, \mathcal{A}, \mathcal{A}^2, \ldots, \mathcal{A}^{n^2} \quad (n = \dim V)$$

in Hom V. These are linearly dependent due to $\dim \operatorname{Hom} V = n^2$. Therefore

$$\gamma_0 \mathcal{I} + \gamma_1 \mathcal{A} + \cdots + \gamma_{n^2} \mathcal{A}^{n^2} = \mathcal{O}$$

for some scalars $\gamma_0, \gamma_1, \ldots, \gamma_{n^2} \in F$ not all zero. This means that $\mathcal{A}$ is a root of the non-zero polynomial

$$\gamma_0 + \gamma_1 x + \cdots + \gamma_{n^2} x^{n^2}. \qquad \blacksquare$$

From the proof, we obtain an upper bound for the degree of the minimal polynomial: $\deg m_{\mathcal{A}} \leq n^2$. This can be improved to $\deg m_{\mathcal{A}} \leq n$ and will follow from Theorems 6.3.5 or 6.5.6. This latter bound is already sharp, e.g., the minimal polynomial of differentiation in the n-dimensional vector space of polynomials of degree less than n is x^n.

With the help of the minimal polynomial, we can characterize all polynomials having $\mathcal{A}$ as a root; these are exactly the (polynomial) multiples of the minimal polynomial.

Theorem 6.3.3.

$$g(\mathcal{A}) = \mathcal{O} \iff m_{\mathcal{A}} \mid g. \clubsuit$$

Proof. If $m_{\mathcal{A}} \mid g$, i.e., $g = t m_{\mathcal{A}}$, then

$$g(\mathcal{A}) = t(\mathcal{A}) m_{\mathcal{A}}(\mathcal{A}) = t(\mathcal{A}) \cdot \mathcal{O} = \mathcal{O},$$

so $\mathcal{A}$ is a root of g.

To prove the converse, we assume $g(\mathcal{A}) = \mathcal{O}$. We perform the long division of g by $m_{\mathcal{A}}$: $g = t m_{\mathcal{A}} + r$, where $\deg r < \deg m_{\mathcal{A}}$ or $r = 0$. Then

$$r(\mathcal{A}) = g(\mathcal{A}) - t(\mathcal{A}) m_{\mathcal{A}}(\mathcal{A}) = \mathcal{O} - t(\mathcal{A})\mathcal{O} = \mathcal{O}.$$

By the definition of the minimal polynomial, $\deg r < \deg m_{\mathcal{A}}$ is impossible, so $r = 0$, i.e., $m_{\mathcal{A}} \mid g$. $\blacksquare$

Theorem 6.3.3 can help find the minimal polynomial: If the transformation is a root of some non-zero polynomial, then the minimal polynomial can only be one of its divisors.

Similar to the characteristic polynomial, the minimal polynomial is closely related to the eigenvalues.

Theorem 6.3.4. *The eigenvalues of a linear transformation are exactly the roots in F of its minimal polynomial.* $\clubsuit$

Proof. We show first that every eigenvalue is a root of the minimal polynomial. Let $m_{\mathcal{A}} = \alpha_0 + \alpha_1 x + \cdots + \alpha_k x^k$, and assume that $\lambda \in F$ is an eigenvalue of $\mathcal{A}$, i.e., $\mathcal{A}\mathbf{u} = \lambda \mathbf{u}$ for some $\mathbf{u} \neq \mathbf{0}$. Then

$$\mathcal{A}^2 \mathbf{u} = \mathcal{A}(\mathcal{A}\mathbf{u}) = \mathcal{A}(\lambda \mathbf{u}) = \lambda(\mathcal{A}\mathbf{u}) = \lambda(\lambda \mathbf{u}) = \lambda^2 \mathbf{u},$$

and we obtain $\mathcal{A}^j \mathbf{u} = \lambda^j \mathbf{u}$ by induction similarly for an arbitrary exponent j.

Applying the transformation $m_{\mathcal{A}}(\mathcal{A}) = \mathcal{O}$ to $\mathbf{u}$ the result is $\mathbf{0}$. So

$$\begin{aligned}\mathbf{0} &= m_{\mathcal{A}}(\mathcal{A})\mathbf{u}\\ &= (\alpha_0\mathcal{I} + \alpha_1\mathcal{A} + \cdots + \alpha_k\mathcal{A}^k)\mathbf{u}\\ &= \alpha_0(\mathcal{I}\mathbf{u}) + \alpha_1(\mathcal{A}\mathbf{u}) + \cdots + \alpha_k(\mathcal{A}^k\mathbf{u})\\ &= \alpha_0\mathbf{u} + \alpha_1(\lambda\mathbf{u}) + \cdots + \alpha_k(\lambda^k\mathbf{u})\\ &= (\alpha_0 + \alpha_1\lambda + \cdots + \alpha_k\lambda^k)\mathbf{u}\\ &= \big(m_{\mathcal{A}}(\lambda)\big)\mathbf{u},\end{aligned}$$

i.e., $m_{\mathcal{A}}(\lambda)\mathbf{u} = \mathbf{0}$. Since $\mathbf{u} \neq \mathbf{0}$, we have $m_{\mathcal{A}}(\lambda) = 0$, so λ is a root of the minimal polynomial.

For the converse, we have to show that every root of the minimal polynomial is an eigenvalue. Let $\lambda \in F$ be a root of $m_{\mathcal{A}}$, then the minimal polynomial is of the form $m_{\mathcal{A}} = (x - \lambda)g$. Substituting $x = \mathcal{A}$, we get

$$\mathcal{O} = m_{\mathcal{A}}(\mathcal{A}) = (\mathcal{A} - \lambda\mathcal{I})g(\mathcal{A}),$$

or equivalently $\operatorname{Ker}(\mathcal{A} - \lambda\mathcal{I}) \supseteq \operatorname{Im} g(\mathcal{A})$. Here $\deg g < \deg m_{\mathcal{A}}$, thus $g(\mathcal{A}) \neq \mathcal{O}$, i.e., $\operatorname{Im} g(\mathcal{A}) \neq \mathbf{0}$. Therefore $\operatorname{Ker}(\mathcal{A} - \lambda\mathcal{I}) \neq \mathbf{0}$. Every non-zero vector in $\operatorname{Ker}(\mathcal{A} - \lambda\mathcal{I})$ is an eigenvector belonging to λ, so λ is an eigenvalue. ■

Now we prove a fundamental theorem about the close relationship of the characteristic and minimal polynomials.

Theorem 6.3.5 (Cayley–Hamilton theorem). *The minimal polynomial is a divisor of the characteristic polynomial.* ♣

By Theorem 6.3.3, an equivalent formulation of the Cayley–Hamilton Theorem is that every linear transformation is a root of its characteristic polynomial.

Theorems 6.2.1, 6.3.4, and 6.3.5 connect the eigenvalues, the characteristic polynomial, and the minimal polynomial, but none of them follows from the other two; see Exercise 6.3.19.

Proof. By Theorem 6.3.3, we have to verify that $\mathcal{A}$ is a root of the characteristic polynomial $k_{\mathcal{A}}(x) = \gamma_0 + \gamma_1 x + \cdots + \gamma_n x^n$.

Applying Lemma 2.2.3 to a matrix $A - xI$ of $\mathcal{A} - x\mathcal{I}$, we obtain

$$(A - xI)B(x) = \big(\det(A - xI)\big)I = k_{\mathcal{A}}(x)I, \tag{6.3.1}$$

where $B(x)$ is the adjugate of $A - xI$, i.e., the transpose of the matrix composed of the cofactors of $A - xI$. The cofactors are polynomials (in x) of degree $n - 1$, so the matrix $B(x)$ is a polynomial

$$B(x) = B_0 + B_1 x + \cdots + B_{n-1}x^{n-1}$$

where the coefficients B_i are $n \times n$ matrices (in this respect we put the powers of the variable behind the matrix coefficients). Then (6.3.1) can be written as

$$(A - Ix)(B_0 + B_1 x + \cdots + B_{n-1}x^{n-1}) = \gamma_0 I + \gamma_1 Ix + \cdots + \gamma_n Ix^n. \tag{6.3.2}$$

Performing the multiplication on the left-hand side of (6.3.2) and comparing the coefficients of the powers of x on the two sides, we get

$$
\begin{aligned}
AB_0 &= \gamma_0 I;\\
AB_1 - B_0 &= \gamma_1 I;\\
&\vdots\\
AB_{n-1} - B_{n-2} &= \gamma_{n-1} I;\\
-B_{n-1} &= \gamma_n I.
\end{aligned}
\tag{6.3.3}
$$

In the system of equations (6.3.3), we multiply the second equation by A, the third one by A^2, etc., and the last one by A^n from the left, and add the resulting equations. We obtain the 0 matrix on the left-hand side and $\gamma_0 I + \gamma_1 A + \cdots + \gamma_n A^n = k_{\mathcal{A}}(A)$ on the right-hand side. This means that the matrix A and so the transformation $\mathcal{A}$ is a root of the characteristic polynomial. ■

We can define the minimal polynomial of a square matrix analogously to Definition 6.3.1 and Theorems 6.3.2–6.3.5 remain valid for matrices, too. The connection between transformations and matrices clearly implies that similar matrices have the same minimal polynomial, but this can be verified directly, too, without referring to transformations (Exercise 6.3.20).

Exercises 6.3

6.3.1. Find the minimal polynomials of the transformations in Exercises 6.1.1, 6.2.2, and 6.2.7.

6.3.2. Which transformations have a linear minimal polynomial?

6.3.3. How can we see on the minimal polynomial whether or not the transformation has an inverse?

6.3.4. Verify that if $\mathcal{A}$ is invertible, then $\mathcal{A}^{-1}$ can be obtained as a polynomial of $\mathcal{A}$, i.e., $\mathcal{A}^{-1} = f(\mathcal{A})$ for some $f \in F[x]$ (depending on $\mathcal{A}$).

6.3.5. For invertible matrices, what is the relation between the minimal polynomials of $\mathcal{A}$ and $\mathcal{A}^{-1}$?

6.3.6. True or false?

(a) The minimal polynomial is irreducible over F.

(b) If a linear transformation is a root of an irreducible polynomial over F, then this polynomial is a minimal polynomial of the transformation.

(c) If $F = \mathbf{C}$ and the characteristic polynomial has no multiple roots, then the minimal and characteristic polynomials are equal.

(d) If $F = \mathbf{C}$ and the minimal and characteristic polynomials are equal, then the characteristic polynomial has no multiple roots.

(e) If a transformation has a diagonal matrix, then its minimal polynomial has no multiple roots.

(f) If $\mathcal{A}$ is a root of a polynomial f, then every eigenvalue of $\mathcal{A}$ is a root of f.

(g) If $F = \mathbf{C}$ and every eigenvalue of $\mathcal{A}$ is a root of f, then $\mathcal{A}$ is a root of f.

6.3.7. Give a new proof to Exercise 6.2.5.

6.3.8. Give a new proof to Exercise 5.6.9.

6.3.9. Does there exist a 2×2 matrix $A \neq I$ of (a) real; (b) rational numbers satisfying $A^5 = I$?

6.3.10. What is the relationship of the minimal polynomials of $\mathcal{AB}$ and $\mathcal{BA}$?

6.3.11. Verify that $\mathcal{A}$ and $\mathcal{B}^{-1}\mathcal{AB}$ have the same minimal polynomial.

6.3.12. Let $\dim V = n$, $\mathcal{A} \in \operatorname{Hom} V$, and let $k \geq n$ be an arbitrary integer. Prove that $\mathcal{A}$ is a root of some polynomial of degree k where the coefficients of all terms of degrees $k-1, k-2, \ldots, n$ are 0.

6.3.13. Determine the dimension of the subspace in Hom V spanned by the transformations $\mathcal{A}^i$, $i = 0, 1, 2, \ldots$ ($\mathcal{A}^0 = \mathcal{I}$)?

*6.3.14. Let $\deg m_{\mathcal{A}} = k$. What are the possible values of $\deg m_{\mathcal{A}^2}$?

****6.3.15.** Prove that, over the complex field, $\mathcal{A}$ and $\mathcal{A}^2$ have the same minimal polynomial if and only if (i) all roots of $m_{\mathcal{A}}$ are 0 or roots of unity of odd order; (ii) 0 is not a multiple root; and (iii) the square of every root is a root with the same multiplicity.

6.3.16. Let h be an arbitrary polynomial. Show that $h(\mathcal{A})$ is invertible if and only if h and $m_{\mathcal{A}}$ are coprime.

6.3.17. Let D be a fixed $n \times n$ matrix and consider the linear transformation $\mathcal{A} \in \operatorname{Hom}(F^{n\times n})$ that multiplies the matrices by D from the left: $\mathcal{A}(B) = DB$ for every $B \in F^{n\times n}$. Describe the connection between the eigenvalues and minimal polynomials of $\mathcal{A}$ and D. Does the same hold for the characteristic polynomials?

*6.3.18. Prove that every non-constant polynomial is a minimal polynomial of some linear transformation.

6.3.19. Find examples of polynomials to demonstrate that none of Theorems 6.2.1, 6.3.4, and 6.3.5 follows from the other two.

6.3.20. Without referring to linear transformations, prove directly that similar matrices have the same minimal polynomial.

6.3.21. A real square matrix A can also be considered as a complex matrix. Demonstrate that the minimal polynomials of A over $\mathbf{R}$ and $\mathbf{C}$ are the same.

6.3.22. (a) Prove that the minimal and characteristic polynomials of a matrix with rational entries contain exactly the same irreducible factors (but their multiplicities can differ).
(b) Give a new solution to Exercise 2.2.14.

6.4. Invariant Subspace

Let $\mathcal{A}$ be a linear transformation of V. Restricting $\mathcal{A}$ to a subspace U, this does not induce a linear transformation of U in general, as U is not necessarily mapped into itself by $\mathcal{A}$. The subspaces mapped into themselves by $\mathcal{A}$ play an important role in the sequel.

Definition 6.4.1. A subspace U is an *invariant subspace* of $\mathcal{A}$ (or is an $\mathcal{A}$-*invariant subspace* or a *subspace invariant with respect to* $\mathcal{A}$) if $\mathbf{u} \in U \Rightarrow \mathcal{A}\mathbf{u} \in U$. ♣

If it is clear which transformation $\mathcal{A}$ is considered, then we can omit the reference to $\mathcal{A}$ and use just the expression invariant subspace. We have to be aware, however, that a subspace can only be invariant with respect to some transformation and without it an invariant subspace makes no sense in itself.

Examples for invariant subspaces. $\operatorname{Im}\mathcal{A}$ and every subspace containing it; $\operatorname{Ker}\mathcal{A}$ and all its subspaces; a subspace spanned by an eigenvector; a subspace spanned by more eigenvectors (not necessarily with the same eigenvalues). Except the first example and the zero subspace, all others are special cases of the last one.

The following general construction provides many invariant subspaces to a given transformation $\mathcal{A}$. We start with an arbitrary vector $\mathbf{u}$, applying $\mathcal{A}$ we get $\mathcal{A}\mathbf{u}$, applying $\mathcal{A}$ again we obtain $\mathcal{A}^2\mathbf{u}$, etc. Then the subspace spanned by these vectors $\mathcal{A}^i\mathbf{u}$, $i = 0, 1, 2, \ldots$ ($\mathcal{A}^0 = \mathcal{I}$) will be $\mathcal{A}$-invariant.

Definition 6.4.2. The subspace $\langle \mathbf{u}, \mathcal{A} \rangle$ spanned by the vector $\mathbf{u}$ and the transformation $\mathcal{A}$ means the span of the vectors $\mathcal{A}^i\mathbf{u}$, $i = 0, 1, 2, \ldots$:

$$\langle \mathbf{u}, \mathcal{A} \rangle = \langle \mathbf{u}, \mathcal{A}\mathbf{u}, \mathcal{A}^2\mathbf{u}, \ldots \rangle. \clubsuit$$

This is the tightest $\mathcal{A}$-invariant subspace containing $\mathbf{u}$ (see Exercise 6.4.12). Therefore it can also be called as the $\mathcal{A}$-invariant subspace spanned by $\mathbf{u}$ and denoted by $\langle \mathbf{u} \rangle_{\mathcal{A}}$. We must not omit here the index about the transformation, otherwise we get just the usual span (which can be considered as the special case $\mathcal{A} = \mathcal{I}$). We shall stay with the names and notation of Definition 6.4.2.

It is easy to see that in the definition of $\langle \mathbf{u}, \mathcal{A} \rangle$, the infinite spanning set $\mathcal{A}^i\mathbf{u}$ can be replaced by a finite one since there are at most $n = \dim V$ linearly independent vectors among them. Moreover, we can restrict ourselves to the *first* n exponents $0 \le i \le n-1$ (see also Theorem 6.5.4 and its proof).

Exercises 6.4

6.4.1. Prove that if U_1 and U_2 are $\mathcal{A}$-invariant subspaces, then so are $U_1 \cap U_2$ and $\langle U_1, U_2 \rangle$.

6.4.2. True or false?
- (a) If U is an invariant subspace of $\mathcal{A}$, then U is an invariant subspace of $\mathcal{A}^3$.
- (b) If U is an invariant subspace of $\mathcal{A}^3$, then U is an invariant subspace of $\mathcal{A}$.
- (c) If U is an invariant subspace of $\mathcal{A}$ and $U \cap \operatorname{Im}\mathcal{A} = \mathbf{0}$, then $U \subseteq \operatorname{Ker}\mathcal{A}$.
- (d) If U is an invariant subspace of $\mathcal{A}$ and $U \cap \operatorname{Ker}\mathcal{A} = \mathbf{0}$, then $U \subseteq \operatorname{Im}\mathcal{A}$.

6.4.3. Consider the matrix of a transformation in a basis $\mathbf{b}_1, \mathbf{b}_2, \ldots, \mathbf{b}_n$ and let $k < n$. How can we see from the matrix whether or not the span $\langle \mathbf{b}_1, \mathbf{b}_2, \ldots, \mathbf{b}_k \rangle$ of the first k basis vectors is an invariant subspace?

6.4.4. Let $\dim V = n$ and let U be a k-dimensional subspace. Consider all transformations $\mathcal{A} \in \operatorname{Hom} V$ such that U is $\mathcal{A}$-invariant.
(a) Verify that these transformations form a subalgebra in $\operatorname{Hom} V$.
(b) Compute the dimension of this subalgebra.

6.4.5. (a) Which transformations satisfy that every subspace is invariant?
(b) Which transformations satisfy that every 13-dimensional subspace is invariant?

6.4.6. Assume $\mathcal{AB} = \mathcal{BA}$. Verify that $\operatorname{Im}\mathcal{A}$, $\operatorname{Ker}\mathcal{A}$, and the eigenspaces of $\mathcal{A}$ are invariant subspaces of $\mathcal{B}$.

6.4.7. (a) Prove that the invariant subspaces of $\mathcal{A}, \lambda\mathcal{A}, \mathcal{A}+\mu\mathcal{I}$, and $\lambda\mathcal{A}+\mu\mathcal{I}$ are the same for every $\lambda \neq 0$ and μ.
(b) Assume that $\mathcal{A}$ and $\mathcal{B}$ have the same invariant subspaces. Does this imply $\mathcal{B} = \lambda\mathcal{A} + \mu\mathcal{I}$ with a suitable $\lambda \neq 0$ and μ?

6.4.8. Find the number of invariant subspaces of differentiation in the usual vector space of polynomials of degree not greater than n.

6.4.9. (a) Verify that $\operatorname{Ker} f(\mathcal{A})$ is $\mathcal{A}$-invariant for any polynomial f.
(b) Prove $\operatorname{Ker} f(\mathcal{A}) = \operatorname{Ker} g(\mathcal{A}) \iff (f, m_{\mathcal{A}}) = (g, m_{\mathcal{A}})$.
(c) Show that the number of invariant subspaces of a transformation is not less than the number of pairwise non-associate divisors of its minimal polynomial.
(d) Which transformations have only the two trivial invariant subspaces ($\mathbf{0}$ and the entire V)?

6.4.10. Solve the analogous problem if kernel is replaced by image in the previous exercise.

6.4.11. Does $m_{\mathcal{A}} = fg$ imply
(a) $\operatorname{Im} f(\mathcal{A}) \subseteq \operatorname{Ker} g(\mathcal{A})$;
(b) $\operatorname{Im} f(\mathcal{A}) = \operatorname{Ker} g(\mathcal{A})$?

6.4.12. Verify that $\langle \mathbf{u}, \mathcal{A} \rangle$ is the tightest $\mathcal{A}$-invariant subspace containing $\mathbf{u}$, i.e., (i) it is an $\mathcal{A}$-invariant subspace; (ii) it contains $\mathbf{u}$; and (iii) it is part of every $\mathcal{A}$-invariant subspace containing $\mathbf{u}$.

6.4.13. For which vectors $\mathbf{u}$ is $\dim\langle \mathbf{u}, \mathcal{A} \rangle = 0$ or 1?

6.4.14. Give an example of an $\mathcal{A}$-invariant subspace not of the form $\langle \mathbf{u}, \mathcal{A} \rangle$.

6.4.15. True or false?
(a) If $\lambda \neq 0$, then $\langle \mathbf{u}, \mathcal{A} \rangle = \langle \lambda\mathbf{u}, \mathcal{A} \rangle$.
(b) $\langle \mathbf{u}+\mathbf{v}, \mathcal{A} \rangle = \langle \langle \mathbf{u}, \mathcal{A} \rangle, \langle \mathbf{v}, \mathcal{A} \rangle \rangle$.
(c) $\langle \mathbf{u}, \mathcal{A}^2 \rangle \subseteq \langle \mathbf{u}, \mathcal{A} \rangle$.
(d) If $\langle \mathbf{u}, \mathcal{A} \rangle = \langle \mathbf{v}, \mathcal{A} \rangle$ for some $\mathcal{A}$, then $\mathbf{u} = \lambda\mathbf{v}$.
(e) If $\langle \mathbf{u}, \mathcal{A} \rangle = \langle \mathbf{v}, \mathcal{A} \rangle$ for every $\mathcal{A}$, then $\mathbf{u} = \lambda\mathbf{v}$.
(f) If $\langle \mathbf{u}, \mathcal{A} \rangle = \langle \mathbf{u}, \mathcal{B} \rangle$ for some $\mathbf{u} \neq \mathbf{0}$, then $\mathcal{A} = \lambda\mathcal{B}$.
(g) If $\langle \mathbf{u}, \mathcal{A} \rangle = \langle \mathbf{u}, \mathcal{B} \rangle$ for every $\mathbf{u} \neq \mathbf{0}$, then $\mathcal{A} = \lambda\mathcal{B}$.

6.5. Order

Let $V \neq \mathbf{0}$ be a finite-dimensional vector space over a commutative field F, $\dim V = n$, and $\mathcal{A} \in \operatorname{Hom} V$.

By the definition of the minimal polynomial, $m_{\mathcal{A}}(\mathcal{A})\mathbf{u} = \mathbf{0}$ for every $\mathbf{u} \in V$. However, to a given vector $\mathbf{u}$, in general, we can find polynomials $f \in F[x]$ of smaller degree, too, satisfying $f(\mathcal{A})\mathbf{u} = \mathbf{0}$. The ones of smallest degree serve as some sort of local or private minimal polynomials for $\mathbf{u}$ and are called the order of the vector.

Definition 6.5.1. The *order* of $\mathbf{u}$ with respect to $\mathcal{A}$ is a (non-zero) *polynomial* h of minimal degree satisfying $h(\mathcal{A})\mathbf{u} = \mathbf{0}$. ♣

We denote the order by $o_{\mathcal{A}}(\mathbf{u})$. If there is no ambiguity concerning the transformation, then we can omit the index referring to it and write just $o(\mathbf{u})$.

Examples. The zero vector is the only one having order 1 (or any non-zero constant), the elements of the kernel have order x. A vector has an order of degree 1 if and only if it is an eigenvector.

The order has many similar properties as the minimal polynomial. We summarize these in the following theorem.

Theorem 6.5.2.

(i) *Every vector has an order and it is unique up to a constant factor.*

(ii) *The degree of the order is less than or equal to* $n(= \dim V)$.

(iii) $g(\mathcal{A})\mathbf{u} = \mathbf{0} \iff o_{\mathcal{A}}(\mathbf{u}) \mid g$. ♣

The proof is analogous to the arguments used for Theorems 6.3.2 and 6.3.3; see Exercise 6.5.2.

Part (iii) in Theorem 6.5.2 implies that the order always divides the minimal polynomial. Now we show how we can get the minimal polynomial from the orders of suitable vectors.

Theorem 6.5.3. *Let* $\mathbf{u}_1, \dots, \mathbf{u}_s$ *be an arbitrary spanning set in* V. *Then* $m_{\mathcal{A}}$ *is the least common multiple of the polynomials* $o_{\mathcal{A}}(\mathbf{u}_i)$. ♣

Proof. Let $h_i = o_{\mathcal{A}}(\mathbf{u}_i)$ and let H be the least common multiple of these polynomials: $H = [h_1, \dots, h_s]$. Since $h_i \mid m_{\mathcal{A}}$, we have $H \mid m_{\mathcal{A}}$. To prove the reverse divisibility, we have to show $H(\mathcal{A}) = \mathcal{O}$. Since $h_i \mid H$, we know that $H(\mathcal{A})\mathbf{u}_i = \mathbf{0}$. The vectors $\mathbf{u}_i$ span V, so every $\mathbf{v} \in V$ is their linear combination: $\mathbf{v} = \sum_{i=1}^{s} \lambda_i \mathbf{u}_i$. Then $H(\mathcal{A})\mathbf{v} = \sum_{i=1}^{s} \lambda_i H(\mathcal{A})\mathbf{u}_i = \mathbf{0}$, so $H(\mathcal{A}) = \mathcal{O}$. ■

The order is suitable to compute the dimension of certain invariant subspaces.

Theorem 6.5.4. *The dimension of the subspace* $\langle \mathbf{u}, \mathcal{A} \rangle$ *spanned by the vector* $\mathbf{u}$ *and transformation* $\mathcal{A}$ *is equal to the degree of the order of* $\mathbf{u}$:

$$\dim\langle \mathbf{u}, \mathcal{A} \rangle = \deg o_{\mathcal{A}}(\mathbf{u}). \text{ ♣}$$

Proof. The statement is clearly true for the zero vector (both values are 0). Let $\mathbf{u} \neq \mathbf{0}$ and $o_{\mathcal{A}}(\mathbf{u}) = h = x^k + \alpha_{k-1}x^{k-1} + \cdots + \alpha_0$. We have to prove that the dimension of $\langle \mathbf{u}, \mathcal{A} \rangle$ is k. We show this by verifying that $\mathbf{u}, \mathcal{A}\mathbf{u}, \ldots, \mathcal{A}^{k-1}\mathbf{u}$ is a basis in $\langle \mathbf{u}, \mathcal{A} \rangle$.

For a proof of linear independence by contradiction, we assume the existence of a non-trivial linear combination $\beta_0\mathbf{u} + \beta_1\mathcal{A}\mathbf{u} + \cdots + \beta_{k-1}\mathcal{A}^{k-1}\mathbf{u} = \mathbf{0}$. Then the non-zero polynomial $f = \beta_0 + \beta_1 x + \cdots + \beta_{k-1}x^{k-1}$ satisfies $f(\mathcal{A})\mathbf{u} = \mathbf{0}$. This, however, contradicts $\deg o(\mathbf{u}) = k$.

Now we show that the span of the vectors $\mathbf{u}, \mathcal{A}\mathbf{u}, \ldots, \mathcal{A}^{k-1}\mathbf{u}$ is the subspace $\langle \mathbf{u}, \mathcal{A} \rangle$. It is enough to verify that every $\mathcal{A}^i\mathbf{u}$ can be written as their linear combination. This is obvious for the exponents $i < k$. For $i = k$ we obtain it by ordering $h(\mathcal{A})\mathbf{u} = \mathbf{0}$:

$$\mathcal{A}^k\mathbf{u} = -\alpha_0\mathbf{u} - \alpha_1(\mathcal{A}\mathbf{u}) - \cdots - \alpha_{k-1}(\mathcal{A}^{k-1}\mathbf{u}). \tag{6.5.1}$$

Consider now the exponent $i = k + 1$. Applying $\mathcal{A}$ to (6.5.1), we obtain that $\mathcal{A}^{k+1}\mathbf{u}$ can be expressed as a linear combination of $\mathcal{A}\mathbf{u}, \ldots, \mathcal{A}^k\mathbf{u}$. Replacing here $\mathcal{A}^k\mathbf{u}$ by its representation (6.5.1), we obtain $\mathcal{A}^{k+1}\mathbf{u}$ as a linear combination of $\mathbf{u}, \mathcal{A}\mathbf{u}, \ldots, \mathcal{A}^{k-1}\mathbf{u}$. We can proceed to higher exponents similarly by induction.

Alternatively, for the exponents $i \geq k$, we can replace x^i by its remainder when divided by the order, and substitute the transformation $\mathcal{A}$ into this remainder. ∎

Theorem 6.5.4 implies the existence of invariant subspaces of certain dimensions.

Theorem 6.5.5. *If the minimal polynomial has a divisor of degree r that is irreducible over F, then $\mathcal{A}$ has an r-dimensional invariant subspace.* ♣

Remarks.

1. The special case $r = 1$ yields a part of Theorem 6.3.4, the proof, too, will be analogous.

2. Later we show (see Theorem 6.5.8) that Theorem 6.5.5 holds for reducible divisors, too.

Proof. Let $m_{\mathcal{A}} = hg$ where h is irreducible and $\deg h = r$. We show that $o_{\mathcal{A}}(\mathbf{u}) = h$ for some $\mathbf{u} \neq \mathbf{0}$. Then the dimension of the subspace $\langle \mathbf{u}, \mathcal{A} \rangle$ is $\deg h = r$ by Theorem 6.5.4.

Since $\deg g < \deg m_{\mathcal{A}}$, we have $g(\mathcal{A}) \neq \mathcal{O}$, i.e., $\operatorname{Im} g(\mathcal{A}) \neq \mathbf{0}$. Substituting $\mathcal{A}$ into $m_{\mathcal{A}}$ yields $\mathcal{O} = m_{\mathcal{A}}(\mathcal{A}) = h(\mathcal{A})g(\mathcal{A})$. Thus $\operatorname{Ker} h(\mathcal{A}) \supseteq \operatorname{Im} g(\mathcal{A})$. Let $\mathbf{u}$ be any non-zero vector in $\operatorname{Im} g(\mathcal{A})$. Then $h(\mathcal{A})\mathbf{u} = \mathbf{0}$, so $o_{\mathcal{A}}(u) \mid h$. Since h is irreducible and $\mathbf{u} \neq \mathbf{0}$, only $o_{\mathcal{A}}(\mathbf{u}) = h$ is possible. ∎

Now we show that the minimal polynomial itself occurs among the orders.

Theorem 6.5.6. *For each transformation, there exists a vector whose order is the minimal polynomial.* ♣

Proof. By Theorem 6.5.3, the minimal polynomial is the least common multiple of the orders of a spanning set. Hence it is sufficient to prove the following lemma.

Lemma 6.5.7. *If the polynomials $h_1, \ldots, h_s$ are orders of vectors $\mathbf{u}_1, \ldots, \mathbf{u}_s$, then their least common multiple is the order of some vector $\mathbf{u}$.* ♣

Proof of Lemma 6.5.7. We arrive at the general result through several special cases. Let H denote the least common multiple $[h_1, h_2, \ldots, h_s]$ of the polynomials h_i.

(i) Two coprime polynomials:

$$(h_1, h_2) = 1 \Longrightarrow o(\mathbf{u}_1 + \mathbf{u}_2) = h_1 h_2 = H.$$

Let $\mathbf{u} = \mathbf{u}_1 + \mathbf{u}_2$ and $K = o(\mathbf{u})$. We verify $H = K$. To prove $K \mid H$, we need $H(\mathcal{A})\mathbf{u} = \mathbf{0}$. Since $o(\mathbf{u}_i) \mid H$, we infer

$$H(\mathcal{A})\mathbf{u} = H(\mathcal{A})(\mathbf{u}_1 + \mathbf{u}_2) = H(\mathcal{A})\mathbf{u}_1 + H(\mathcal{A})\mathbf{u}_2 = \mathbf{0} + \mathbf{0} = \mathbf{0}.$$

To prove the reverse divisibility $H \mid K$, we have to verify $h_i \mid K$ $(i = 1, 2)$. Since

$$(Kh_1)(\mathcal{A})\mathbf{u}_2 = (Kh_1)(\mathcal{A})\mathbf{u} - (Kh_1)(\mathcal{A})\mathbf{u}_1 = \mathbf{0} - \mathbf{0} = \mathbf{0},$$

we have $o(\mathbf{u}_2) = h_2 \mid Kh_1$. By $(h_1, h_2) = 1$, we infer $h_2 \mid K$. We obtain $h_1 \mid K$ similarly.

(ii) Pairwise coprime polynomials:

$$(h_i, h_j) = 1, \quad 1 \le i < j \le s \Longrightarrow o(\mathbf{u}_1 + \cdots + \mathbf{u}_s) = h_1 \ldots h_s = H.$$

This follows from (i) by induction.

(iii) Each divisor of an order is an order itself: If $f = gh$ and $f = o(\mathbf{v})$, then $g = o[h(\mathcal{A})\mathbf{v}]$. The proof is set in Exercise 6.5.5.

(iv) If two polynomials h_1 and h_2 are orders, then their least common multiple is an order.

We write h_1 and h_2 as a product of powers of irreducible factors:

$$h_1 = \alpha_1 p_1^{k_{11}} \ldots p_r^{k_{r1}},$$
$$h_2 = \alpha_2 p_1^{k_{12}} \ldots p_r^{k_{r2}},$$

where α_i are constants, p_j are pairwise non-associate irreducible polynomials, and the exponents k_{ji} are non-negative integers. Then the factorization of $H = [h_1, h_2]$ is

$$H = p_1^{k_{13}} \ldots p_r^{k_{r3}}$$

where $k_{j3} = \max(k_{j1}, k_{j2})$. Since $p_j^{k_{j3}}$ divides h_1 or h_2 for any j, therefore $p_j^{k_{j3}}$ is an order, by (iii). The factors $p_j^{k_{j3}}$ are pairwise coprime, so their product H is an order, by (ii).

(v) The general statement for an arbitrary number of polynomials follows from (iv) by induction. ∎

Theorem 6.5.6 has several important consequences. We mentioned in Section 6.3 that the degree of the minimal polynomial is not greater than the dimension of the vector space. This follows immediately as the degree of the order cannot be greater than the dimension of the space (see Theorem 6.5.2). Another important corollary is a generalization of Theorem 6.5.5.

Theorem 6.5.8. *If the minimal polynomial has a divisor of degree r, then $\mathcal{A}$ has an r-dimensional invariant subspace.* ♣

Proof. The minimal polynomial is an order, by Theorem 6.5.6. By part (iii) in the proof of Lemma 6.5.7 (see Exercise 6.5.5), every divisor of the minimal polynomial is an order. Finally, the degree of each order is the dimension of some invariant subspace, by Theorem 6.5.4. ∎

Exercises 6.5

6.5.1. Given the order of $\mathbf{u}$, compute the order of

(a) $\lambda\mathbf{u}$;

(b) $\mathcal{A}\mathbf{u}$;

(c) $f(\mathcal{A})\mathbf{u}$ where f is an arbitrary polynomial.

6.5.2. Prove Theorem 6.5.2.

6.5.3. Consider the restriction of $\mathcal{A}$ to the invariant subspace $U = \langle \mathbf{u}, \mathcal{A} \rangle$. What is the minimal polynomial of this $U \to U$ transformation?

6.5.4. Prove that $o_{\mathcal{A}}(\mathbf{u})$ has a root in F if and only if $\mathcal{A}$ has an eigenvector in the subspace $\langle \mathbf{u}, \mathcal{A} \rangle$.

6.5.5. Verify part (iii) in the proof of Lemma 6.5.7: Each divisor of an order is an order itself, namely if $f = gh$ and $f = o(\mathbf{v})$, then $g = o[h(\mathcal{A})\mathbf{v}]$.

6.5.6. True or false?

(a) If $\langle \mathbf{u}, \mathcal{A} \rangle = \langle \mathbf{v}, \mathcal{A} \rangle$, then the orders of $\mathbf{u}$ and $\mathbf{v}$ are the same.

(b) If the orders of $\mathbf{u}$ and $\mathbf{v}$ are the same, then $\langle \mathbf{u}, \mathcal{A} \rangle = \langle \mathbf{v}, \mathcal{A} \rangle$.

(c) If $o(\mathbf{u})$ and $o(\mathbf{v})$ are coprime and none of them is a constant, then $\mathbf{u}$ and $\mathbf{v}$ are linearly independent.

(d) If $\mathbf{u}$ and $\mathbf{v}$ are linearly independent, then $o(\mathbf{u})$ and $o(\mathbf{v})$ are coprime.

(e) If $\mathbf{u}_1, \dots, \mathbf{u}_s$ are linearly independent, then

$$o(\mathbf{u}_1 + \dots + \mathbf{u}_s) = [o(\mathbf{u}_1), \dots, o(\mathbf{u}_s)].$$

6.5.7. Prove that the order of a vector is at most by one greater than the dimension of the image. What follows from this for the degree of the minimal polynomial?

*6.5.8. Let $\dim V = n$ and assume that $\mathcal{A}$ has n distinct eigenvalues. Prove that $\mathcal{A}$ has exactly 2^n invariant subspaces.

6.5.9. Give a new proof to Exercise 6.4.9(c): the number of invariant subspaces cannot be less than the number of pairwise non-associate divisors of the minimal polynomial.

6.5.10. Verify

$$\frac{[o(\mathbf{u}), o(\mathbf{v})]}{(o(\mathbf{u}), o(\mathbf{v}))} | o(\mathbf{u} + \mathbf{v}) | [o(\mathbf{u}), o(\mathbf{v})].$$

What is the connection between this statement and part (i) in the proof of Lemma 6.5.7?

6.5.11. What is the relation of the degrees of $o_{\lambda\mathcal{A}}(\mathbf{u})$ and $o_{\mathcal{A}^2}(\mathbf{u})$ to the degree of $o_{\mathcal{A}}(\mathbf{u})$?

6.5.12. Let $\mathcal{A}, \mathcal{B} \in \operatorname{Hom} V$ and assume that $o_{\mathcal{A}}(\mathbf{v}) = o_{\mathcal{B}}(\mathbf{v})$ holds for every $\mathbf{v} \in V$.
(a) Prove that $\mathcal{A}$ and $\mathcal{B}$ have the same eigenvalues, eigenvectors, and minimal polynomial.
(b) If $\mathcal{A}$ has an eigenbasis, then $\mathcal{A} = \mathcal{B}$.
(c) Show examples over $\mathbf{R}$ and $\mathbf{C}$ for $\mathcal{A} \neq \mathcal{B}$.

6.6. Nice Matrices of Transformations

A matrix of a transformation is "nice" if it is close to the diagonal form and the (eventual) non-zero elements are concentrated around the main diagonal. E.g., we can raise such matrices to powers easily.

We saw in Theorem 6.1.4 that a transformation has a diagonal matrix if and only if it has an eigenbasis. Another interpretation is that the vector space is the direct sum of its one-dimensional invariant subspaces in this case.

We describe first the structure of matrices belonging to bases taken from two invariant subspaces (i.e., when the vector space is the direct sum of two invariant subspaces). We next show how to find such invariant subspaces. Finally, we determine the "nicest possible" matrix of a transformation over the complex field (this is the so-called *Jordan normal form*).

Theorem 6.6.1. *Let $\mathbf{b}_1, \ldots, \mathbf{b}_n$ be a basis of the vector space V such that $U_1 = \langle \mathbf{b}_1, \ldots, \mathbf{b}_k \rangle$ and $U_2 = \langle \mathbf{b}_{k+1}, \ldots, \mathbf{b}_n \rangle$ are invariant subspaces of the transformation $\mathcal{A}$. Then every element in the matrix $A = [\mathcal{A}]_b$ is zero except for the top left $k \times k$ and bottom right $(n-k) \times (n-k)$ squares:*

$$\begin{pmatrix} A_1 & 0 \\ 0 & A_2 \end{pmatrix}$$

where A_1 is a $k \times k$, A_2 is an $(n-k) \times (n-k)$ matrix, the bottom left $(n-k) \times k$ and the top right $k \times (n-k)$ parts are zero matrices. ♣

Theorem 6.6.1 immediately follows from U_1 and U_2 being invariant subspaces. Observe that matrices A_i $(i = 1, 2)$ are just the matrices of the restriction of $\mathcal{A}$ onto U_i (in the corresponding bases). We say that A is decomposed into *blocks* A_1 and A_2, or A is the *direct sum* of matrices A_1 and A_2.

Theorem 6.6.2. *Assume $m_{\mathcal{A}} = g_1 g_2$ where $(g_1, g_2) = 1$. Then $V = U_1 \oplus U_2$ where U_i are invariant subspaces of $\mathcal{A}$, and the minimal polynomial of the restriction of $\mathcal{A}$ onto U_i is g_i.* ♣

Proof. We show that $U_i = \operatorname{Ker} g_i(\mathcal{A})$ meet the requirements.

(a) U_i are invariant subspaces by Exercise 6.4.9(a).

(b) To prove $V = U_1 \oplus U_2$, we have to verify

$$\text{(b1)} \quad U_1 \cap U_2 = \mathbf{0} \qquad \text{and} \qquad \text{(b2)} \quad \langle U_1, U_2 \rangle = V.$$

(b1) Assume $\mathbf{u} \in U_1 \cap U_2$, i.e., $g_i(\mathcal{A})\mathbf{u} = \mathbf{0}$, $i = 1, 2$. This means $o_{\mathcal{A}}(\mathbf{u}) \mid g_i$, hence, $o_{\mathcal{A}}(\mathbf{u}) \mid (g_1, g_2) = 1$, i.e., $\mathbf{u} = \mathbf{0}$.

(b2) Since g_1 and g_2 are coprime, $1 = g_1h_1 + g_2h_2$ with suitable polynomials h_1 and h_2. Therefore $\mathcal{I} = g_1(\mathcal{A})h_1(\mathcal{A}) + g_2(\mathcal{A})h_2(\mathcal{A})$. Applying this to an arbitrary vector $\mathbf{v}$, we get $\mathbf{v} = \mathbf{v}_1 + \mathbf{v}_2$ where $\mathbf{v}_i = g_i(\mathcal{A})h_i(\mathcal{A})\mathbf{v}$. Here $\mathbf{v}_1 \in U_2$ as

$$g_2(\mathcal{A})\mathbf{v}_1 = (g_2g_1h_1)(\mathcal{A})\mathbf{v} = (m_{\mathcal{A}}h_1)(\mathcal{A})\mathbf{v} = \mathbf{0}.$$

We obtain $\mathbf{v}_2 \in U_1$ similarly.

(c) Finally, let r_i be the minimal polynomial of the restriction of $\mathcal{A}$ onto U_i. We have to show $r_i = g_i$. Since $U_1 = \operatorname{Ker} g_1(\mathcal{A})$, $\mathbf{u} \in U_1$ implies $g_1(\mathcal{A})\mathbf{u} = \mathbf{0}$, so $r_1 \mid g_1$.

To prove the converse divisibility $g_1 \mid r_1$, consider the decomposition $\mathbf{v} = \mathbf{v}_1 + \mathbf{v}_2$ in (b2) for any vector $\mathbf{v} \in V$. Here $\mathbf{v}_1 \in U_2$, $\mathbf{v}_2 \in U_1$. We see that $g_2(\mathcal{A})\mathbf{v}_1 = \mathbf{0}$, and $r_1(\mathcal{A})\mathbf{v}_2 = \mathbf{0}$ by the definition of r_1. Let $s = r_1g_2$, then $s(\mathcal{A})\mathbf{v} = s(\mathcal{A})\mathbf{v}_1 + s(\mathcal{A})\mathbf{v}_2 = \mathbf{0}$ for every $\mathbf{v} \in V$. Thus $m_{\mathcal{A}} = g_1g_2 \mid s = r_1g_2$, i.e., $g_1 \mid r_1$.

This proves $r_1 = g_1$, and we get $r_2 = g_2$ the same way. ∎

Theorem 6.6.2 implies by induction that if we factor the minimal polynomial into the product of (pairwise non-associate) irreducible polynomials, $m_{\mathcal{A}} = p_1^{k_1} \dots p_t^{k_t}$, then the vector space V is the direct sum of invariant subspaces U_i where $p_i^{k_i}$ is the minimal polynomial of the restriction $\mathcal{A}_i$ of $\mathcal{A}$ onto U_i. By Theorem 6.6.1, the matrix of $\mathcal{A}$ in a basis taken from the subspaces U_i is decomposed into blocks $A_i = [\mathcal{A}_i]$.

By the above, it is sufficient to find "nice" matrices of transformations where the minimal polynomial is a power of an irreducible polynomial. This is a hard task in general. The situation is simpler over the complex field, since an irreducible polynomial can have only degree one.

Theorem 6.6.3. *Let $\mathcal{B} \in \operatorname{Hom} V$, $m_{\mathcal{B}} = (x - \lambda)^k$. Then, the matrix of $\mathcal{B}$ in a suitable basis consists of blocks B_j (this may mean just one block, too) where*

(i) *the maximal size of the blocks B_j is k, i.e., each B_j is at most $k \times k$ with equality for at least one of them; and*

(ii) *every element in the main diagonal of each B_j is λ, every element just below the main diagonal is 1, and all other elements are 0.*

(A block can be of size 1, too, then it consists of a single λ. If $k = 1$, then every block has size 1, and we get the diagonal matrix λI.)

Expressing this in formulae,

$$[\mathcal{B}] = \begin{pmatrix} B_1 & 0 & \dots & 0 \\ 0 & B_2 & \dots & 0 \\ \vdots & \vdots & \ddots & \vdots \\ 0 & 0 & \dots & B_r \end{pmatrix}, \qquad B_j = \begin{pmatrix} \lambda & 0 & 0 & \dots & 0 \\ 1 & \lambda & 0 & \dots & 0 \\ 0 & 1 & \lambda & \dots & 0 \\ \vdots & \vdots & \vdots & \ddots & \vdots \\ 0 & 0 & 0 & \dots & \lambda \end{pmatrix},$$

where the size of B_j is $q_j \times q_j$, $\sum_{j=1}^{r} q_j = \dim V$ and $\max_{1 \le j \le r} q_j = k$. ♣

Proof. Let $\mathcal{C} = \mathcal{B} - \lambda\mathcal{I}$; then the minimal polynomial of $\mathcal{C}$ is x^k, and we have to show that the matrix of $\mathcal{C}$ in a suitable basis consists of blocks C_j with a maximum of sizes k, every element just below the main diagonal in each C_j is 1, and all other elements are 0 (including the main diagonal).

This means that a basis belonging to a block of size q is of the form

$$\mathbf{u}_1,\ \mathbf{u}_2 = \mathcal{C}\mathbf{u}_1,\ \mathbf{u}_3 = \mathcal{C}\mathbf{u}_2 = \mathcal{C}^2\mathbf{u}_1,\ \ldots,\ \mathbf{u}_q = \mathcal{C}\mathbf{u}_{q-1} = \mathcal{C}^{q-1}\mathbf{u}_1,$$

where

$$\mathcal{C}\mathbf{u}_q = \mathcal{C}^q\mathbf{u}_1 = \mathbf{0}.$$

In other words, the q-dimensional invariant subspace belonging to a block of size q is $\langle \mathbf{u}_1, \mathcal{C}\rangle$ where q is the degree of the order x^q of $\mathbf{u}_1$ with respect to $\mathcal{C}$.

This leads us to the following definition.

Definition 6.6.4. Let $\mathcal{C} \in \operatorname{Hom} V$, $m_{\mathcal{C}} = x^k$. The *orbit* of a vector $\mathbf{u} \in V$ of order x^q is the system of vectors

$$\mathbf{u}, \quad \mathcal{C}\mathbf{u}, \quad \mathcal{C}^2\mathbf{u}, \quad \ldots, \quad \mathcal{C}^{q-1}\mathbf{u}$$

where q is the smallest positive integer satisfying

$$\mathcal{C}^q\mathbf{u} = \mathbf{0}.$$

The integer q is the *length* of the orbit. ♣

Theorem 6.5.4 yields that the vectors in an orbit are linearly independent.

By the above, Theorem 6.6.3 is equivalent to the following lemma.

Lemma 6.6.5. *Let $\mathcal{C} \in \operatorname{Hom} V$, $m_{\mathcal{C}} = x^k$. Then V has a basis consisting of orbits with a maximum of lengths k.* ♣

We give two proofs for Lemma 6.6.5 (which is a rephrased form of Theorem 6.6.3).

First proof. We proceed by induction on the degree of the minimal polynomial. For $k = 1$, $\mathcal{C} = \mathcal{O}$, so $\mathcal{C}\mathbf{u} = \mathbf{0}$ for every vector $\mathbf{u}$. This means that the orbit of every non-zero vector has length 1, thus any basis meets the requirements.

Assume now that the statement holds for $k-1$ and let $\mathcal{C} \in \operatorname{Hom} V$, $m_{\mathcal{C}} = x^k$. Consider the restriction $\mathcal{C}'$ of the transformation $\mathcal{C}$ onto the invariant subspace $W = \operatorname{Im} \mathcal{C}$. We claim that the minimal polynomial of $\mathcal{C}'$ is x^{k-1}. On the one hand, for every vector $\mathbf{w} = \mathcal{C}\mathbf{u} \in \operatorname{Im} \mathcal{C}$, we have $\mathcal{C}^{k-1}\mathbf{w} = \mathcal{C}^k\mathbf{u} = \mathbf{0}$. On the other hand, if $\mathcal{C}^j\mathbf{w} = \mathbf{0}$ holds for every $\mathbf{w} \in \operatorname{Im} \mathcal{C}$, then for every $\mathbf{u} \in V$, we get $\mathcal{C}^{j+1}\mathbf{u} = \mathcal{C}^j(\mathcal{C}\mathbf{u}) = \mathbf{0}$, so $j \geq k-1$.

By the induction hypothesis, W has a basis composed of orbits with maximum length $k-1$. Let $\mathbf{w}_i = \mathcal{C}\mathbf{v}_i$ be the starting element of the ith orbit of length q_i, $i = 1, 2, \ldots, r$ where $q_1 = k-1$. Extending the orbit of $\mathbf{w}_i$ by $\mathbf{v}_i$, we get the orbit O_i of $\mathbf{v}_i$ in V having length $q_i + 1$. Let U be the subspace in V spanned by all vectors in the orbits O_i, $H = U \cap \operatorname{Ker} \mathcal{C}$, and let Z be an arbitrary basis in any direct complement H_1 satisfying $H \oplus H_1 = \operatorname{Ker} \mathcal{C}$. We prove that

(i) the union of the orbits O_i form a basis in U;

(ii) the last elements $\mathcal{C}^{q_i}\mathbf{v}_i$ in O_i form a basis in H; and

(iii) the union of Z and the orbits O_i form a basis in V.

As the elements of Z are in the kernel, they mean orbits of length 1, so the basis of V given in (iii) is the union of orbits, and we are done.

Turning to the proof of (i), we have to show that the vectors in the union of the orbits O_i are linearly independent, i.e.,

$$\sum_{i=1}^{r}\sum_{j=0}^{q_i}\lambda_{ij}\mathcal{C}^j\mathbf{v}_i = \mathbf{0} \tag{6.6.1}$$

implies $\lambda_{ij} = 0$ for every i, j. We cut the linear combination on the left-hand side of (6.6.1) into two parts: $\mathbf{s}_1$ consists of the terms with $j = q_i$, and $\mathbf{s}_2$ comprises the terms with $j < q_i$ (so $\mathbf{s}_1$ is the linear combination of the last elements of the orbits, $\mathbf{s}_2$ joins the others, $\mathbf{s}_1+\mathbf{s}_2 = \mathbf{0}$). Applying $\mathcal{C}$ to (6.6.1), the terms in $\mathbf{s}_1$ are mapped to $\mathbf{0}$ implying $\mathcal{C}\mathbf{s}_2 = \mathbf{0}$. But $\mathcal{C}\mathbf{s}_2$ is a linear combination of vectors in the orbits of $\mathcal{C}\mathbf{v}_i = \mathbf{w}_i$ instead of $\mathbf{v}_i$. These vectors are linearly independent by the induction hypothesis, so every λ_{ij} in $\mathbf{s}_2$ is 0. This reduces (6.6.1) to $\mathbf{s}_1 = \mathbf{0}$. The vectors in $\mathbf{s}_1$ are linearly independent by the induction hypothesis, hence each coefficient λ_{ij} is 0 in $\mathbf{s}_1$, too. This proves (i).

Passing on (ii), we examine which elements of U are in the kernel. We know by (i) that every $\mathbf{v} \in U$ has a unique representation as $\mathbf{s}_1 + \mathbf{s}_2$ on the left-hand side of (6.6.1). Thus, $\mathcal{C}\mathbf{u} = \mathbf{0}$ means $\mathcal{C}\mathbf{s}_1 + \mathcal{C}\mathbf{s}_2 = \mathbf{0} + \mathcal{C}\mathbf{s}_2 = \mathbf{0}$. As seen in (i), the linear independence of the vectors in $\mathcal{C}\mathbf{s}_2$ implies that every $\lambda_{ij} = 0$ in $\mathbf{s}_2$, so $\mathbf{s}_2 = \mathbf{0}$, i.e., $\mathbf{u} = \mathbf{s}_1$. Since the vectors in $\mathbf{s}_1$ are linearly independent, therefore they form a basis in the subspace H spanned by them.

Finally, to verify (iii), we have $U \cap H_1 = \mathbf{0}$ by (ii), thus the vectors given in (iii) are linearly independent. They are

$$\dim U + \dim H_1 = \dim \operatorname{Im} \mathcal{C} + r + \dim H_1 = \dim \operatorname{Im} \mathcal{C} + \dim \operatorname{Ker} \mathcal{C} = \dim V$$

in number, hence, they form a basis in V. ∎

Second proof. This nice idea is due to Terence Tao. (For Tao's original argument, see https://terrytao.wordpress.com/2007/10/12/the-jordan-normal-form-and-the-euclidean-algorithm/#more-202.) The vector space V has a spanning set of orbits, e.g., the union of orbits starting from an arbitrary basis of V has this property. We show that if the elements of such a spanning set are linearly dependent, then a suitable modification leads to a spanning set of orbits of smaller total length. Repeating this process, finally we get a linearly independent spanning set, i.e., a basis.

For clarity, we illustrate every step through a concrete example, too.

We represent $\mathbf{0}$ as a non-trivial linear combination of the spanning set and keep only the terms with non-zero coefficients. In our example let the lengths of the orbits be 3, 4, 5, and 5, and, say,

$$6\mathbf{v}_1 + 7\mathcal{C}^2\mathbf{v}_1 + 8\mathcal{C}\mathbf{v}_2 + 9\mathcal{C}^2\mathbf{v}_2 + 10\mathcal{C}^3\mathbf{v}_3 + 11\mathcal{C}^2\mathbf{v}_4 = \mathbf{0}.$$

Applying a suitable power of $\mathcal{C}$ to this linear combination, we achieve that at most one term will remain from each orbit (the others are mapped to $\mathbf{0}$). In our example, we need $\mathcal{C}^2$, and the result is

$$6\mathcal{C}^2\mathbf{v}_1 + 8\mathcal{C}^3\mathbf{v}_2 + 11\mathcal{C}^4\mathbf{v}_4 = \mathbf{0}.$$

We now take the smallest integer m occurring as an exponent of $\mathcal{C}$ and write the linear combination in the form $\mathcal{C}^m\mathbf{v} = \mathbf{0}$. In the example, $m = 2$ and

$$\mathcal{C}^2(6\mathbf{v}_1 + 8\mathcal{C}\mathbf{v}_2 + 11\mathcal{C}^2\mathbf{v}_4) = \mathbf{0},$$

so

$$\mathbf{v} = 6\mathbf{v}_1 + 8\mathcal{C}\mathbf{v}_2 + 11\mathcal{C}^2\mathbf{v}_4.$$

As we worked with the largest possible power $\mathcal{C}^m$, at least one $\mathbf{v}_i$ occurs "itself" (i.e., **not** its image under a power of $\mathcal{C}$) in $\mathbf{v}$; in our example $i = 1$. Therefore, this $\mathbf{v}_i$ can be expressed as a linear combination of $\mathbf{v}$ and elements from the other orbits. In the example,

$$\mathbf{v}_1 = \frac{1}{6}\mathbf{v} - \frac{4}{3}\mathcal{C}\mathbf{v}_2 - \frac{11}{6}\mathcal{C}^2\mathbf{v}_4.$$

Hence, replacing the orbit of $\mathbf{v}_i$ by the orbit of $\mathbf{v}$, we still have a spanning set. In our example, the orbit $\mathbf{v}_1, \mathcal{C}\mathbf{v}_1, \mathcal{C}^2\mathbf{v}_1$ is replaced by the orbit $\mathbf{v}, \mathcal{C}\mathbf{v}$. The length of the new orbit is m, which is less than the length of the deleted orbit O_i was, since O_i contained $\mathcal{C}^m\mathbf{v}_i$ as its $(m+1)$st element. ■

Theorems 6.6.1–6.6.3 immediately imply the "nicest" matrix of an arbitrary transformation over the complex field.

Theorem 6.6.6 (Jordan normal form). *Let V be a finite-dimensional vector space over the complex field, $\mathcal{A} \in \operatorname{Hom} V$, $m_{\mathcal{A}} = (x-\lambda_1)^{k_1}\dots(x-\lambda_t)^{k_t}$. Then, in a suitable basis, the matrix of $\mathcal{A}$ consists of blocks A_i with minimal polynomial $m_{\mathcal{A}_i} = (x-\lambda_i)^{k_i}$, $i = 1, 2, \dots, t$, according to Theorems* 6.6.1–6.6.2, *and a block A_i is composed of sub-blocks A_{ij} conforming to Theorem* 6.6.3. *For a fixed i, the maximal size of the sub-blocks A_{ij} is k_i, every element in the main diagonal of a sub-block A_{ij} is λ_i, every element just below the main diagonal is 1, and all other elements are* 0. ♣

The matrix $[\mathcal{A}]$ described in Theorem 6.6.6 is called the *Jordan normal form* or briefly the *Jordan form* belonging to the transformation $\mathcal{A}$. If we consider (any) matrix of a transformation, then the Jordan form of this matrix means the Jordan form belonging to the given transformation.

We can add to the theorem that the Jordan form of a transformation (or matrix) is uniquely determined apart from the order of the blocks and the order of the sub-blocks within a block.

The Jordan form yields that over the complex field, every transformation has a "nearly diagonal matrix": non-zero elements can appear only in the main diagonal and just below it; in the main diagonal, we have the eigenvalues, and the elements just below the main diagonal are 1 (within a sub-block) or 0 (at the border of sub-blocks or blocks).

Exercises 6.6

6.6.1. Prove that $\mathcal{A}$ has a diagonal matrix if and only if $m_{\mathcal{A}}$ is the product of distinct root factors.

6.6.2. Let $\dim V = n$, and assume that $\mathcal{A}$ has n distinct eigenvalues. Verify:
(a) If $\mathcal{A}\mathcal{B} = \mathcal{B}\mathcal{A}$, then $\mathcal{B}$ has a diagonal matrix.
(b) $\mathcal{A}\mathcal{B} = \mathcal{B}\mathcal{A}$ if and only if $\mathcal{B} = f(\mathcal{A})$ for some polynomial f.

6.6.3. Let $\mathcal{A} \in \operatorname{Hom} V$ and let U be an invariant subspace of $\mathcal{A}$. What is the relation between the minimal polynomial of the restriction of $\mathcal{A}$ onto U and the original minimal polynomial? Investigate the analogous question for characteristic polynomials, too.

6.6.4. Let U_1 and U_2 be invariant subspaces of $\mathcal{A}$. Consider the restrictions of $\mathcal{A}$ onto the (invariant) subspaces U_1, U_2, $U_1 \cap U_2$, and $\langle U_1, U_2\rangle$. Let the corresponding minimal polynomials be m_1, m_2, $m_\cap$, and $m_{\langle\rangle}$. Verify
(a) $m_{\langle\rangle} = [m_1, m_2]$;
(b) $m_\cap \mid (m_1, m_2)$, but equality does not hold in general.

6.6.5. Using the notation of the previous exercise, let the corresponding characteristic polynomials be k_1, k_2, $k_\cap$, and $k_{\langle\rangle}$. Prove
(a) $[k_1, k_2] \mid k_{\langle\rangle}$, but equality does not hold in general;
(b) $k_\cap \mid (k_1, k_2)$, but equality does not hold in general;
(c) $k_1 \cdot k_2 = k_\cap \cdot k_{\langle\rangle}$.

*6.6.6. Prove that over an infinite field, a transformation has only finitely many invariant subspaces if and only if the degree of the minimal polynomial equals the dimension of the vector space. In this case, the number of invariant subspaces is equal to the number of pairwise non-associate divisors of the minimal polynomial.

6.6.7. Show that the following conditions are equivalent for $\mathcal{A} \in \operatorname{Hom} V$.
(i) Every invariant subspace of $\mathcal{A}$ is of the form $\operatorname{Ker} f(\mathcal{A})$ with a suitable polynomial $f \in F[x]$.
(ii) Every invariant subspace of $\mathcal{A}$ is of the form $\operatorname{Im} g(\mathcal{A})$ with a suitable polynomial $g \in F[x]$.
(iii) Every invariant subspace of $\mathcal{A}$ is of the form $\langle \mathbf{u}, \mathcal{A}\rangle$ with a suitable vector $\mathbf{u} \in V$.
(iv) The degree of the minimal polynomial equals the dimension of V.
(v) The minimal polynomial is equal to the characteristic polynomial.

6.6.8. Analogously to matrices, we can define similarity for linear transformations: $\mathcal{A}$ and $\mathcal{B} \in \operatorname{Hom} V$ are *similar* if they have a "common matrix", i.e., $[\mathcal{B}]_b = [\mathcal{A}]_a$ for some bases $\mathbf{a}_1, \dots, \mathbf{a}_n$ and $\mathbf{b}_1, \dots, \mathbf{b}_n$. We denote similarity by $\mathcal{A} \sim \mathcal{B}$.
(a) Which transformations are similar just to themselves?
(b) Prove that $\mathcal{A}$ and $\mathcal{B}$ are similar if and only if $\mathcal{B} = \mathcal{C}^{-1}\mathcal{A}\mathcal{C}$ for some invertible matrix $\mathcal{C}$.
(c) Verify that similarity is an equivalence relation.
(d) Does $\mathcal{A} \sim \mathcal{B}$ imply $\mathcal{A} + \mathcal{D} \sim \mathcal{B} + \mathcal{D}$ and $\mathcal{A}\mathcal{D} \sim \mathcal{B}\mathcal{D}$?

(e) Prove that $\mathcal{A} \sim \mathcal{B}$ implies $f(\mathcal{A}) \sim f(\mathcal{B})$ for any polynomial f.

(f) Show that similar transformations have the same minimal polynomial and the same characteristic polynomial. Is the converse true?

6.6.9. For a given V, which transformations are uniquely determined by the minimal polynomial?

6.6.10. Find the Jordan form of the following 3×3 matrices:

$$\text{(a)} \begin{pmatrix} 1 & 0 & 1 \\ 0 & 2 & 0 \\ 1 & 0 & 1 \end{pmatrix}; \quad \text{(b)} \begin{pmatrix} 0 & 1 & 0 \\ 0 & 0 & 1 \\ 1 & 0 & 0 \end{pmatrix}; \quad \text{(c)} \begin{pmatrix} 1 & 1 & 1 \\ 0 & 2 & 0 \\ 1 & 1 & 1 \end{pmatrix}; \quad \text{(d)} \begin{pmatrix} 1 & 0 & 0 \\ 2 & 1 & 2 \\ 0 & 0 & 1 \end{pmatrix}.$$

6.6.11. Find the Jordan form of the following $n \times n$ matrices $A = (\alpha_{ij})$:

(a) $\alpha_{ij} = \begin{cases} 1, & \text{if } j = i + 1; \\ 0, & \text{otherwise.} \end{cases}$

(The elements just above the main diagonal are 1, all other elements are 0.)

(b) $\alpha_{ij} = \begin{cases} 1, & \text{if } j \equiv i + 1 \pmod{n}; \\ 0, & \text{otherwise.} \end{cases}$

(The elements just above the main diagonal and in the bottom left corner are 1, all other elements are 0.)

(c) $\alpha_{ij} = \begin{cases} 1, & \text{if } j = i - 2; \\ 0, & \text{otherwise.} \end{cases}$

(The elements in the second diagonal below the main diagonal are 1, all other elements are 0.)

(d) $\alpha_{ij} = \begin{cases} 1, & \text{if } j < i; \\ 0, & \text{otherwise.} \end{cases}$

(The elements below the main diagonal are 1, all other elements are 0.)

(e) $\alpha_{ij} = 1$. (Every element is 1.)

(f) $\alpha_{ij} = \begin{cases} 1, & \text{if } i + j = n + 1; \\ 0, & \text{otherwise.} \end{cases}$

(The elements in the diagonal connecting the bottom left and upper right corners are 1, all other elements are 0.)

6.6.12. How can we raise a matrix in Jordan form to a(n arbitrarily large) power?

6.6.13. How can we see the minimal polynomial and the characteristic polynomial from the Jordan form?

6.6.14. Let V be an n-dimensional vector space over the complex field, $\mathcal{A} \in \operatorname{Hom} V$, and $0 \le k \le n$. Show that $\mathcal{A}$ has a k-dimensional invariant subspace.

6.6.15. Let V be a finite-dimensional vector space over the complex field. Which transformations are determined uniquely up to similarity by their

(a) minimal polynomial;

(b) characteristic polynomial;

(c) minimal and characteristic polynomials together?

*6.6.16. Prove that every square matrix is similar to its transpose.

*6.6.17. Compute the 2000th power of

$$(a)\, A = \begin{pmatrix} 3 & 1 \\ 2 & 4 \end{pmatrix};$$

$$(b)\, B = \begin{pmatrix} 3 & 1 \\ -1 & 1 \end{pmatrix}.$$

Chapter 7

Bilinear Functions

The theory of real bilinear functions and quadratic forms emerged from the geometry of quadratic curves and surfaces (conics and quadrics). Perpendicular vectors induced a general notion of orthogonality and this is a major tool for classifying bilinear functions. Their perhaps most important type is the inner product giving rise to Euclidean spaces (see the next chapter). We also discuss briefly the modifications necessary to preserve the good properties when switching from real to complex bilinear functions.

7.1. Real Bilinear Function

Definition 7.1.1. Let V be a vector space over $\mathbf{R}$. An $\mathbf{A}: V \times V \to \mathbf{R}$ map is a (real) *bilinear function* if it is linear in each variable, i.e., fixing any value of one variable, $\mathbf{A}$ is linear in the other variable.

In detail ($\mathbf{u}, \mathbf{u}', \mathbf{v}, \mathbf{v}' \in V, \lambda \in \mathbf{R}$):

(i) $\mathbf{A}$ assigns to every pair of vectors $(\mathbf{u}, \mathbf{v})$ a real number;

(ii) $\mathbf{A}(\mathbf{u} + \mathbf{u}', \mathbf{v}) = \mathbf{A}(\mathbf{u}, \mathbf{v}) + \mathbf{A}(\mathbf{u}', \mathbf{v})$;

(iii) $\mathbf{A}(\lambda\mathbf{u}, \mathbf{v}) = \lambda\mathbf{A}(\mathbf{u}, \mathbf{v})$;

(iv) $\mathbf{A}(\mathbf{u}, \mathbf{v} + \mathbf{v}') = \mathbf{A}(\mathbf{u}, \mathbf{v}) + \mathbf{A}(\mathbf{u}, \mathbf{v}')$;

(v) $\mathbf{A}(\mathbf{u}, \lambda\mathbf{v}) = \lambda\mathbf{A}(\mathbf{u}, \mathbf{v})$. ♣

We denote bilinear functions by uppercase bold Roman letters. The properties of linear maps imply the laws

(vi) $\mathbf{A}(\mathbf{u}, \mathbf{0}) = \mathbf{A}(\mathbf{0}, \mathbf{v}) = 0$;

(vii) $\mathbf{A}(-\mathbf{u}, \mathbf{v}) = \mathbf{A}(\mathbf{u}, -\mathbf{v}) = -\mathbf{A}(\mathbf{u}, \mathbf{v})$;

(viii) $\mathbf{A}\left(\sum_{i=1}^{k} \lambda_i \mathbf{u}_i, \sum_{j=1}^{m} \mu_j \mathbf{v}_j\right) = \sum_{i=1}^{k} \sum_{j=1}^{m} \lambda_i \mu_j \mathbf{A}(\mathbf{u}_i, \mathbf{v}_j)$.

Examples of bilinear functions

E1 Let V be the usual vector space of plane or space vectors and let $\mathbf{A}$ be the *inner product* (or *dot product*) assigning to two vectors the product of their lengths multiplied by the cosine of their angle. (If at least one of the vectors is $\mathbf{0}$, then their angle makes no sense but the length of $\mathbf{0}$ is 0, so the inner product is 0.) The inner product can also be obtained as the sum of the products of the corresponding coordinates (with respect to perpendicular unit vectors).

E2 We can generalize the second characterization of the inner product (or *dot product*) to any $\mathbf{R}^k$: If

$$\mathbf{u} = \begin{pmatrix} u_1 \\ \vdots \\ u_k \end{pmatrix}, \mathbf{v} = \begin{pmatrix} v_1 \\ \vdots \\ v_k \end{pmatrix},$$

then $\mathbf{A}(\mathbf{u}, \mathbf{v}) = \sum_{i=1}^{k} u_i v_i$.

This makes it possible to define an inner product on any k-dimensional real vector space: we fix a basis and take the sum of the products of the corresponding coordinates.

Please note that, adopting the general convention, the components or coordinates of the vectors are denoted by lower case Roman letters instead of Greek ones until the end of the next chapter.

E3 Using the notations of the previous example, e.g., $u_1v_2+2u_2v_1$ or $u_1v_1-3u_2v_2$ are bilinear functions on $\mathbf{R}^k$.

E4 Assigning the zero number to every pair of vectors we obtain the $\mathbf{0}$ bilinear function.

E5 Finally, we mention a few infinite-dimensional examples. Let V be the usual vector space of polynomials with real coefficients, and we assign to two polynomials the value of their product assumed in a given point. This can be generalized, e.g., for continuous functions. Another important example is the integral of the product on a given interval.

In the rest of the chapter, we only deal with *finite-dimensional* vector spaces. Let $\dim V = n$ and we fix a basis $\mathbf{b}_1, \dots, \mathbf{b}_n$.

We verify that, similar to linear maps, bilinear functions can be characterized by the images of the basis vectors and, consequently, by matrices.

Theorem 7.1.2. *Let $\mathbf{b}_1, \dots, \mathbf{b}_n$ be a basis in V and let α_{ij}, $i, j = 1, 2, \dots, n$ be arbitrary real numbers. Then there exists exactly one bilinear function $\mathbf{A}$ satisfying*

$$\mathbf{A}(\mathbf{b}_i, \mathbf{b}_j) = \alpha_{ij}, \quad i, j = 1, 2, \dots, n. \clubsuit$$

Proof. We follow the proof of Theorem 5.3.1. Any two vectors $\mathbf{u}$ and $\mathbf{v}$ have a unique representation $\mathbf{u} = u_1\mathbf{b}_1 + \cdots + u_n\mathbf{b}_n$ and $\mathbf{v} = v_1\mathbf{b}_1 + \cdots + v_n\mathbf{b}_n$. *If* there exists a bilinear function **A** with the prescribed property, then, by (viii),

$$\begin{aligned}\mathbf{A}(\mathbf{u},\mathbf{v}) &= \mathbf{A}(u_1\mathbf{b}_1 + \cdots + u_n\mathbf{b}_n, v_1\mathbf{b}_1 + \cdots + v_n\mathbf{b}_n)\\ &= \sum_{i,j=1}^{n} u_i v_j \mathbf{A}(\mathbf{b}_i,\mathbf{b}_j)\\ &= \sum_{i,j=1}^{n} u_i v_j \alpha_{ij}.\end{aligned}$$

This shows that $\mathbf{A}(\mathbf{u},\mathbf{v})$ is uniquely determined, so at most one such **A** can exist. Moreover, *only* the function defined by $\mathbf{A}(\mathbf{u},\mathbf{v}) = \sum_{i,j=1}^{n} u_i v_j \alpha_{ij}$ can work. We have to verify that this is bilinear by checking properties (ii)–(v). We leave these simple computations to the reader. ■

Definition 7.1.3. The *matrix* of a bilinear function **A** in the basis $\mathbf{b}_1,\dots,\mathbf{b}_n$ is an $n\times n$ matrix where the jth entry in the ith row is $\alpha_{ij} = \mathbf{A}(\mathbf{b}_i,\mathbf{b}_j)$. We denote this matrix by $[\mathbf{A}]_b$. ♣

Do not forget that, similar to linear maps, the matrix of a bilinear function depends heavily on the basis and choosing another basis the matrix will be quite different in general.

Theorem 7.1.4. *For a fixed basis, there is a bijection between the bilinear functions on V and the $n \times n$ (real) matrices. If the coordinates of* **u** *and* **v** *are $u_1,\dots,u_n$ and $v_1,\dots,v_n$, then*

$$\mathbf{A}(\mathbf{u},\mathbf{v}) = \sum_{i,j=1}^{n} \alpha_{ij} u_i v_j \tag{7.1.1}$$

or

$$\mathbf{A}(\mathbf{u},\mathbf{v}) = [\mathbf{u}]^T[\mathbf{A}][\mathbf{v}] = (u_1,u_2,\dots,u_n)\begin{pmatrix}\alpha_{11} & \alpha_{12} & \cdots & \alpha_{1n}\\ \alpha_{21} & \alpha_{22} & \cdots & \alpha_{2n}\\ \vdots & \vdots & \ddots & \vdots\\ \alpha_{n1} & \alpha_{n2} & \cdots & \alpha_{nn}\end{pmatrix}\begin{pmatrix}v_1\\ v_2\\ \vdots\\ v_n\end{pmatrix}. \clubsuit \tag{7.1.2}$$

Proof. Bijection follows from Theorem 7.1.2. Decomposition (7.1.1) was verified in the proof of Theorem 7.1.2. Performing the multiplication of matrices in (7.1.2), we obtain the equivalence of formulas (7.1.1) and (7.1.2). ■

The use of matrices and formulas (7.1.1) and (7.1.2) provide a convenient handling of bilinear functions.

Exercises 7.1

V denotes a vector space over $\mathbf{R}$.

7.1.1. Let V be the usual vector space of polynomials with real coefficients of degree not greater than four (including the 0 polynomial). Select the bilinear functions and write their matrices in the usual basis. Let the image of the pair f, g be

(a) fg; (b) $f(1) + g(1)$;

(c) $f(1)g(2)$; (d) $f'(1)g(2)$;

(e) the coefficient of x^2 in fg.

7.1.2. Compute the matrices of the bilinear functions of Examples E2, E3, and E4 (in suitable bases).

7.1.3. Which sets can occur as a range of a real bilinear function?

7.1.4. Formulate the analogues of Theorem 7.1.2 if, instead of a basis, the values of the bilinear function are prescribed on a

(a) spanning set;

(b) linearly independent system.

7.1.5. Define a natural addition and multiplication by scalars among the bilinear functions on V, show that this creates a vector space, and determine its dimension.

7.1.6. Let $\mathbf{b}_1, \dots, \mathbf{b}_n$ be a basis in V. Describe the changes in the matrix of the bilinear function $\mathbf{A}$ corresponding to the modification of the basis.

(a) We swap $\mathbf{b}_1$ and $\mathbf{b}_2$.

(b) We replace $\mathbf{b}_3$ by $\lambda\mathbf{b}_3$ (where $\lambda \neq 0$).

(c) We replace $\mathbf{b}_3$ by $\mathbf{b}_3 + \mu\mathbf{b}_2$.

7.1.7. Find all bilinear functions having the same matrix in every basis.

7.1.8. Let $\mathbf{b}_1, \dots, \mathbf{b}_n$ be a fixed basis in V and consider the inner product defined in Example E2. We denote the inner product of $\mathbf{c}$ and $\mathbf{d}$ by $\mathbf{c} \cdot \mathbf{d}$.

(a) Let $\mathcal{A} \in \operatorname{Hom} V$ be an arbitrary linear transformation. Verify that $\mathbf{A}(\mathbf{u}, \mathbf{v}) = \mathbf{u} \cdot \mathcal{A}\mathbf{v}$ is a bilinear function.

(b) Show that the linear map $\mathcal{A}$ in (a) and the bilinear function $\mathbf{A}$ have the same matrix in the basis $\mathbf{b}_1, \dots, \mathbf{b}_n$: $[\mathcal{A}]_b = [\mathbf{A}]_b$.

(c) Prove that every bilinear function has a form described in (a) with a suitable $\mathcal{A} \in \operatorname{Hom} V$.

7.1.9. The $V \to \mathbf{R}$ linear maps, i.e., the elements of $\operatorname{Hom}(V, \mathbf{R})$ are called *linear functionals*.

(a) Let Φ and Ψ be linear functionals. Show that $\mathbf{A}(\mathbf{u}, \mathbf{v}) = \Phi(\mathbf{u})\Psi(\mathbf{v})$ is a bilinear function and determine the rank of the matrix of $\mathbf{A}$.

***(b)** Prove that every bilinear function $\mathbf{A}$ can be decomposed as $\mathbf{A}(\mathbf{u}, \mathbf{v}) = \sum_{m=1}^{r} \Phi_m(\mathbf{u})\Psi_m(\mathbf{v})$ where Φ_m and Ψ_m are linear functionals ($m = 1, 2, \dots, r$). Given $\mathbf{A}$, what is the smallest possible value of r?

7.2. Orthogonalization

Let V be a finite-dimensional vector space over $\mathbf{R}$. In this section we investigate which bilinear functions $\mathbf{A}$ have a diagonal matrix. It turns out that the situation is much more favorable than it was for linear transformations. It is easy to see that $\mathbf{A}$ has to be symmetric (see below). Surprisingly, this is not only necessary but also sufficient for diagonalization.

Definition 7.2.1. A bilinear function $\mathbf{A}$ is *symmetric* if $\mathbf{A}(\mathbf{u}, \mathbf{v}) = \mathbf{A}(\mathbf{v}, \mathbf{u})$ for every $\mathbf{u}, \mathbf{v} \in V$. ♣

Among the examples in the previous section, the inner product, the zero function, and the second function in E3 are symmetric but the first function in E3 is not.

Theorem 7.2.2. *A bilinear function $\mathbf{A}$ is symmetric if and only if its matrix (in any basis) is symmetric (i.e., it equals its transpose).* ♣

Proof. If the bilinear function $\mathbf{A}$ is symmetric, then applying symmetry to the basis elements $\mathbf{b}_i$, we get $\alpha_{ij} = \mathbf{A}(\mathbf{b}_i, \mathbf{b}_j) = \mathbf{A}(\mathbf{b}_j, \mathbf{b}_i) = \alpha_{ji}$, so the matrix is symmetric, too. The converse follows, e.g., from formula (7.1.1) in Theorem 7.1.4. ∎

Now we turn to the question of diagonalization. Assume first that $\mathbf{A}$ has a diagonal matrix. A diagonal matrix is symmetric, so $\mathbf{A}$ has to be symmetric by Theorem 7.2.2. As indicated at the beginning of this section, the converse also holds.

Theorem 7.2.3. *Let $\mathbf{A}$ be a symmetric bilinear function on V. Then the matrix of $\mathbf{A}$ is diagonal in a suitable basis.* ♣

Before turning to the proof, we introduce an important notion. In a diagonal matrix $\mathbf{A}(\mathbf{c}_i, \mathbf{c}_j) = 0$ for $i \neq j$ for the given basis elements $\mathbf{c}_1, \dots, \mathbf{c}_n$. If the bilinear function is the usual inner product of plane or space vectors, then $\mathbf{A}(\mathbf{u}, \mathbf{v}) = 0$ means that $\mathbf{u}$ and $\mathbf{v}$ are perpendicular or, with another expression, orthogonal. This is generalized in the following definition.

Definition 7.2.4. Let A be a symmetric bilinear function. The vectors $\mathbf{u}, \mathbf{v} \in V$ are $\mathbf{A}$-*orthogonal* if $\mathbf{A}(\mathbf{u}, \mathbf{v}) = 0$. ♣

Using this notion, Theorem 7.2.3 asserts that there exist $\mathbf{A}$-orthogonal bases to every symmetric bilinear function $\mathbf{A}$.

We give three proofs of this theorem. The first one is not complete, as it works only for a certain – but as we shall see in the next chapter, perhaps the most important – class of bilinear functions. A great advantage of this proof is that it yields a natural and easily implementable construction of an $\mathbf{A}$-orthogonal basis. The second and third proofs work for an arbitrary symmetric bilinear function and these are also constructive. The basic idea of the second proof is similar to the argument of the first one but needs some technical modifications to handle the general case. The third proof uses Gaussian elimination for the diagonalization of the matrix of $\mathbf{A}$. After each proof, we illustrate the method on the same concrete example.

First proof of Theorem 7.2.3. (Gram–Schmidt orthogonalization). We assume that $\mathbf{A}(\mathbf{u},\mathbf{u}) \neq 0$ for every $\mathbf{u} \neq \mathbf{0}$. Let $\mathbf{b}_1,\dots,\mathbf{b}_n$ be any basis in V. We construct an $\mathbf{A}$-orthogonal basis $\mathbf{c}_1,\dots,\mathbf{c}_n$ recursively:

$$\begin{aligned}
\mathbf{c}_1 &= \mathbf{b}_1,\\
\mathbf{c}_2 &= \mathbf{b}_2 + \rho_{21}\mathbf{c}_1,\\
\mathbf{c}_3 &= \mathbf{b}_3 + \rho_{31}\mathbf{c}_1 + \rho_{32}\mathbf{c}_2,\\
&\vdots\\
\mathbf{c}_n &= \mathbf{b}_n + \rho_{n1}\mathbf{c}_1 + \rho_{n2}\mathbf{c}_2 + \cdots + \rho_{n,n-1}\mathbf{c}_{n-1},
\end{aligned}$$

with suitably chosen scalars ρ_{qs}.

We show first that $\mathbf{c}_1,\dots,\mathbf{c}_n$ is a basis in V for arbitrary values of the scalars ρ_{qs}. Since $n = \dim V$, it suffices to verify that $\mathbf{c}_1,\dots,\mathbf{c}_n$ is a spanning set. This follows if every $\mathbf{b}_j$ is a linear combination of the vectors $\mathbf{c}_i$. We obtain such a representation by expressing $\mathbf{b}_j$ with the help of $\mathbf{c}_1,\dots,\mathbf{c}_j$ from the equation defining $\mathbf{c}_j$.

Now we prove that the vectors $\mathbf{c}_i$ are $\mathbf{A}$-orthogonal under a suitable choice of the coefficients ρ_{qs}. We determine the vectors $\mathbf{c}_j$ successively so that $\mathbf{c}_j$ is $\mathbf{A}$-orthogonal to each of the previous vectors $\mathbf{c}_1,\dots,\mathbf{c}_{j-1}$.

We start with $\mathbf{c}_2$. Here the only requirement is $\mathbf{A}(\mathbf{c}_2,\mathbf{c}_1) = 0$. Using the definition of $\mathbf{c}_2$, this means

$$0 = \mathbf{A}(\mathbf{c}_2,\mathbf{c}_1) = \mathbf{A}(\mathbf{b}_2 + \rho_{21}\mathbf{c}_1,\mathbf{c}_1) = \mathbf{A}(\mathbf{b}_2,\mathbf{c}_1) + \rho_{21}\mathbf{A}(\mathbf{c}_1,\mathbf{c}_1).$$

So $\rho_{21} = -\mathbf{A}(\mathbf{b}_2,\mathbf{c}_1)/\mathbf{A}(\mathbf{c}_1,\mathbf{c}_1)$. (By assumption, the denominator is not zero.)

We can argue in general similarly. We assume that we used the first $m-1$ equations to prove the $\mathbf{A}$-orthogonality of $\mathbf{c}_1,\dots,\mathbf{c}_{m-1}$. Now we consider the mth equation and check the $\mathbf{A}$-orthogonality of $\mathbf{c}_m$ and $\mathbf{c}_j$ for $j < m$:

$$\begin{aligned}
0 &= \mathbf{A}(\mathbf{c}_m,\mathbf{c}_j)\\
&= \mathbf{A}(\mathbf{b}_m + \rho_{m1}\mathbf{c}_1 + \rho_{m2}\mathbf{c}_2 + \cdots + \rho_{mj}\mathbf{c}_j + \cdots + \rho_{m,m-1}\mathbf{c}_{m-1},\mathbf{c}_j)\\
&= \mathbf{A}(\mathbf{b}_m,\mathbf{c}_j) + \rho_{m1}\mathbf{A}(\mathbf{c}_1,\mathbf{c}_j) + \cdots + \rho_{mj}\mathbf{A}(\mathbf{c}_j,\mathbf{c}_j) + \cdots + \rho_{m,m-1}\mathbf{A}(\mathbf{c}_{m-1},\mathbf{c}_j).
\end{aligned}$$

For $i,j < m$ and $i \neq j$, we know that $\mathbf{c}_i$ and $\mathbf{c}_j$ are $\mathbf{A}$-orthogonal, meaning, $\mathbf{A}(\mathbf{c}_i,\mathbf{c}_j) = 0$. So, our condition is reduced to $0 = \mathbf{A}(\mathbf{b}_m,\mathbf{c}_j) + \rho_{mj}\mathbf{A}(\mathbf{c}_j,\mathbf{c}_j)$. Thus $\rho_{mj} = -\mathbf{A}(\mathbf{b}_m,\mathbf{c}_j)/\mathbf{A}(\mathbf{c}_j,\mathbf{c}_j)$. Hereby we have verified that the coefficients ρ_{mj} in the mth equation can be chosen so that $\mathbf{c}_m$ is $\mathbf{A}$-orthogonal to each of $\mathbf{c}_1,\dots,\mathbf{c}_{m-1}$. ■

Remarks. We only used the condition $\mathbf{A}(\mathbf{u},\mathbf{u}) \neq 0$ for $\mathbf{c}_1,\dots,\mathbf{c}_{n-1}$. So if we are lucky, then the procedure might work even if $\mathbf{A}(\mathbf{u},\mathbf{u}) = 0$ for some $\mathbf{u} \neq \mathbf{0}$ but these do not show up during the orthogonalization (see the example below).

Given $\mathbf{A}$ in a vector space, there are many $\mathbf{A}$-orthogonal bases. This is clear from the proof as starting from another basis $\mathbf{b}_i$ we generally obtain a different basis $\mathbf{c}_i$.

We emphasize once again that the proof yields a practical algorithm to construct an $\mathbf{A}$-orthogonal basis. We illustrate it with an example.

Example. Let $V = \mathbf{R}^3$,

$$\mathbf{u} = \begin{pmatrix} u_1 \\ u_2 \\ u_3 \end{pmatrix}, \text{ and } \mathbf{v} = \begin{pmatrix} v_1 \\ v_2 \\ v_3 \end{pmatrix}.$$

We want to find an $\mathbf{A}$-orthogonal basis and a diagonal matrix for the symmetric bilinear function

$$(*) \qquad \mathbf{A}(\mathbf{u}, \mathbf{v}) = 4u_1v_1 + 2u_1v_2 + 2u_2v_1 + 2u_1v_3 + 2u_3v_1 + 2u_2v_3 + 2u_3v_2.$$

We take the usual unit vectors

$$\mathbf{b}_1 = \begin{pmatrix} 1 \\ 0 \\ 0 \end{pmatrix}, \mathbf{b}_2 = \begin{pmatrix} 0 \\ 1 \\ 0 \end{pmatrix}, \mathbf{b}_3 = \begin{pmatrix} 0 \\ 0 \\ 1 \end{pmatrix}$$

as an initial basis. By the construction, $\mathbf{c}_1 = \mathbf{b}_1$. Next, $\mathbf{c}_2 = \mathbf{b}_2 + \rho_{21}\mathbf{c}_1$. We get ρ_{21} from the condition $\mathbf{A}(\mathbf{c}_2, \mathbf{c}_1) = 0$:

$$0 = \mathbf{A}(\mathbf{c}_2, \mathbf{c}_1) = \mathbf{A}\left(\begin{pmatrix} 0 \\ 1 \\ 0 \end{pmatrix}, \begin{pmatrix} 1 \\ 0 \\ 0 \end{pmatrix}\right) + \rho_{21}\mathbf{A}\left(\begin{pmatrix} 1 \\ 0 \\ 0 \end{pmatrix}, \begin{pmatrix} 1 \\ 0 \\ 0 \end{pmatrix}\right) = 2 + 4\rho_{21},$$

thus $\rho_{21} = -1/2$. So

$$\mathbf{c}_2 = \begin{pmatrix} 0 \\ 1 \\ 0 \end{pmatrix} - \frac{1}{2}\begin{pmatrix} 1 \\ 0 \\ 0 \end{pmatrix} = \begin{pmatrix} -1/2 \\ 1 \\ 0 \end{pmatrix}.$$

(If we prefer integers, we can multiply $\mathbf{c}_2$ by a non-zero scalar as that does not affect $\mathbf{A}$-orthogonality.) Next we consider

$$\mathbf{c}_3 = \mathbf{b}_3 + \rho_{31}\mathbf{c}_1 + \rho_{32}\mathbf{c}_2 = \begin{pmatrix} 0 \\ 0 \\ 1 \end{pmatrix} + \rho_{31}\begin{pmatrix} 1 \\ 0 \\ 0 \end{pmatrix} + \rho_{32}\begin{pmatrix} -1/2 \\ 1 \\ 0 \end{pmatrix}.$$

Conditions $\mathbf{A}(\mathbf{c}_3, \mathbf{c}_1) = \mathbf{A}(\mathbf{c}_3, \mathbf{c}_2) = 0$ imply $\rho_{31} = -1/2$ and $\rho_{32} = 1$. So

$$\mathbf{c}_3 = \begin{pmatrix} -1 \\ 1 \\ 1 \end{pmatrix}.$$

Thus we obtain an $\mathbf{A}$-orthogonal basis $\mathbf{c}_1, \mathbf{c}_2, \mathbf{c}_3$. The main diagonal of the corresponding diagonal matrix consists of the values $\mathbf{A}(\mathbf{c}_i, \mathbf{c}_i)$, so

$$[\mathbf{A}]_c = \begin{pmatrix} 4 & 0 & 0 \\ 0 & -1 & 0 \\ 0 & 0 & 0 \end{pmatrix}.$$

Second proof of Theorem 7.2.3. First we verify that if $\mathbf{A}(\mathbf{u}, \mathbf{u}) = 0$ for every $\mathbf{u} \in V$, then $\mathbf{A}$ is the (identically) zero bilinear function, i.e., $\mathbf{A}(\mathbf{u}, \mathbf{v}) = 0$ for every $\mathbf{u}, \mathbf{v} \in V$. This follows from

$$\begin{aligned} \mathbf{A}(\mathbf{u} + \mathbf{v}, \mathbf{u} + \mathbf{v}) &= \mathbf{A}(\mathbf{u}, \mathbf{u}) + \mathbf{A}(\mathbf{u}, \mathbf{v}) + \mathbf{A}(\mathbf{v}, \mathbf{u}) + \mathbf{A}(\mathbf{v}, \mathbf{v}) \\ &= \mathbf{A}(\mathbf{u}, \mathbf{u}) + 2\mathbf{A}(\mathbf{u}, \mathbf{v}) + \mathbf{A}(\mathbf{v}, \mathbf{v}). \end{aligned}$$

Turning to the proof of the theorem, we note that if $\mathbf{A}$ is the zero function, then every basis is $\mathbf{A}$-orthogonal (the matrix of $\mathbf{A}$ is the zero matrix). Otherwise choose any vector $\mathbf{d}$ satisfying $\mathbf{A}(\mathbf{d},\mathbf{d}) \neq 0$ (such a vector must exist by the preliminary remark).

Consider the set W of all vectors $\mathbf{A}$-orthogonal to $\mathbf{d}$:

$$W = \{\mathbf{w} \in V \mid \mathbf{A}(\mathbf{d},\mathbf{w}) = 0\}.$$

We show that W is a subspace and every $\mathbf{v} \in V$ has a unique representation

$$\mathbf{v} = \lambda\mathbf{d} + \mathbf{w} \tag{7.2.1}$$

where $\mathbf{w} \in W$ (i.e., V is the direct sum of the subspaces $\langle\mathbf{d}\rangle$ and W).

$W \neq \emptyset$ as $\mathbf{0} \in W$. A simple calculation shows that W is closed under addition and multiplication by a scalar, so it is a subspace. Equivalent forms of (7.2.1) are

$$\mathbf{v} - \lambda\mathbf{d} \in W \iff \mathbf{A}(\mathbf{d},\mathbf{v}-\lambda\mathbf{d}) = 0 \iff \mathbf{A}(\mathbf{d},\mathbf{v}) - \lambda\mathbf{A}(\mathbf{d},\mathbf{d}) = 0.$$

Hence (7.2.1) holds if and only if $\lambda = \mathbf{A}(\mathbf{d},\mathbf{v})/\mathbf{A}(\mathbf{d},\mathbf{d})$.

To construct an $\mathbf{A}$-orthogonal basis, let $\mathbf{d}$ be its first element, then repeat the above procedure for the subspace W (having a smaller dimension), etc. This yields an $\mathbf{A}$-orthogonal basis in not more than $\dim V - 1$ steps. (It terminates earlier if $\mathbf{A}$ is the zero function on some intermediate subspace of dimension greater than one.) ∎

Remarks. Comparing the first and second proofs, the first one constructs an $\mathbf{A}$-orthogonal system via an increasing chain of subspaces whereas the second one uses a decreasing chain of subspaces.

Concerning the second proof, note that the properties of W can also be verified from the matrix equation $[\mathbf{d}]^T[\mathbf{A}][\mathbf{w}] = 0$ and we can obtain the elements of W explicitly. This leads to an effective construction of an $\mathbf{A}$-orthogonal basis. We present this method on the same example $(*)$ used after the first proof.

The matrix of $\mathbf{A}$ in the original basis $\mathbf{b}_i$ is

$$\begin{pmatrix} 4 & 2 & 2 \\ 2 & 0 & 2 \\ 2 & 2 & 0 \end{pmatrix}.$$

We start with $\mathbf{d} = \mathbf{b}_1$. This works as $\mathbf{A}(\mathbf{b}_1,\mathbf{b}_1) = 4 \neq 0$. We determine the vectors $\mathbf{w}$ $\mathbf{A}$-orthogonal to $\mathbf{d}$ from the equation $\mathbf{A}(\mathbf{d},\mathbf{w}) = [\mathbf{d}]^T[\mathbf{A}][\mathbf{w}] = 0$:

$$0 = \begin{pmatrix} 1 & 0 & 0 \end{pmatrix}\begin{pmatrix} 4 & 2 & 2 \\ 2 & 0 & 2 \\ 2 & 2 & 0 \end{pmatrix}\begin{pmatrix} w_1 \\ w_2 \\ w_3 \end{pmatrix} = 4w_1 + 2w_2 + 2w_3.$$

This is equivalent to

$$\mathbf{w} = \begin{pmatrix} \alpha \\ \beta \\ -2\alpha - \beta \end{pmatrix}.$$

We continue the orthogonalization among these vectors (as these are $\mathbf{A}$-orthogonal to the vector

$$\begin{pmatrix} 1 \\ 0 \\ 0 \end{pmatrix}$$

selected first). We can take, e.g.,

$$\mathbf{d}_2 = \begin{pmatrix} 1 \\ 0 \\ -2 \end{pmatrix}$$

as a second vector (obtained from $\alpha = 1, \beta = 0$) since $\mathbf{A}(\mathbf{d}_2, \mathbf{d}_2) = -4 \neq 0$. The vectors in W $\mathbf{A}$-orthogonal to $\mathbf{d}_2$ are determined by

$$0 = \begin{pmatrix} 1 & 0 & -2 \end{pmatrix} \begin{pmatrix} 4 & 2 & 2 \\ 2 & 0 & 2 \\ 2 & 2 & 0 \end{pmatrix} \begin{pmatrix} \alpha \\ \beta \\ -2\alpha - \beta \end{pmatrix} = -4\alpha - 4\beta.$$

This means $\beta = -\alpha$, so these vectors are of the form

$$\begin{pmatrix} \alpha \\ -\alpha \\ -\alpha \end{pmatrix}.$$

($\mathbf{A}$ is already zero on this one-dimensional subspace.) So, e.g., $\alpha = 1$ gives

$$\mathbf{d}_3 = \begin{pmatrix} 1 \\ -1 \\ -1 \end{pmatrix}$$

as the last element of an $\mathbf{A}$-orthogonal basis. The matrix of the bilinear function in this basis is

$$\begin{pmatrix} 4 & 0 & 0 \\ 0 & -4 & 0 \\ 0 & 0 & 0 \end{pmatrix}.$$

Third proof of Theorem 7.2.3. We start from the matrix of $\mathbf{A}$ in an arbitrary basis $\mathbf{b}_1, \dots, \mathbf{b}_n$. We shall apply the steps given in Exercise 7.1.6. These are the well-known elementary equivalent transformations, but we have to perform them simultaneously for rows and columns. Note that such a combined step does not violate the symmetry of the matrix.

We summarize these steps and the induced changes in the matrix. (a) Swapping two basis elements we have to swap the corresponding rows and columns of the matrix. (b) Multiplying the basis vector $\mathbf{b}_i$ by a scalar $\lambda \neq 0$, the ith row and column are multiplied by λ, so α_{ii} is multiplied by λ^2. (c) If $\mathbf{b}_i$ is replaced by $\mathbf{b}_i + \mu\mathbf{b}_j$ (where $j \neq i$), then we add μ times the jth row/column to the ith row/column, so the ith element in the new ith row will be $\alpha'_{ii} = \alpha_{ii} + \mu\alpha_{ji} + \mu\alpha_{ij} + \mu^2\alpha_{jj}$.

Let us implement the Gaussian elimination using these steps. If the element in the upper left corner is not zero, then subtracting the suitable multiples of the first row/column from the other rows/columns, the other elements in the first row and column will become zero. If the element in the upper left corner is zero but some other element in the main diagonal is not zero, then swapping the corresponding rows/columns will put this non-zero element into the upper left corner. If every element in the main diagonal is zero, but (say) the third element in the first row/column $\alpha_{13} = \alpha_{31} \neq 0$, then we add the third row/column to the first row/column yielding

$\alpha'_{11} = \alpha_{11} + 2\alpha_{31} + \alpha_{33} = 2\alpha_{31} \neq 0$ in the upper left corner. If every element in the first row/column is zero, then we do nothing (temporarily).

Thus we have achieved that every element in the first row and column is zero except (perhaps) the first element. We do not perform further changes with this row and column.

Now we repeat the procedure with the symmetric matrix of the other rows and columns (these steps do not affect the first row and column). We continue similarly until finally we obtain a diagonal matrix. ■

Also the third proof provides a simple algorithm to construct a diagonal matrix. We get the corresponding **A**-orthogonal basis, too, if we keep track of the changes in the basis after each step. We illustrate this again on the example (∗) discussed after the first two proofs.

The initial basis is $\mathbf{b}_1, \mathbf{b}_2, \mathbf{b}_3$, and the initial matrix is

$$\begin{pmatrix} 4 & 2 & 2 \\ 2 & 0 & 2 \\ 2 & 2 & 0 \end{pmatrix}.$$

We subtract the half of the first row from the second row and subtract the half of the new(!) first column from the second column.

New basis: $\mathbf{b}_1, \mathbf{b}_2 - \frac{1}{2}\mathbf{b}_1, \mathbf{b}_3$, new matrix:

$$\begin{pmatrix} 4 & 0 & 2 \\ 0 & -1 & 1 \\ 2 & 1 & 0 \end{pmatrix}.$$

Now we subtract the half of the first row/column from the third row/column.

New basis: $\mathbf{b}_1, \mathbf{b}_2 - \frac{1}{2}\mathbf{b}_1, \mathbf{b}_3 - \frac{1}{2}\mathbf{b}_1$, new matrix:

$$\begin{pmatrix} 4 & 0 & 0 \\ 0 & -1 & 1 \\ 0 & 1 & -1 \end{pmatrix}.$$

Finally, we add the second row/column to the third row/column.

New basis: $\mathbf{b}_1, \mathbf{b}_2 - \frac{1}{2}\mathbf{b}_1, \mathbf{b}_3 + \mathbf{b}_2 - \mathbf{b}_1$, new matrix:

$$\begin{pmatrix} 4 & 0 & 0 \\ 0 & -1 & 0 \\ 0 & 0 & 0 \end{pmatrix}.$$

We obtain the same **A**-orthogonal basis as with the method of the first proof.

We can bring the matrix of a bilinear function to an even nicer form: multiplying the **A**-orthogonal basis vectors by suitable scalars, we can achieve that the elements in the main diagonal of the diagonal matrix are ± 1 and 0. Using this basis, the bilinear function is reduced to a very simple form similar to the inner product. The precise formulation is the following theorem.

Theorem 7.2.5. *Every symmetric bilinear function* $\mathbf{A}$ *has a matrix where the elements of the main diagonal are* 1, −1, *or* 0, *and every other element is* 0.

The bilinear function in the corresponding basis is $\mathbf{A}(\mathbf{u},\mathbf{v}) = \sum_{i=1}^{n} \gamma_i u_i v_i$ *where* $\gamma_i = \pm 1$ *or* 0 *and* $u_1, \dots, u_n$ *and* $v_1, \dots, v_n$ *denote the coordinates of* $\mathbf{u}$ *and* $\mathbf{v}$. ♣

The inner product is the special case when every $\gamma_i = 1$.

Proof. Let $\mathbf{c}_1, \dots, \mathbf{c}_n$ be an $\mathbf{A}$-orthogonal basis and let $\mathbf{A}(\mathbf{c}_i, \mathbf{c}_i) = \mu_i$. We normalize $\mathbf{c}_i$ by multiplying it with a suitable scalar: if $\mu_i = 0$, then $\mathbf{e}_i = \mathbf{c}_i$; if $\mu_i \neq 0$, then $\mathbf{e}_i = (1/\sqrt{|\mu_i|})\mathbf{c}_i$. (This is a transformation of type (b) indicated in the third proof but not used there.) Also the vectors $\mathbf{e}_i$ are $\mathbf{A}$-orthogonal and $\mathbf{A}(\mathbf{e}_i, \mathbf{e}_i) = \pm 1$ or 0 according to μ_i being positive, negative, or zero. So the matrix of $\mathbf{A}$ in the basis $\mathbf{e}_i$ has the desired form.

The second statement of the theorem immediately follows from Theorem 7.1.4. ∎

As mentioned before, a bilinear function has many $\mathbf{A}$-orthogonal bases. The diagonal matrix, however, is pretty much determined.

Theorem 7.2.6 (Law of inertia). *In the main diagonal of a diagonal matrix of a bilinear function the numbers of positive, negative, and zero elements are uniquely determined.* ♣

Proof. Consider two diagonal matrices of $\mathbf{A}$. We denote the number of positive, negative, and zero elements in the main diagonals by r, s, t and r', s', t'. (Any number may be zero, and $r + s + t = r' + s' + t' = \dim V$). Let

$$\mathbf{p}_1, \dots, \mathbf{p}_r, \mathbf{n}_1, \dots, \mathbf{n}_s, \mathbf{o}_1, \dots, \mathbf{o}_t,$$

be the $\mathbf{A}$-orthogonal basis belonging to the first matrix where

$$\mathbf{A}(\mathbf{p}_i, \mathbf{p}_i) > 0,\ \mathbf{A}(\mathbf{n}_j, \mathbf{n}_j) < 0,\ \mathbf{A}(\mathbf{o}_k, \mathbf{o}_k) = 0.$$

The other basis is formed similarly from the vectors $\mathbf{m}'_i$, $\mathbf{n}'_j$, $\mathbf{o}'_k$.

We show first that the $r + t + s'$ vectors $\mathbf{p}_i$, $\mathbf{o}_k$, and $\mathbf{n}'_j$ are linearly independent. Consider $\mathbf{0}$ as their linear combination and order it as

$$\pi_1\mathbf{p}_1 + \cdots + \pi_r\mathbf{p}_r + \omega_1\mathbf{o}_1 + \cdots + \omega_t\mathbf{o}_t = \nu_1\mathbf{n}'_1 + \cdots + \nu_{s'}\mathbf{n}'_{s'}. \tag{7.2.2}$$

It suffices to show that both sides are $\mathbf{0}$. Then the linear independence of the vectors $\mathbf{p}_i, \mathbf{o}_k$ implies that every coefficient is 0 on the left-hand side and the linear independence of the vectors $\mathbf{n}'_j$ guarantees the same on the right-hand side.

Let $\mathbf{u}$ be the common value of the two sides in (7.2.2). Since $\mathbf{u}$ is a linear combination of the vectors $\mathbf{p}_i$ and $\mathbf{o}_k$, the formula of Theorem 7.1.4 implies

$$\mathbf{A}(\mathbf{u},\mathbf{u}) = \sum_{i=1}^{r} \pi_i^2 \mathbf{A}(\mathbf{p}_i, \mathbf{p}_i) + \sum_{k=1}^{t} \omega_k^2 \mathbf{A}(\mathbf{o}_k, \mathbf{o}_k) \geq 0. \tag{7.2.3}$$

We obtain similarly

$$\mathbf{A}(\mathbf{u},\mathbf{u}) = \sum_{j=1}^{s'} \nu_j^2 \mathbf{A}(\mathbf{n}'_j, \mathbf{n}'_j) \leq 0. \tag{7.2.4}$$

Comparing (7.2.3) and (7.2.4), we have $\mathbf{A}(\mathbf{u},\mathbf{u}) = 0$. So (7.2.4) implies $\mathbf{u} = \mathbf{0}$.

Herewith we have proved that the vectors $\mathbf{p}_i$, $\mathbf{o}_k$, and $\mathbf{n}'_j$ are linearly independent.

The number of independent vectors is not greater than the dimension, so $r+t+s' \leq \dim V = r + t + s$. Hence $s' \leq s$. Swapping roles we get $s \leq s'$; thus $s = s'$.

Swapping roles of the vectors $\mathbf{p}$ and $\mathbf{n}$, we obtain $r = r'$. Finally, $t = \dim V - r - s = \dim V - r' - s' = t'$. ■

Exercises 7.2

7.2.1. A bilinear function is *antisymmetric* (or *skew-symmetric*) if $\mathbf{A}(\mathbf{u},\mathbf{v}) = -\mathbf{A}(\mathbf{v},\mathbf{u})$ for every $\mathbf{u},\mathbf{v} \in V$. Verify that $\mathbf{A}$ is antisymmetric if and only if $\mathbf{A}(\mathbf{x},\mathbf{x}) = 0$ for every $\mathbf{x} \in V$.

7.2.2. Prove that every bilinear function has a unique representation as the sum of a symmetric and an antisymmetric bilinear function.

7.2.3. Let B be the vector space of all bilinear functions defined on V (see Exercise 7.1.5).
(a) Show that the symmetric and antisymmetric functions form subspaces S and A in B.
(b) Determine the dimensions of S and A.
(c) Prove that B is the direct sum of S and A.

7.2.4. If $\mathbf{A}$ is a bilinear function on V, then it(s restriction) is bilinear on every subspace of V. Let U and W be subspaces in V. True or false?
(a) If $\mathbf{A}$ is symmetric on $\langle U, W\rangle$, then it is symmetric on U and W.
(b) If $\mathbf{A}$ is symmetric on U and W, then it is symmetric on $\langle U, W\rangle$.

7.2.5. Let $\mathbf{A}$ be the usual inner product on $\mathbf{R}^4$. We denote the components of a general $\mathbf{u} \in \mathbf{R}^4$ by u_1, u_2, u_3, u_4. Construct an $\mathbf{A}$-orthogonal basis on the subspaces U_i of $\mathbf{R}^4$:
(a) $U_1 = \{\mathbf{u} \mid u_1 = u_4\}$;
(b) $U_2 = \{\mathbf{u} \mid u_1, \ldots, u_4$ is an arithmetic progression$\}$;
(c) $U_3 = \{\mathbf{u} \mid \sum_{i=1}^4 u_i = 0\}$.

7.2.6. Let V be the usual vector space of the polynomials with real coefficients of degree at most 4 (including the 0 polynomial). Find diagonal matrices and $\mathbf{A}$-orthogonal bases for the bilinear functions mapping f and g into
(a) $f(1)g(1)$;
(b) the coefficient of x^2 in fg;
(c) $f(1)g(1) + f(2)g(2) + f(3)g(3) + f(4)g(4) + f(5)g(5)$.

7.2.7. Let $\dim V = 3$ with a basis $\mathbf{b}_1, \mathbf{b}_2, \mathbf{b}_3$, and let u_1, u_2, u_3 and v_1, v_2, v_3 denote the coordinates of $\mathbf{u}$ and $\mathbf{v}$. Find a diagonal matrix and an $\mathbf{A}$-orthogonal basis to the bilinear function
(a) $\sum_{i,j=1}^3 iju_iv_j$;
(b) $\sum_{i,j=1}^3 (i+j-1)u_iv_j$.

7.2.8. Let $\mathbf{A}$ be a symmetric bilinear function and assume that $\mathbf{A}(\mathbf{u},\mathbf{u}) > 0$ and $\mathbf{A}(\mathbf{v},\mathbf{v}) < 0$ holds for some $\mathbf{u},\mathbf{v} \in V$. Prove that $\mathbf{A}(\mathbf{w},\mathbf{w}) = 0$ for some non-zero $\mathbf{w} \in V$.

7.2.9. Let $\dim V = n$, $\mathbf{A}$ a symmetric bilinear function on V, and $\mathbf{v} \in V$. Verify that the vectors $\mathbf{A}$-orthogonal to $\mathbf{v}$ form a subspace in V. What can be its dimension?

7.2.10. Let us call the bilinear functions $\mathbf{A}$ and $\mathbf{B}$ defined on the same vector space V equivalent if they have a common matrix, i.e., $[\mathbf{B}]_b = [\mathbf{A}]_a$ in some bases $\mathbf{a}_1,\dots,\mathbf{a}_n$ and $\mathbf{b}_1,\dots,\mathbf{b}_n$. Find the number of pairwise non-equivalent bilinear functions on V.

7.3. Quadratic Form

As seen in the previous section the values $\mathbf{A}(\mathbf{u},\mathbf{u})$ play an important role in studying a bilinear function $\mathbf{A}$.

Definition 7.3.1. The function $V \to \mathbf{R}$ defined by $\tilde{\mathbf{A}}(\mathbf{x}) = \mathbf{A}(\mathbf{x},\mathbf{x})$ is called the *quadratic form* of the bilinear function $\mathbf{A}$. ♣

A quadratic form is thus the restriction of a bilinear function when we write the same vector into both variables. So, every quadratic form comes from some bilinear function.

Clearly, a bilinear function defines its quadratic form uniquely. The converse is false, different bilinear functions can yield the same quadratic form. However, there is a bijection between the quadratic forms and the *symmetric* bilinear functions (see Exercise 7.3.1). Therefore, we shall always consider a quadratic form as derived from a symmetric bilinear function.

By the formulas of Theorem 7.1.4,

$$\tilde{\mathbf{A}}(\mathbf{x}) = [\mathbf{x}]^T[\mathbf{A}][\mathbf{x}] = \sum_{i,j=1}^{n} \alpha_{ij}x_ix_j$$

where $x_1,\dots,x_n$ are the coordinates of $\mathbf{x}$ in the given basis. This expression is a (homogeneous) polynomial of second degree in the variables x_i thus justifying the name quadratic form.

Consider now a diagonal matrix of the bilinear function with entries ± 1 and 0. Then the quadratic form will become a *signed sum of squares* by Theorem 7.2.5: $\tilde{\mathbf{A}}(\mathbf{x}) = \sum_{i=1}^{n} \gamma_i x_i^2$ where $\gamma_i = \pm 1$ or 0. As an important special case, the quadratic form of an inner product is the sum of squares of the coordinates.

We show how to obtain this sum of signed squares on the example of the symmetric bilinear function

$$(*) \qquad \mathbf{A}(\mathbf{u},\mathbf{v}) = 4u_1v_1 + 2u_1v_2 + 2u_2v_1 + 2u_1v_3 + 2u_3v_1 + 2u_2v_3 + 2u_3v_2$$

studied to illustrate the various proofs of Theorem 7.2.3. The corresponding quadratic form is

$$\tilde{\mathbf{A}}(\mathbf{x}) = 4x_1^2 + 4x_1x_2 + 4x_1x_3 + 4x_2x_3.$$

Here x_1, x_2, x_3 are the coordinates of $\mathbf{x}$ in the original basis $\mathbf{b}_i$. Consider now a diagonal matrix with entries $\gamma_i = \pm 1$ or 0 in the main diagonal. If the coordinates of $\mathbf{x}$ in this basis are $\hat{x}_1, \hat{x}_2, \hat{x}_3$, then $\tilde{\mathbf{A}}(\mathbf{x}) = \gamma_1 \hat{x}_1^2 + \gamma_2 \hat{x}_2^2 + \gamma_3 \hat{x}_3^2$. So we have to derive these new coordinates $\hat{x}_1, \hat{x}_2, \hat{x}_3$ from the original coordinates x_1, x_2, x_3. There exist relatively simple general methods but we confine ourselves to the concrete case.

The most comfortable process is to follow the third proof. Here we perform simultaneous and symmetric row/column equivalent transformations, and keep track of the changes in the basis. At the same time, we can also monitor the changes in the coordinates. So we get the desired coordinates $\hat{x}_1, \hat{x}_2, \hat{x}_3$ at the end of the procedure automatically.

Let us see the concrete case. In the first step, we transformed the basis $\mathbf{b}_1, \mathbf{b}_2, \mathbf{b}_3$ into the basis $\mathbf{b}_1, \mathbf{b}_2 - (1/2)\mathbf{b}_1, \mathbf{b}_3$. This means that only x_1 has to be modified with the help of x_2 in a reverse manner compared to the modification of $\mathbf{b}_2$ by $\mathbf{b}_1$. So the new coordinates are $x_1 + (1/2)x_2$, x_2, x_3 since

$$x_1\mathbf{b}_1 + x_2\mathbf{b}_2 + x_3\mathbf{b}_3 = [x_1 + (1/2)x_2]\mathbf{b}_1 + x_2[\mathbf{b}_2 - (1/2)\mathbf{b}_1] + x_3\mathbf{b}_3.$$

Similarly, the next two transformations modify the coordinates the following way: $x_1 + (1/2)x_2 + (1/2)x_3$, x_2, x_3 and then $x_1 + (1/2)x_2 + (1/2)x_3$, $x_2 - x_3$, x_3. Finally we have to normalize the diagonal matrix

$$\begin{pmatrix} 4 & 0 & 0 \\ 0 & -1 & 0 \\ 0 & 0 & 0 \end{pmatrix}:$$

halving the first row and column, we get

$$\begin{pmatrix} 1 & 0 & 0 \\ 0 & -1 & 0 \\ 0 & 0 & 0 \end{pmatrix}.$$

This means that the (actual) first basis element is multiplied by 1/2, thus the first coordinate is doubled yielding $2x_1 + x_2 + x_3$, $x_2 - x_3$, x_3. These are the desired coordinates $\hat{x}_i$ as the matrix is diagonal with entries $1, -1, 0$ in the main diagonal. Thus, we get the signed sum of squares as

$$\tilde{\mathbf{A}}(\mathbf{x}) = \hat{x}_1^2 - \hat{x}_2^2 = (2x_1 + x_2 + x_3)^2 - (x_2 - x_3)^2.$$

If we do not want to follow the changes in the coordinates step by step, then we can consider immediately the equality $\mathbf{x} = \sum x_i \mathbf{b}_i = \sum \hat{x}_i \hat{\mathbf{b}}_i$. We express the new basis vectors $\hat{\mathbf{b}}_i$ with the help of the original basis vectors $\mathbf{b}_i$ and then compare the coefficients of each $\mathbf{b}_i$ on the two sides. This means a system of linear equations for the coordinates $\hat{x}_i$ and can be solved (e.g., by Gaussian elimination).

The representation of a quadratic form as a signed sum of squares can have various realizations as there can be many suitable $\mathbf{A}$-orthogonal bases. But the number of terms with positive, negative, and zero signs must always be the same by the law of inertia (Theorem 7.2.6).

We underline that the signed sums of squares cannot be arbitrary (cf. Exercise 7.3.8), they must be obtained from a suitable **A**-orthogonal basis; this means the linear independence of the terms in the sum (in a precisely definable sense).

Now we survey the various types of quadratic forms. $\tilde{\mathbf{A}}(\mathbf{0}) = 0$ for every quadratic form. We classify the non-identically zero quadratic forms (and so the symmetric bilinear functions) depending on their values assumed at the non-zero vectors.

Definition 7.3.2. The quadratic form $\tilde{\mathbf{A}}$ of the symmetric bilinear function $\mathbf{A} \neq \mathbf{0}$ is

(PD) *positive definite* if $\tilde{\mathbf{A}}(\mathbf{x}) > 0$ for every $\mathbf{x} \neq \mathbf{0}$;

(ND) *negative definite* if $\tilde{\mathbf{A}}(\mathbf{x}) < 0$ for every $\mathbf{x} \neq \mathbf{0}$;

(PS) *positive semidefinite* if $\tilde{\mathbf{A}}(\mathbf{x}) \geq 0$ for every $\mathbf{x}$ and $\tilde{\mathbf{A}}(\mathbf{x}) = 0$ for some $\mathbf{x} \neq \mathbf{0}$;

(NS) *negative semidefinite* if $\tilde{\mathbf{A}}(\mathbf{x}) \leq 0$ for every $\mathbf{x}$ and $\tilde{\mathbf{A}}(\mathbf{x}) = 0$ for some $\mathbf{x} \neq \mathbf{0}$;

(I) *indefinite* if $\tilde{\mathbf{A}}(\mathbf{x})$ assumes both positive and negative values. ♣

The type of a quadratic form can be easily determined from the diagonal matrix of the bilinear function.

Theorem 7.3.3. *Consider a diagonal matrix of the symmetric bilinear function* **A**. *Then* **A** *(or* $\tilde{\mathbf{A}}$*) is*

- *identically zero if every entry in the main diagonal is* 0;
- *positive definite if every entry in the main diagonal is positive;*
- *negative definite if every entry in the main diagonal is negative;*
- *positive semidefinite if there are both positive and zero entries in the main diagonal, but no negative entries occur;*
- *negative semidefinite if there are both negative and zero entries in the main diagonal, but no positive entries occur;*
- *indefinite if there are both positive and negative elements in the main diagonal.* ♣

Proof. Each statement follows immediately from the representation of the quadratic form as a sum of signed squares. ∎

Note that some indefinite forms do, some others do not, contain a 0 entry in the main diagonal of their diagonal matrix.

We often have to determine the type of a quadratic form from its *arbitrary* matrix (without diagonalization). We have a really good criterion only for definite forms; we state it without proof.

Theorem 7.3.4. *Let* $A \in \mathbf{R}^{n \times n}$ *be any matrix of the symmetric bilinear function* **A**.

A *is positive definite if and only if the* $k \times k$ *minors in the upper left corner of* A *are positive for every* $k \leq n$.

A *is negative definite if and only if the* $k \times k$ *minors in the upper left corner of* A *are positive if* k *is even, and are negative if* k *is odd.* ♣

As mentioned in the introduction of the chapter, quadratic forms arise naturally in geometry at the characterization of quadratic curves and surfaces but also have many other applications in various fields of mathematics.

Finally we note that positive definite forms play a key role in the next chapter.

Exercises 7.3

7.3.1. (a) Prove that the (not necessarily symmetric) bilinear functions $\mathbf{A}$ and $\mathbf{B}$ determine the same quadratic form if and only if $\mathbf{A}-\mathbf{B}$ is antisymmetric (see the definition in Exercise 7.2.1).

(b) Verify that there is a bijection between symmetric bilinear functions and quadratic forms.

7.3.2. Determine the types of the symmetric bilinear functions (or of the induced quadratic forms) in Exercises 7.2.5–7.2.7.

7.3.3. What are the ranges of the quadratic forms of various types? What happens if we only consider the values assumed at non-zero vectors?

7.3.4. Let $\tilde{\mathbf{A}}$ be the quadratic form of the symmetric bilinear function $\mathbf{A}$.

(a) How can we express $\tilde{\mathbf{A}}(\lambda\mathbf{x})$ from $\tilde{\mathbf{A}}(\mathbf{x})$?

(b) Find a necessary and sufficient condition for vectors $\mathbf{x}$ and $\mathbf{z}$ to satisfy $\tilde{\mathbf{A}}(\mathbf{x}+\mathbf{z}) = \tilde{\mathbf{A}}(\mathbf{x}) + \tilde{\mathbf{A}}(\mathbf{z})$.

7.3.5. (a) Determine the type of the symmetric bilinear function $\lambda\mathbf{A}$ depending on λ and the type of $\mathbf{A}$.

(b) What are the possible types of $\mathbf{A}+\mathbf{B}$ depending on the types of $\mathbf{A}$ and $\mathbf{B}$?

7.3.6. Represent the three-dimensional quadratic forms as signed sums of squares.

(a) x_1x_2;

(b) $x_1x_2 + x_2x_3$;

(c) $x_1x_2 + x_2x_3 + x_3x_1$;

(d) $x_1^2 - 3x_3^2 - 2x_1x_2 + 2x_1x_3 - 6x_2x_3$;

(e) $x_1^2 + x_2^2 + 3x_3^2 + 4x_1x_2 + 2x_1x_3 + 2x_2x_3$.

7.3.7. Represent the four-dimensional quadratic forms as signed sums of squares.

(a) $\sum_{i,j=1}^4 x_ix_j$;

(b) $x_1x_2 + x_2x_3 + x_3x_4$;

(c) $x_1x_2 + x_2x_3 + x_3x_4 + x_4x_1$;

(d) $\sum_{i<j} x_ix_j$.

7.3.8. The quadratic form $x_1^2 + x_2^2 + (x_1+x_2)^2 = (x_1\sqrt{3})^2 + (x_2\sqrt{3})^2 - (x_1-x_2)^2$ consists of signed sums on both sides All three coefficients are positive on the left-hand side, but there is a negative coefficient on the right-hand side. This seems to contradict the law of inertia. What is wrong?

7.3.9. Show that switching bases, the determinant of a symmetric bilinear function generally changes, but its sign (positive, negative, or zero) does not depend on the choice of the basis.

7.3.10. True or false?

(a) If $\mathbf{A}$ is positive or negative definite, then $\det[\mathbf{A}] \neq 0$.

(b) If $\det[\mathbf{A}] \neq 0$, then $\mathbf{A}$ is positive or negative definite.

7.3.11. Determine the number of pairwise non-equivalent symmetric bilinear functions of various types on an n-dimensional vector space. (The notion of equivalence was defined in Exercise 7.2.10.)

7.3.12. Which symmetric bilinear functions $\mathbf{A}$ have the property that if non-zero vectors are pairwise $\mathbf{A}$-orthogonal, then they are necessarily linearly independent?

***7.3.13.** For which symmetric bilinear functions $\mathbf{A}$ can any vector $\mathbf{v} \neq \mathbf{0}$ be extended to an $\mathbf{A}$-orthogonal basis?

***7.3.14.** We define the *kernel* of a quadratic form $\tilde{\mathbf{A}}$ as the set of its roots:

$$\operatorname{Ker} \tilde{\mathbf{A}} = \{\mathbf{v} \in V \mid \tilde{\mathbf{A}}(\mathbf{v}) = 0\}.$$

(a) For which quadratic forms is the kernel a subspace in V?

(b) For which quadratic forms does the kernel contain a basis of V?

(c) What is the maximal number of linearly independent vectors in the kernel?

(d) What is the maximal dimension of a subspace contained in the kernel?

7.4. Complex Bilinear Function

In this section V denotes a finite-dimensional vector space over the complex field. In defining the bilinear functions on V we want to preserve the properties of orthogonalization and quadratic forms for the complex case.

If keeping the original Definition 7.1.1, $\mathbf{A}(i\mathbf{x}, i\mathbf{x}) = i^2\mathbf{A}(\mathbf{x},\mathbf{x}) = -\mathbf{A}(\mathbf{x},\mathbf{x})$ would imply the non-existence of definite or semidefinite quadratic forms. For this and other similar reasons we modify the definition of a bilinear function so that multiplying the *first* variable by λ, the value of the function gets multiplied by the *conjugate* of λ. All other requirements for the sums and multiplying the second variable by a scalar remain unchanged.

Definition 7.4.1. A map $\mathbf{A}: V \times V \to \mathbf{C}$ is a (complex) *bilinear function* if (using the numbering of formulas in Definition 7.1.1)

(ii) $\mathbf{A}(\mathbf{u}+\mathbf{u}', \mathbf{v}) = \mathbf{A}(\mathbf{u},\mathbf{v}) + \mathbf{A}(\mathbf{u}',\mathbf{v})$;

!(iii)! $\mathbf{A}(\lambda\mathbf{u}, \mathbf{v}) = \overline{\lambda}\mathbf{A}(\mathbf{u},\mathbf{v})$ ←**ATTENTION!** We have the *conjugate* of lambda on the right-hand side;

(iv) $\mathbf{A}(\mathbf{u}, \mathbf{v}+\mathbf{v}') = \mathbf{A}(\mathbf{u},\mathbf{v}) + \mathbf{A}(\mathbf{u},\mathbf{v}')$;

(v) $\mathbf{A}(\mathbf{u}, \lambda\mathbf{v}) = \lambda\mathbf{A}(\mathbf{u},\mathbf{v})$. ♣

Similar to the real case, a complex bilinear function can be characterized by its values on the basis vectors (Theorem 7.1.2, but there is a slight change in the formula, see below) and we define the matrix of a bilinear function the same way (Definition 7.1.3). In the representations (7.1.1) and (7.1.2) of Theorem 7.1.4 we have to replace the coordinates of the *first* vector by their complex conjugates.

Theorem 7.4.2. *Let α_{ij} be the elements of the matrix of a complex bilinear function* **A** *and let $u_1, \dots, u_n$ and $v_1, \dots, v_n$ be the coordinates of the vectors* **u** *and* **v**. *Then*

$$\mathbf{A}(\mathbf{u},\mathbf{v}) = \sum_{i,j=1}^{n} \alpha_{ij}\overline{u_i}v_j = [\mathbf{u}]^*[\mathbf{A}][\mathbf{v}] = (\overline{u_1}, \overline{u_2}, \dots, \overline{u_n}) \begin{pmatrix} \alpha_{11} & \alpha_{12} & \dots & \alpha_{1n} \\ \alpha_{21} & \alpha_{22} & \dots & \alpha_{2n} \\ \vdots & \vdots & \ddots & \vdots \\ \alpha_{n1} & \alpha_{n2} & \dots & \alpha_{nn} \end{pmatrix} \begin{pmatrix} v_1 \\ v_2 \\ \vdots \\ v_n \end{pmatrix}$$

where * *means the adjoint (i.e., the conjugate of the transpose) of the matrix (see Definition* 2.1.6). ♣

The proof is the same as in the real case. (We only present proofs if a significant difference arises compared to real bilinear functions.)

Before turning to orthogonalization, we consider the quadratic forms. Their Definition 7.3.1 and notation $\tilde{\mathbf{A}}$ remain the same. We show first that, in contrast to the real case, the quadratic form determines the bilinear function uniquely. This means that there is a bijection between the quadratic forms and *all* bilinear functions in the complex case.

Theorem 7.4.3. *Every quadratic form originates from exactly one bilinear function.* ♣

Proof. We express $\mathbf{A}(\mathbf{u},\mathbf{v})$ with values $\tilde{\mathbf{A}}(\mathbf{x}) = \mathbf{A}(\mathbf{x},\mathbf{x})$. Expanding the expressions $\tilde{\mathbf{A}}(\mathbf{u}+\mathbf{v})$ and $\tilde{\mathbf{A}}(\mathbf{u}+i\mathbf{v})$ we get

$$\mathbf{A}(\mathbf{u}+\mathbf{v},\mathbf{u}+\mathbf{v}) = \mathbf{A}(\mathbf{u},\mathbf{u}) + \mathbf{A}(\mathbf{u},\mathbf{v}) + \mathbf{A}(\mathbf{v},\mathbf{u}) + \mathbf{A}(\mathbf{v},\mathbf{v}),$$
$$\mathbf{A}(\mathbf{u}+i\mathbf{v},\mathbf{u}+i\mathbf{v}) = \mathbf{A}(\mathbf{u},\mathbf{u}) + i\mathbf{A}(\mathbf{u},\mathbf{v}) - i\mathbf{A}(\mathbf{v},\mathbf{u}) + \mathbf{A}(\mathbf{v},\mathbf{v}).$$

Multiplying the first equality by i and adding it to the second one, $\mathbf{A}(\mathbf{v},\mathbf{u})$ gets canceled and $\mathbf{A}(\mathbf{u},\mathbf{v})$ can (uniquely) be expressed with the values of the quadratic form at $\mathbf{u}, \mathbf{v}, \mathbf{u}+\mathbf{v}$, and $\mathbf{u}+i\mathbf{v}$. ■

A quadratic form derived from a complex bilinear function assumes complex values in general. To define the definite, semidefinite, and indefinite characters, we need a quadratic form assuming only real values. A simple necessary and sufficient condition is the following theorem.

Theorem 7.4.4. *A quadratic form* $\tilde{\mathbf{A}}$ *belonging to a complex bilinear function* **A** *assumes exclusively real numbers if and only if* $\mathbf{A}(\mathbf{u},\mathbf{v}) = \overline{\mathbf{A}(\mathbf{v},\mathbf{u})}$ *for every vector* **u**, **v**. ♣

Bilinear functions **A** with this property are called *Hermitian* (or *self-adjoint*; see explanation below at the matrix of such bilinear functions).

Proof. If $\mathbf{A}(\mathbf{u},\mathbf{v}) = \overline{\mathbf{A}(\mathbf{v},\mathbf{u})}$, then applying this to $\mathbf{u}=\mathbf{v}=\mathbf{x}$ we obtain $\mathbf{A}(\mathbf{x},\mathbf{x}) = \overline{\mathbf{A}(\mathbf{x},\mathbf{x})}$. Thus every $\tilde{\mathbf{A}}(\mathbf{x})$ is equal to its conjugate, so it is a real number.

Conversely, assume that all values of the quadratic form are real numbers. The two equalities in the proof of Theorem 7.4.3 yield that both $\mathbf{A}(\mathbf{u},\mathbf{v}) + \mathbf{A}(\mathbf{v},\mathbf{u})$ and $i[\mathbf{A}(\mathbf{u},\mathbf{v}) - \mathbf{A}(\mathbf{v},\mathbf{u})]$ are real. This implies that $\mathbf{A}(\mathbf{u},\mathbf{v})$ and $\mathbf{A}(\mathbf{v},\mathbf{u})$ are conjugates of each other. ■

In the sequel we only consider Hermitian bilinear functions. We can adapt the notions and theorems of the real case mutatis mutandis as summarized below.

A bilinear function is Hermitian if and only if its matrix (in an arbitrary basis) is *self-adjoint*, i.e., it equals its conjugate. (This is the complex analogue of Theorem 7.2.2.) Due to self-adjointness, every entry in the main diagonal of the matrix is a real number.

Every Hermitian bilinear function has a diagonal matrix. (This corresponds to Theorem 7.2.3. Each of the three proofs can be easily adapted to the complex case, see Exercise 7.4.5.) This diagonal matrix is self-adjoint, so all its entries are real numbers.

Let A be an Hermitian bilinear function. Similar to Definition 7.2.4, the vectors $\mathbf{u}, \mathbf{v} \in V$ are $\mathbf{A}$-*orthogonal* if $\mathbf{A}(\mathbf{u}, \mathbf{v}) = 0$.

The analogue of Theorem 7.2.5 states that every Hermitian bilinear function has a diagonal matrix where the elements of the main diagonal are 1, -1, or 0. In a basis belonging to such a matrix, the bilinear function has a simple form $\mathbf{A}(\mathbf{u}, \mathbf{v}) = \sum_{j=1}^{n} \gamma_j \overline{u_j} v_j$ where $\gamma_j = \pm 1$ or 0 and $u_1, \dots, u_n$ and $v_1, \dots, v_n$ denote the coordinates of the vectors $\mathbf{u}$ and $\mathbf{v}$. Accordingly, the signed sum representation of the quadratic form is

$$\tilde{\mathbf{A}}(\mathbf{x}) = \sum_{j=1}^{n} \gamma_j \overline{x_j} x_j = \sum_{j=1}^{n} \gamma_j |x_j|^2.$$

The numbers of positive, negative, and zero entries in (the main diagonal of) a diagonal matrix are uniquely determined. (Law of inertia, Theorem 7.2.6. As mentioned before, all entries are real numbers in a diagonal matrix.)

Finally, the definition and properties of positive/negative definite/semidefinite or indefinite characters of a quadratic form conform to the real case. (These are the analogues of Definition 7.3.2 and Theorems 7.3.3–7.3.4; "symmetric" is replaced by "Hermitian", and $\mathbf{R}$ is replaced by $\mathbf{C}$ in the last theorem.)

Exercises 7.4

The exercises refer to complex bilinear functions in a finite-dimensional vector space over $\mathbf{C}$.

7.4.1. How should we modify the definition of the inner product for the complex case?

7.4.2. Which complex bilinear functions are symmetric, i.e., $\mathbf{A}(\mathbf{u}, \mathbf{v}) = \mathbf{A}(\mathbf{v}, \mathbf{u})$ for every $\mathbf{u}, \mathbf{v}$?

7.4.3. True or false? (A is a square matrix with complex entries.)
   (a) If A is self-adjoint, then $\det A \in \mathbf{R}$.
   (b) If $\det A \in \mathbf{R}$, then A is self-adjoint.

7.4.4. Solve Exercise 7.1.6 for complex bilinear functions.

7.4.5. Adapt each of the three proofs of Theorem 7.2.3 to the complex case.

7.4.6. The matrices below define Hermitian bilinear functions. In each case, find a diagonal matrix and an **A**-orthogonal basis, determine the character of the quadratic form, and write the quadratic form as a signed sum of squares.

(a) $\begin{pmatrix} 1 & i \\ -i & 1 \end{pmatrix}$; (b) $\begin{pmatrix} 0 & i \\ -i & 0 \end{pmatrix}$; (c) $\begin{pmatrix} 1 & \rho & \rho^2 \\ \rho^2 & 1 & \rho \\ \rho & \rho^2 & 1 \end{pmatrix}$

where $\rho = \cos 120° + i \sin 120°$.

7.4.7. (a) What can be the range of an Hermitian quadratic form?

(b) Show examples demonstrating that there is a much greater variety of general bilinear functions: the range in the complex plane of a quadratic form can be any line containing the origin; a half-plane bordered by such a line; a half-line starting from the origin; the entire plane; or any angle with its vertex in the origin.

7.4.8. We call a bilinear function $\mathbf{S}$ *skew-Hermitian* if $\mathbf{S}(\mathbf{u},\mathbf{v}) = -\overline{\mathbf{S}(\mathbf{v},\mathbf{u})}$ for every $\mathbf{u}, \mathbf{v}$.

(a) How can we recognize a skew-Hermitian function from its matrix (in any basis)?

(b) Prove that the skew-Hermitian functions are exactly the Hermitian functions multiplied by i.

(c) How can we recognize a skew-Hermitian function from its quadratic form?

(d) Show that every complex bilinear function has a unique representation as the sum of an Hermitian and a skew-Hermitian function.

7.4.9. True or false?

(a) If $\mathbf{A}$ has a diagonal matrix, then $\mathbf{A}$ is Hermitian.

(b) The $\mathbf{A}$-orthogonality of two vectors is a symmetric relation for every $\mathbf{A}$, i.e., $\mathbf{A}(\mathbf{u},\mathbf{v}) = 0 \iff \mathbf{A}(\mathbf{v},\mathbf{u}) = 0$.

(c) If $\mathbf{A}$ is Hermitian, then $\mathbf{A}(\mathbf{u},\mathbf{v}) = 0 \iff \mathbf{A}(\mathbf{v},\mathbf{u}) = 0$.

(d) If $\mathbf{A}(\mathbf{u},\mathbf{v}) = 0 \iff \mathbf{A}(\mathbf{v},\mathbf{u}) = 0$, then $\mathbf{A}$ is Hermitian.

*7.4.10. Prove that $\mathbf{A}(\mathbf{u},\mathbf{v}) = 0 \iff \mathbf{A}(\mathbf{v},\mathbf{u}) = 0$ holds exactly for the scalar multiples of Hermitian functions.

Chapter 8

Euclidean Spaces

The Euclidean space is the most direct generalization of the usual plane and space: it means a vector space with an inner product. Besides orthogonality, the inner product makes possible to define length, distance, and — in the real case — angles preserving most properties from geometry. The generalization of the usual orthogonal unit vectors is the orthonormal basis. The adjoint transformation plays an important role in studying the linear transformations of a Euclidean space. We discuss some types of transformations having a special relation with their adjoints. We determine which transformations have an orthonormal eigenbasis, i.e., which transformations have a diagonal matrix belonging to orthogonal unit vectors. The answers differ in the real and complex cases.

8.1. Real Euclidean Space

This section deals with finite-dimensional vector spaces over **R**, though some results can be generalized to infinite dimension, too (see Exercises 8.1.15–8.1.17).

The starting point is the inner product from geometry assigning to two plane or space vectors the sum of the products of their Descartes coordinates. This is generalized in the following definition.

Definition 8.1.1. Let $\mathbf{e}_1, \dots, \mathbf{e}_n$ be a *fixed* basis in V. The *inner product* (or *dot product*) related to this basis is the function $\mathbf{D}: V \times V \to \mathbf{R}$ assigning to two vectors the sum of the products of their coordinates:

$$\mathbf{D}(\mathbf{x}, \mathbf{z}) = \mathbf{x} \cdot \mathbf{z} = \begin{pmatrix} x_1 \\ \vdots \\ x_n \end{pmatrix} \cdot \begin{pmatrix} z_1 \\ \vdots \\ z_n \end{pmatrix} = \sum_{j=1}^{n} x_j z_j. \clubsuit$$

We shall denote the inner product with a dot between the two vectors.

The inner product was introduced already in Examples E1 and E2 of Section 7.1.

Sometimes the name scalar product occurs as a synonym for the inner product. We shall not use it to avoid confusion with the multiplication by a scalar. Inner product assigns a scalar to two vectors, whereas multiplication by a scalar assigns a vector (matrix, map, etc.) to a scalar and a vector (matrix, map, etc.).

The inner product is a symmetric bilinear function with a positive definite quadratic form:

$$\mathbf{x}\cdot\mathbf{z}=\mathbf{z}\cdot\mathbf{x};\quad (\mathbf{x}+\mathbf{x}')\cdot\mathbf{z}=\mathbf{x}\cdot\mathbf{z}+\mathbf{x}'\cdot\mathbf{z};\quad (\lambda\mathbf{x})\cdot\mathbf{z}=\lambda(\mathbf{x}\cdot\mathbf{z});\quad \mathbf{x}\neq\mathbf{0}\Rightarrow\mathbf{x}\cdot\mathbf{x}>0.$$

The properties concerning the sum and scalar multiple are also valid for the second variable. We omitted listing them separately due to symmetry.

The orthogonalization Theorem 7.2.3 (combined with Theorem 7.2.5) implies that the converse is also true: every positive definite symmetric bilinear function becomes an inner product in a suitable basis. Indeed, it has a diagonal matrix where every entry is ±1 or 0. But 0 and -1 cannot occur due to positive definiteness. So, the function is $[\mathbf{x}]^T I[\mathbf{z}]=\sum_{j=1}^n x_j z_j$, i.e., the inner product $\mathbf{x}\cdot\mathbf{z}$.

This equivalence makes it possible to define the inner product without referring to a basis. It is very useful, e.g., when extending the inner product to infinite dimension, but even in the finite-dimensional case it is often more convenient than the form using the coordinates.

Due to its importance, we formulate it as a theorem.

Theorem 8.1.2. *The inner product can also be defined as a positive definite bilinear function.* ♣

We are ready to define Euclidean spaces.

Definition 8.1.3. A *Euclidean space* is a vector space with an inner product. ♣

So, we obtain a Euclidean space if we fix an inner product in a (real, finite-dimensional) vector space. Then the dot $(\cdot)$ refers to this fixed inner product. By Theorem 8.1.2, there are two ways to select an inner product: we either fix a basis, or indicate a positive definite symmetric bilinear function.

A basis determines the inner product uniquely by assigning to two vectors the sum of the products of their coordinates. The converse is false; different bases can create the same inner product. By the argument above this happens exactly if the matrix of the (positive definite symmetric bilinear) function is the identity matrix in the given basis. Such a basis consists of orthogonal unit vectors: the inner product of a basis vector with itself is 1 (it is the corresponding element of the main diagonal), and the inner product of two distinct basis vectors is 0. We introduce a special name for such vector systems in a Euclidean space.

Definition 8.1.4. The vectors $\mathbf{e}_1,\dots,\mathbf{e}_n$ form an *orthonormal system* if $\mathbf{e}_i\cdot\mathbf{e}_j=0$ for $i\neq j$ and 1 for $i=j$.

If the vectors $\mathbf{e}_i$ also form a basis, then we speak about an *orthonormal basis.* ♣

If the inner product in a Euclidean space was specified by a basis, then this is an orthonormal basis, but there are many other orthonormal bases, too. Moreover, every orthonormal system can be extended to an orthonormal basis by the Gram–Schmidt orthogonalization (see the first proof of Theorem 7.2.3).

Any subspace of a Euclidean space is a Euclidean space with (the restriction of) the same inner product, even if the basis originally defining the inner product has no element in this subspace. This follows from Theorem 8.1.2: also the restriction is a positive definite symmetric bilinear function, so it defines an inner product.

Next, we introduce two fundamental notions: orthogonality and orthogonal complement.

Definition 8.1.5. The vectors $\mathbf{a}$ and $\mathbf{b}$ in a Euclidean space are *orthogonal* (or *perpendicular*) if their inner product is zero: $\mathbf{a} \cdot \mathbf{b} = 0$.

We denote this relation by $\mathbf{a} \perp \mathbf{b}$. ♣

Note that orthogonality depends heavily on the selection of the inner product. If we define two different inner products on the same vector space (yielding two different Euclidean spaces, of course), then (generally) there will arise other pairs of orthogonal vectors in the two spaces.

Definition 8.1.6. Let H be a subset in a Euclidean space V. Its *orthogonal complement* $H^\perp$ is the set of vectors orthogonal to every element in H, i.e.,

$$H^\perp = \{\mathbf{x} \in V \mid (\mathbf{h} \in H \Rightarrow \mathbf{h} \cdot \mathbf{x} = 0)\}. \clubsuit$$

A simple calculation shows that $H^\perp$ is a subspace in V. If H itself is a subspace, then we have a much stronger result.

Theorem 8.1.7. *If U is a subspace, then every vector in the Euclidean space V has a unique representation as the sum of a vector in U and a vector in $U^\perp$.* ♣

In other words, V is the direct sum of the subspaces U and $U^\perp$ (see Theorem 4.3.6, Definition 4.3.7): $V = U \oplus U^\perp$. If V is the (usual three-dimensional) space (with the usual inner product) and U is, e.g., a plane (containing the origin), then this means the well-known geometric fact that every vector in the space is a unique sum of its perpendicular projection in U and a vector perpendicular to the plane U. Generalizing this to an arbitrary Euclidean space V and its subspace U, we say that $\mathbf{u} \in U$ is the *orthogonal projection* of $\mathbf{v} \in V$ if $\mathbf{v} = \mathbf{u} + \mathbf{u}^\perp$ where $\mathbf{u}^\perp \in U^\perp$.

First proof. Let $\mathbf{b}_1, \dots, \mathbf{b}_k$ be an orthonormal basis in U and extend it to an orthonormal basis of V with the vectors $\mathbf{b}_{k+1}, \dots, \mathbf{b}_n$. (As mentioned above, this can be performed, e.g., via a Gram–Schmidt orthogonalization.) Then every $\mathbf{v} \in V$ can be written as $\mathbf{v} = \sum_{j=1}^{n} \lambda_j \mathbf{b}_j$. Here the linear combination of the first k vectors is in U and that of the remaining $n - k$ vectors is in $U^\perp$. This provides a required representation of $\mathbf{v}$.

To show the uniqueness we have to verify $U \cap U^\perp = \mathbf{0}$ by Theorem 4.3.6. Assume $\mathbf{a} \in U \cap U^\perp$. Then $\mathbf{a} \perp \mathbf{a}$ also holds, i.e., $\mathbf{a} \cdot \mathbf{a} = 0$ implying $\mathbf{a} = \mathbf{0}$. ∎

We see from the proof that $\mathbf{b}_{k+1}, \ldots, \mathbf{b}_n$ is a basis in $U^\perp$. Based on this, we can even construct the actual representation of a vector, though this procedure is rather complicated. The next proof yields a much simpler and better implementable algorithm.

Second proof. Let $\mathbf{b}_1, \ldots, \mathbf{b}_k$ be an orthonormal basis in U, $\mathbf{v} \in V$, and we are looking for a representation $\mathbf{v} = \mathbf{u} + \mathbf{u}^\perp$ where $\mathbf{u} \in U$ and $\mathbf{u}^\perp \in U^\perp$. Substitute $\mathbf{u} = \sum_{j=1}^{k} \lambda_j \mathbf{b}_j$ and form the inner product of the resulting equality with the vectors $\mathbf{b}_m$ $(m = 1, 2, \ldots, k)$:

$$\mathbf{v} \cdot \mathbf{b}_m = \sum_{j=1}^{k} \lambda_j (\mathbf{b}_j \cdot \mathbf{b}_m) + \mathbf{u}^\perp \cdot \mathbf{b}_m. \tag{8.1.1}$$

On the right-hand side, the products $\mathbf{b}_j \cdot \mathbf{b}_m = 0$ for $j \neq m$. Also $\mathbf{u}^\perp \cdot \mathbf{b}_m = 0$ as $\mathbf{b}_m \in U$ and $\mathbf{u}^\perp \in U^\perp$. Finally, $\mathbf{b}_m \cdot \mathbf{b}_m = 1$. So (8.1.1) is equivalent to $\mathbf{v} \cdot \mathbf{b}_m = \lambda_m$. This determines the only possible value of the coefficient λ_m. Therefore the only candidate is $\mathbf{u} = \sum_{j=1}^{k} (\mathbf{v} \cdot \mathbf{b}_j) \mathbf{b}_j$.

We show that this $\mathbf{u}$ satisfies the requirements, i.e., $\mathbf{u}^\perp = \mathbf{v} - \sum_{j=1}^{k} (\mathbf{v} \cdot \mathbf{b}_j) \mathbf{b}_j \in U^\perp$. It is easy to see that a vector is orthogonal to every element in U if and only if it is orthogonal to an arbitrary basis in U. So, it is enough to verify that the inner product of the above $\mathbf{u}^\perp$ with every $\mathbf{b}_j$ is zero. This follows from a simple computation. ■

Exercises 8.1

8.1.1. Verify that the usual unit vectors

$$\begin{pmatrix}1\\0\\0\\0\end{pmatrix}, \begin{pmatrix}0\\1\\0\\0\end{pmatrix}, \begin{pmatrix}0\\0\\1\\0\end{pmatrix}, \begin{pmatrix}0\\0\\0\\1\end{pmatrix}$$

in $\mathbf{R}^4$ and the vectors

$$\begin{pmatrix}1/2\\1/2\\1/2\\1/2\end{pmatrix}, \begin{pmatrix}1/2\\-1/2\\1/2\\-1/2\end{pmatrix}, \begin{pmatrix}1/2\\-1/2\\-1/2\\1/2\end{pmatrix}, \begin{pmatrix}1/2\\1/2\\-1/2\\-1/2\end{pmatrix}$$

define the same inner product.

8.1.2. Prove that pairwise orthogonal non-zero vectors in a Euclidean space are necessarily linearly independent.

8.1.3. Show that there are infinitely many orthonormal bases in every Euclidean space of dimension greater than one.

8.1.4. Let V be the usual vector space of the polynomials with real coefficients of degree at most two. Verify that the functions define an inner product and construct an orthonormal basis in each case. (f and g denote arbitrary polynomials in V, and f' and f'' are the first two derivatives of f.)

(a) $\int_{-1}^{+1} f(t)g(t)dt$; (b) $f(1)g(1) + f'(1)g'(1) + f''(1)g''(1)$;

(c) $\sum_{|j|<5} f(j)g(j)$; (d) $\sum_{j=1}^{9} f(j)g(j)$.

8.1.5. Show that in a Euclidean space $\mathbf{b}$ is orthogonal to each of the vectors $\mathbf{a}_1, \ldots, \mathbf{a}_k$ if and only if $\mathbf{b} \in \langle \mathbf{a}_1, \ldots, \mathbf{a}_k \rangle^\perp$.

8.1.6. Let U be a subspace in a Euclidean space V. Prove $\dim U + \dim U^\perp = \dim V$.

8.1.7. Consider the Euclidean space $V = \mathbf{R}^5$ with the usual inner product. Let $v_1, \ldots, v_5$ be the components of a general vector $\mathbf{v} \in V$. Find the orthogonal complements of the subspaces.

(a) $W_1 = \{\mathbf{v} \mid v_1 = v_2 = 0\}$;

(b) $W_2 = \{\mathbf{v} \mid v_1 = v_2 = v_3 = v_4 = v_5\}$;

(c) $W_3 = \left\{\mathbf{v} \mid \sum_{j=1}^{5} v_j = 0,\ 2v_1 + v_2 = v_4 + 2v_5\right\}$.

8.1.8. Let U and W be two subspaces in a Euclidean space V satisfying $\dim U + \dim W \geq \dim V$ and $\mathbf{u} \cdot \mathbf{w} = 0$ for every $\mathbf{u} \in U$ and $\mathbf{w} \in W$. Prove $W = U^\perp$.

8.1.9. Let U_1 and U_2 be subspaces in a Euclidean space. Prove:

(a) $U_1 \subseteq U_2 \iff U_1^\perp \supseteq U_2^\perp$;

(b) $\langle U_1, U_2 \rangle^\perp = U_1^\perp \cap U_2^\perp$;

(c) $(U_1 \cap U_2)^\perp = \langle U_1^\perp, U_2^\perp \rangle$.

8.1.10. (a) Find infinitely many vectors in the usual Euclidean space $\mathbf{R}^2$ such that any two are linearly independent but no two are orthogonal.

*(b) Let V be an n-dimensional Euclidean space ($n > 0$). Find infinitely many vectors in V such that any n are linearly independent but no two are orthogonal.

*8.1.11 Let V be an n-dimensional Euclidean space and let $\mathbf{b}_1, \ldots, \mathbf{b}_n$ be an arbitrary basis. Prove that there exists exactly one basis $\mathbf{c}_1, \ldots, \mathbf{c}_n$ satisfying $\mathbf{b}_i \cdot \mathbf{c}_j = 0$ if $i \neq j$, and 1 if $i = j$.

8.1.12. Let the finite-dimensional vector space V be the direct sum of its subspaces U and W. Define an inner product on V so that $W = U^\perp$ in the resulting Euclidean space. How many inner products have this property?

8.1.13. Let V be an n-dimensional Euclidean space. The $V \to \mathbf{R}$ linear maps, i.e., the elements of $\mathrm{Hom}(V, \mathbf{R})$ are called *linear functionals* (this notion was already introduced in Exercise 7.1.9, too).

(a) Let $\mathbf{c} \in V$ be a fixed vector. Verify that $\Phi_\mathbf{c}(\mathbf{x}) = \mathbf{c} \cdot \mathbf{x}$ is a linear functional.

(b) Let Ψ be an arbitrary linear functional. Prove that there exists exactly one vector $\mathbf{c} \in V$ satisfying $\Psi(\mathbf{x}) = \mathbf{c} \cdot \mathbf{x}$.

Remark. The two parts of the exercise express that the linear functionals are obtained as the inner products of vectors in V with a fixed vector. The above map $\mathbf{c} \to \Phi_\mathbf{c}$ is an isomorphism between the vector spaces V and the *dual space* $\mathrm{Hom}(V, \mathbf{R})$ of the linear functionals.

8.1.14. Two Euclidean spaces are *isomorphic* if there is a bijection between them also preserving the inner product (besides addition and multiplication by a scalar). Prove that two Euclidean spaces are isomorphic if and only if they have the same dimension.

*8.1.15. We can define infinite-dimensional Euclidean spaces by specifying the inner product (without relying on a basis) as a positive definite symmetric bilinear function. Let V be the set of all infinite real sequences $\mathbf{a} = (\alpha_1, \alpha_2, \dots, \alpha_k, \dots)$ where the infinite series of the squares of the terms is convergent:

$$V = \{\mathbf{a} = (\alpha_1, \alpha_2, \dots, \alpha_k, \dots) \mid \sum_{j=1}^{\infty} \alpha_j^2 < \infty\}.$$

(a) Verify that V is a vector space under the usual addition and multiplication by a scalar of sequences.

(b) Show that $\mathbf{a} \cdot \mathbf{b} = (\alpha_1, \alpha_2, \dots, \alpha_k, \dots) \cdot (\beta_1, \beta_2, \dots, \beta_k, \dots) = \sum_{j=1}^{\infty} \alpha_j \beta_j$ defines an inner product on V.

(c) Let U be the subspace of sequences having at most finitely many non-zero elements. Determine $U^{\perp}$.

*8.1.16. Let U be a subspace in an infinite-dimensional Euclidean space V.

(a) Verify $U^{\perp} = V \iff U = \mathbf{0}$.

(b) Is $U^{\perp} = \mathbf{0} \iff U = V$ true?

(c) Prove $(U^{\perp})^{\perp} \supseteq U$ but equality does not hold in general.

(d) Show $((U^{\perp})^{\perp})^{\perp} = U^{\perp}$.

*8.1.17. Investigate the statements of Theorem 8.1.7 and Exercise 8.1.9 in an infinite-dimensional Euclidean space.

8.2. Length, Distance, and Angles

We rely on the inner product to develop the geometry of Euclidean spaces. We use the relations in the usual plane and space when extending the notions in the title to an arbitrary Euclidean space.

Definition 8.2.1. In a Euclidean space, the *length* (or *norm* or *magnitude* or *absolute value*) of a vector $\mathbf{x}$ is the square root of its inner product with itself, i.e., the square root of the sum of squares of its coordinates in an orthonormal basis. We denote the length of $\mathbf{x}$ by $||\mathbf{x}||$. So

$$||\mathbf{x}|| = \sqrt{\mathbf{x} \cdot \mathbf{x}} = \sqrt{\sum_{j=1}^{n} x_j^2}$$

where $x_1, \dots, x_n$ are the coordinates of $\mathbf{x}$ in an orthonormal basis. ♣

Theorem 8.2.2. *Properties of length are:*

(N1) $||\mathbf{x}|| \geq 0$ *and* $||\mathbf{x}|| = 0 \iff \mathbf{x} = \mathbf{0}$.

(N2) $||\lambda \mathbf{x}|| = |\lambda| \cdot ||\mathbf{x}||$.

(N3) $||\mathbf{x} + \mathbf{z}|| \leq ||\mathbf{x}|| + ||\mathbf{z}||$. ♣

Proof. (N1) and (N2) immediately follow from the positive definite and bilinear properties of the inner product. We shall prove the *triangle inequality* (N3) after Theorem 8.2.8. ∎

Definition 8.2.3. A real vector space V is a *normed (vector) space* if a *norm* $\|\cdot\| : V \to \mathbf{R}$ is defined on V satisfying properties (N1), (N2), and (N3). ♣

Theorem 8.2.2 states that every Euclidean space is a normed space. The converse is false, see Exercises 8.2.4–8.2.5.

Now we define distance with the help of length.

Definition 8.2.4. The *distance* of two vectors in a normed space is the length of their difference. The distance of vectors $\mathbf{x}$ and $\mathbf{z}$ is denoted by $\delta(\mathbf{x}, \mathbf{z})$. So, $\delta(\mathbf{x}, \mathbf{z}) = \|\mathbf{x}-\mathbf{z}\|$. ♣

Theorem 8.2.5. *Properties of distance are:*

(M1) $\delta(\mathbf{x}, \mathbf{z}) \geq 0$ *and* $\delta(\mathbf{x}, \mathbf{z}) = 0 \iff \mathbf{x} = \mathbf{z}$.

(M2) $\delta(\mathbf{x}, \mathbf{z}) = \delta(\mathbf{z}, \mathbf{x})$.

(M3) $\delta(\mathbf{x}, \mathbf{z}) \leq \delta(\mathbf{x}, \mathbf{w}) + \delta(\mathbf{w}, \mathbf{z})$. ♣

Proof. Each property (M) immediately follows from property (N) with the same serial number. We need (N2) only for $\lambda = -1$. ∎

Definition 8.2.6. A set H is a *metric space* if a *distance* (or *metric*) $\delta : H \times H \to \mathbf{R}$ satisfying properties (M1), (M2), and (M3) is defined on H. ♣

Theorem 8.2.5 states that every normed space (consequently every Euclidean space) is a metric space. The converse is false; see Exercises 8.2.6–8.2.7.

Finally, we define angles. In the usual plane or space, the inner product of two non-zero vectors is the product of their lengths multiplied by the cosine of the angle between them: $\mathbf{x} \cdot \mathbf{z} = \|\mathbf{x}\| \cdot \|\mathbf{z}\| \cdot \cos\varphi$. Expressing $\cos\varphi$, we get $\cos\varphi = (\mathbf{x} \cdot \mathbf{z})/(\|\mathbf{x}\| \cdot \|\mathbf{z}\|)$. (The factors $\|\mathbf{x}\|$ and $\|\mathbf{z}\|$ in the denominator are not zero since $\mathbf{x}$ and $\mathbf{z}$ are non-zero vectors.) So, the cosine of the angle can be expressed with the inner product. This can be generalized to define an angle in an arbitrary Euclidean space:

Definition 8.2.7. If $\mathbf{x}$ and $\mathbf{z}$ are non-zero vectors in a Euclidean space, then the *angle* between them is $0 \leq \varphi \leq \pi$ satisfying

$$\cos\varphi = \frac{\mathbf{x} \cdot \mathbf{z}}{\|\mathbf{x}\| \cdot \|\mathbf{z}\|} = \frac{\mathbf{x} \cdot \mathbf{z}}{\sqrt{\mathbf{x} \cdot \mathbf{x}}\sqrt{\mathbf{z} \cdot \mathbf{z}}}. \clubsuit$$

The definition only makes sense if the expression for $\cos\varphi$ is between -1 and $+1$. This follows from the following theorem.

Theorem 8.2.8 (Cauchy–Bunyakovsky–Schwarz inequality). *Any two vectors* $\mathbf{x}$ *and* $\mathbf{z}$ *in a Euclidean space satisfy*

$$|\mathbf{x} \cdot \mathbf{z}| \leq \|\mathbf{x}\| \cdot \|\mathbf{z}\|.$$

Equality holds if and only if $\mathbf{x}$ *and* $\mathbf{z}$ *are linearly dependent (i.e., one of them is a scalar multiple of the other).* ♣

If at least one of $\mathbf{x}$ and $\mathbf{z}$ is the zero vector, then both sides of the inequality are 0. So, it is sufficient to deal with the case when none of $\mathbf{x}$ and $\mathbf{z}$ is $\mathbf{0}$.

First proof. Consider the subspace spanned by $\mathbf{x}$ and $\mathbf{z}$. This is a Euclidean space of dimension at most 2, so it is isomorphic (as a Euclidean space, see Exercise 8.1.14) with the usual plane or its subspace. The inequality is valid in the plane (this was our starting point). Further, equality holds if and only if the vectors are parallel, i.e., linearly dependent. ■

Second proof. As both sides are non-negative, squaring leads to an equivalent inequality. Expressing both sides using the coordinates related to an orthonormal basis, we obtain

$$(x_1z_1 + \cdots + x_nz_n)^2 \leq (x_1^2 + \cdots + x_n^2)(z_1^2 + \cdots + z_n^2).$$

Performing the operations and ordering leads to

$$0 \leq \sum_{1\leq i<j\leq n} (x_iz_j - x_jz_i)^2.$$

The sum of squares on the right-hand side is non-negative (proving the inequality) and it is zero if and only if every term in it is zero. The latter means that the coordinates of $\mathbf{x}$ and $\mathbf{z}$ are proportional, so one vector is a scalar multiple of the other. ■

Note that the second proof verified an elementary inequality for real numbers using high-school methods. As a special case, we get the inequality of arithmetic and quadratic means (see Exercise 8.2.8).

Third proof. Let λ be an arbitrary scalar and consider the inner product

$$||\lambda\mathbf{x} + \mathbf{z}||^2 = (\lambda\mathbf{x} + \mathbf{z}) \cdot (\lambda\mathbf{x} + \mathbf{z}) = \lambda^2(\mathbf{x} \cdot \mathbf{x}) + 2\lambda(\mathbf{x} \cdot \mathbf{z}) + \mathbf{z} \cdot \mathbf{z}.$$

This is a polynomial of second degree of λ (as $\mathbf{x} \neq \mathbf{0}$) and is non-negative for every real number λ. Therefore its discriminant is non-positive:

$$(\mathbf{x} \cdot \mathbf{z})^2 - (\mathbf{x} \cdot \mathbf{x})(\mathbf{z} \cdot \mathbf{z}) \leq 0.$$

This is just the squared form of the inequality in question.

Equality holds if and only if the discriminant is zero. This happens exactly if the polynomial has a real root (in contrast to the case of negative discriminant). Therefore $||\lambda\mathbf{x} + \mathbf{z}|| = 0$ for some λ, i.e., $\lambda\mathbf{x} + \mathbf{z} = \mathbf{0}$, so $\mathbf{z} = -\lambda\mathbf{x}$. ■

The Cauchy–Bunyakovsky–Schwarz inequality (we abbreviate it as CBS) has many applications. Now we use it to prove the triangle inequality (N3) in Theorem 8.2.2.

Proof of the triangle inequality, (N3), Theorem 8.2.2. We have to verify $||\mathbf{x}+\mathbf{z}|| \leq ||\mathbf{x}|| + ||\mathbf{z}||$. Both sides are non-negative, so squaring leads to an equivalent inequality. The square of the left-hand side is $||\mathbf{x}||^2 + 2(\mathbf{x} \cdot \mathbf{z}) + ||\mathbf{z}||^2$ and the square of the right-hand side is $||\mathbf{x}||^2 + 2 \cdot ||\mathbf{x}|| \cdot ||\mathbf{z}|| + ||\mathbf{z}||^2$. They differ only in the middle terms where CBS guarantees the inequality required. ■

We obtained from the proof that (in harmony with our geometric knowledge) the triangle inequality holds with equality exactly if the two vectors have the same direction, i.e., one of them is a *non-negative* scalar multiple of the other.

Exercises 8.2

8.2.1. What is the distance of two elements in an orthonormal basis?

8.2.2. Compute the angle between $\mathbf{x}$ and $\mathbf{z}$ if $||\mathbf{x}|| = ||\mathbf{z}|| = ||\mathbf{x} - \mathbf{z}|| \neq 0$.

8.2.3. Prove the statements for an arbitrary Euclidean space. Which well-known geometric theorems are generalized?
(a) $\mathbf{x} \perp \mathbf{z} \iff ||\mathbf{x} + \mathbf{z}||^2 = ||\mathbf{x}||^2 + ||\mathbf{z}||^2$.
(b) $||\mathbf{x}|| = ||\mathbf{z}|| \iff \mathbf{x} + \mathbf{z} \perp \mathbf{x} - \mathbf{z}$.
(c) $||\mathbf{x} + \mathbf{z}||^2 + ||\mathbf{x} - \mathbf{z}||^2 = 2||\mathbf{x}||^2 + 2||\mathbf{z}||^2$.

8.2.4. Which $\mathbf{R}^n \to \mathbf{R}$ functions define a norm on $\mathbf{R}^n$? (x_j denotes the jth component of $\mathbf{x}$.)
(a) $\max_{j=1}^n x_j$; (b) $\max_{j=1}^n |x_j|$; (c) $|x_1|$;
(d) $\sum_{j=1}^n |x_j|$; **(e) $(\sum_{j=1}^n |x_j|^3)^{1/3}$.

8.2.5. (a) Give an example of a normed space that is not a Euclidean space, i.e., the norm does not originate from an inner product.
**(b) Prove that a normed space is a Euclidean space, i.e., there can be defined an inner product satisfying $||\mathbf{x}||^2 = \mathbf{x} \cdot \mathbf{x}$ if and only if $||\mathbf{x}+\mathbf{z}||^2 + ||\mathbf{x}-\mathbf{z}||^2 = 2||\mathbf{x}||^2 + 2||\mathbf{z}||^2$ holds for every $\mathbf{x}$ and $\mathbf{z}$.

8.2.6. Which functions define a distance on $\mathbf{R}^n$, i.e., in which cases do we obtain a metric space? (x_j and z_j denote the components of $\mathbf{x}$ and $\mathbf{z}$.)
(a) $|x_1 - z_1|$;
(b) $\sum_{j=1}^n |x_j - z_j|$;
(c) The number of components where $\mathbf{x}$ and $\mathbf{z}$ differ, i.e., $x_j \neq z_j$.

8.2.7. Give an example of a metric vector space that is not a normed space, i.e., the metric does not originate from a norm. This means that one cannot define a norm satisfying $\delta(\mathbf{x}, \mathbf{z}) = ||\mathbf{x} - \mathbf{z}||$.

8.2.8. How does CBS imply the inequality of arithmetic and quadratic means

$$\sqrt{\frac{x_1^2 + \cdots + x_n^2}{n}} \geq \frac{x_1 + \cdots + x_n}{n}?$$

When do we have equality?

8.2.9. True or false? (k is an arbitrary positive integer and $\mathbf{c}_i$ are vectors in a Euclidean space.)
(a) If $\mathbf{c}_1, \ldots, \mathbf{c}_k$ are pairwise orthogonal, then $||\sum_{j=1}^k \mathbf{c}_j||^2 = \sum_{j=1}^k ||\mathbf{c}_j||^2$.
(b) If $||\sum_{j=1}^k \mathbf{c}_j||^2 = \sum_{j=1}^k ||\mathbf{c}_j||^2$, then $\mathbf{c}_1, \ldots, \mathbf{c}_k$ are pairwise orthogonal.

8.2.10. Compute the angle between the vectors in the usual Euclidean space $\mathbf{R}^4$.

(a) $\begin{pmatrix}1\\1\\1\\1\end{pmatrix}$ and $\begin{pmatrix}1\\1\\1\\0\end{pmatrix}$;

(b) $\begin{pmatrix}1\\1\\1\\1\end{pmatrix}$ and $\begin{pmatrix}1\\-1\\-1\\-1\end{pmatrix}$;

(c) $\begin{pmatrix}1\\\sqrt{2}\\1\\0\end{pmatrix}$ and $\begin{pmatrix}0\\1\\\sqrt{2}\\1\end{pmatrix}$.

8.2.11. Consider a unit cube in the usual Euclidean space $\mathbf{R}^4$.
(a) Determine the number of vertices, edges, and body diagonals.
(b) How long are the body diagonals?
(c) What is the angle between a body diagonal and an edge?
(d) What is the angle between two body diagonals?
(e) What are the radii of the (four-dimensional) circumsphere and insphere of the cube?

8.2.12. Define the distance of a subspace U and a vector $\mathbf{v}$ in a Euclidean space and compute it for

$$U = \{\begin{pmatrix}u_1\\u_2\\u_3\\u_4\end{pmatrix} \mid u_1 + u_2 + u_3 + u_4 = 0\} \text{ and } \mathbf{v} = \begin{pmatrix}1\\2\\3\\4\end{pmatrix}$$

in the usual Euclidean space $\mathbf{R}^4$.

8.2.13. Assume that the real system of linear equations $A\mathbf{x} = \mathbf{b}$ has no solution. We want to find an approximate solution $\mathbf{z}$ such that $A\mathbf{z}$ is as close to $\mathbf{b}$ as possible. How can we determine $\mathbf{z}$ and in which sense will it be the best approximate solution? Illustrate the method on the system of equations

$$\begin{aligned} x_1 + x_2 + x_3 + x_4 &= 1 \\ x_1 + 2x_2 + 3x_3 + 4x_4 &= 1 \\ x_1 + 3x_2 + 5x_3 + 7x_4 &= 7. \end{aligned}$$

8.2.14. Let $\mathbf{e}_1, \ldots, \mathbf{e}_n$ be an orthonormal basis in a Euclidean pace. Prove:
(a) $\mathbf{x} = \sum_{j=1}^{n} (\mathbf{e}_j \cdot \mathbf{x})\mathbf{e}_j$;
(b) $\mathbf{x} \cdot \mathbf{z} = \sum_{j=1}^{n} (\mathbf{x} \cdot \mathbf{e}_j)(\mathbf{e}_j \cdot \mathbf{z})$;
(c) *Parseval formula*: $\|\mathbf{x}\|^2 = \sum_{j=1}^{n} |\mathbf{x} \cdot \mathbf{e}_j|^2$.

8.2.15. (*Bessel inequality.*) Show that if $\mathbf{c}_1, \ldots, \mathbf{c}_k$ is an orthonormal system, then $\|\mathbf{x}\|^2 \geq \sum_{j=1}^{k} |\mathbf{x} \cdot \mathbf{c}_j|^2$ for every $\mathbf{x}$. When does equality hold?

8.2.16. Verify that CBS remains valid if the positive definite property of the inner product is replaced by positive semidefiniteness: If $\mathbf{A}$ is a positive semidefinite symmetric bilinear function, then $|\mathbf{A}(\mathbf{x},\mathbf{z})|^2 \leq \mathbf{A}(\mathbf{x},\mathbf{x}) \cdot \mathbf{A}(\mathbf{z},\mathbf{z})$ for every $\mathbf{x}$ and $\mathbf{z}$.

***8.2.17.** At most how many non-zero vectors are there in an n-dimensional Euclidean space so that the angle between any two is (a) 60; (b) 120 degrees?

*8.2.18. Prove that CBS remains valid in infinite-dimensional Euclidean spaces, too (see the definition of such spaces in Exercise 8.1.15).

8.3. Complex Euclidean Space

We adapt the notions and result of the previous two sections to finite-dimensional vector spaces over the complex field.

Definition 8.3.1. Let $\mathbf{e}_1, \dots, \mathbf{e}_n$ be a *fixed* basis in V. The *inner product* belonging to this basis is the $\mathbf{S}\colon V \times V \to \mathbf{C}$ function

$$\mathbf{S}(\mathbf{x},\mathbf{z}) = \mathbf{x}\cdot\mathbf{z} = \begin{pmatrix} x_1 \\ \vdots \\ x_n \end{pmatrix} \cdot \begin{pmatrix} z_1 \\ \vdots \\ z_n \end{pmatrix} = \sum_{j=1}^{n} \overline{x_j} z_j. \clubsuit$$

The difference compared to the real case is that we need the *complex conjugates* of the coordinates of the *first* vector.

This inner product is a positive definite *Hermitian* bilinear function, so

$$!!!\ \mathbf{x}\cdot\mathbf{z} = \overline{\mathbf{z}\cdot\mathbf{x}}\ !!!\ ; \quad (\mathbf{x}+\mathbf{x}')\cdot\mathbf{z} = \mathbf{x}\cdot\mathbf{z} + \mathbf{x}'\cdot\mathbf{z}; \quad !!!\ (\lambda\mathbf{x})\cdot\mathbf{z} = \overline{\lambda}(\mathbf{x}\cdot\mathbf{z})\ !!!\ ;$$
$$\mathbf{x}\cdot(\mathbf{z}+\mathbf{z}') = \mathbf{x}\cdot\mathbf{z} + \mathbf{x}\cdot\mathbf{z}'; \quad \mathbf{x}\cdot(\lambda\mathbf{z}) = \lambda(\mathbf{x}\cdot\mathbf{z}); \quad \mathbf{x} \neq \mathbf{0} \Rightarrow \mathbf{x}\cdot\mathbf{x} > 0.$$

There are two changes compared to the real case: swapping the two factors carries the inner product into its complex conjugate, and multiplying the *first* factor by λ the inner product gets multiplied by $\overline{\lambda}$. Note that the condition $\mathbf{x} \neq \mathbf{0} \Rightarrow \mathbf{x}\cdot\mathbf{x} > 0$ contains that $\mathbf{x}\cdot\mathbf{x}$ is a *real* number for every $\mathbf{x}$ (this also follows from $\mathbf{x}\cdot\mathbf{z} = \overline{\mathbf{z}\cdot\mathbf{x}}$, cf. Theorem 7.4.4).

By the complex version of the orthogonalization theorem, also the converse holds. Given a positive definite Hermitian bilinear function, there exists a basis such that the function is just the inner product defined by this basis. So, the analogue of Theorem 8.1.2 remains valid.

Theorem 8.3.2. *A complex inner product can also be defined as a positive definite Hermitian bilinear function.* $\clubsuit$

The definitions of Euclidean spaces, orthonormal systems, orthonormal bases, orthogonality, and orthogonal complements are the same as in the real case (Definitions 8.1.3–8.1.6). Also, Theorem 8.1.7 remains valid.

The length of a vector is the square root of the inner product of the vector with itself (Definition 8.2.1). Due to the positive definite property, this is a non-negative real number. It is the square root of the sum of squares of the *absolute values* of the coordinates in an orthonormal basis:

$$\|\mathbf{x}\| = \sqrt{\mathbf{x} \cdot \mathbf{x}} = \sqrt{\sum_{j=1}^{n} \overline{x_j} x_j} = \sqrt{\sum_{j=1}^{n} |x_j|^2}.$$

The length (or norm) satisfies the statements (N1)–(N3) in Theorem 8.2.2, so a complex Euclidean space is a (complex) normed space.

The definition induced by the norm and the properties (M1)–(M3) of distance are identical to the real case (Definition 8.2.4 and Theorem 8.2.5), so we get a metric space.

We do not define angles (just orthogonality) as Definition 8.2.7 would yield a complex value for $\cos\varphi$ in general.

The Cauchy–Bunyakovsky–Schwarz inequality (Theorem 8.2.8), however, remains valid. We can adapt the second and third proofs though the latter requires a small trick (see Exercise 8.3.4).

Exercises 8.3

8.3.1. Prove that the statements of Exercises 8.1.2, 8.1.3, 8.1.5, 8.1.6, 8.1.8, 8.1.9, 8.1.11, 8.1.14, 8.2.14, and 8.2.15 also remain valid in a complex Euclidean space.

8.3.2. Let V be a complex Euclidean space. Verify:

(a) $\mathbf{x} - i\mathbf{z}$ and $i\mathbf{x} + \mathbf{z}$ are orthogonal if and only if $\mathbf{x} = i\mathbf{z}$;

(b) $\mathbf{x} + i\mathbf{z}$ and $i\mathbf{x} + \mathbf{z}$ are orthogonal if and only if $\|\mathbf{x}\| = \|\mathbf{z}\|$ and the inner product $\mathbf{x} \cdot \mathbf{z}$ is an imaginary number.

8.3.3. Investigate the statements of Exercise 8.2.3 in a complex Euclidean space.

8.3.4. Prove the Cauchy–Bunyakovsky–Schwarz inequality for a complex Euclidean space.

8.3.5. In the usual Euclidean space $\mathbf{C}^n$ we obtain the conjugate of a vector by conjugating its coordinates:

$$\text{If } \mathbf{x} = \begin{pmatrix} x_1 \\ \vdots \\ x_n \end{pmatrix}, \text{ then } \overline{\mathbf{x}} = \begin{pmatrix} \overline{x_1} \\ \vdots \\ \overline{x_n} \end{pmatrix}.$$

For $H \subseteq \mathbf{C}^n$ let $\overline{H}$ denote the set of conjugates of the vectors in H. Finally, we say that a vector is real if it only has real coordinates. Prove:

(a) $\mathbf{z}$ and $\overline{\mathbf{z}}$ are linearly dependent if and only if $\mathbf{z}$ is a scalar multiple of a real vector.

(b) $\overline{H}$ is a subspace if and only if H is a subspace.

(c) $\overline{U}^{\perp} = \overline{U^{\perp}}$ for any subspace U.

(d) A subspace $U \neq \mathbf{0}$ satisfies $\overline{U} = U$ if and only if U has a basis of real vectors.

(e) There exists a subspace U satisfying $U^{\perp} = \overline{U}$ if and only if n is even.

8.4. The Adjoint of a Transformation

Let V be an n-dimensional (real or complex) Euclidean space.

Theorem 8.4.1. *For every linear transformation $\mathcal{A} \in \operatorname{Hom} V$, there exists exactly one $\mathcal{A}^* \in \operatorname{Hom} V$ satisfying $(\mathcal{A}\mathbf{x}) \cdot \mathbf{z} = \mathbf{x} \cdot (\mathcal{A}^*\mathbf{z})$ for every $\mathbf{x}, \mathbf{z} \in V$.*

This transformation $\mathcal{A}^*$ is called the adjoint of $\mathcal{A}$. ♣

Note that the adjoint depends not only on $\mathcal{A}$ but also on the inner product. If we consider another inner product on the same vector space V (yielding another Euclidean space), then the adjoint of the transformation will be different in general.

Proof. Let $\mathbf{e}_1, \dots, \mathbf{e}_n$ be an orthonormal basis in V.

We consider the real case first. Using the matrices of $\mathcal{A}$, $\mathbf{x}$, and $\mathbf{z}$ in the basis $\mathbf{e}_j$, we can express the inner product $(\mathcal{A}\mathbf{x}) \cdot \mathbf{z}$ as

$$(\mathcal{A}\mathbf{x}) \cdot \mathbf{z} = [\mathcal{A}\mathbf{x}]^T[\mathbf{z}] = [\mathbf{x}]^T[\mathcal{A}]^T[\mathbf{z}].$$

Similarly, the inner product $\mathbf{x} \cdot (\mathcal{A}^*\mathbf{z})$ is

$$\mathbf{x} \cdot (\mathcal{A}^*\mathbf{z}) = [\mathbf{x}]^T[\mathcal{A}^*][\mathbf{z}].$$

Each inner product is a bilinear function (of $\mathbf{x}$ and $\mathbf{z}$). They are equal exactly if their matrices (in the same basis) are equal, i.e., $[\mathcal{A}]^T = [\mathcal{A}^*]$. Thus, the bijection between matrices and linear maps yields that there exists exactly one such transformation $\mathcal{A}^*$.

In the complex case, the only difference is that, instead of the transpose, we have to take the adjoint matrix (the conjugate of the transpose). ∎

We obtain from the proof that the matrix of $\mathcal{A}^*$ in an orthonormal basis is the transpose or the adjoint of the matrix of $\mathcal{A}$ according to the real and complex cases. We formulate this important result as a theorem. Since the transpose and the adjoint of a real matrix are the same, we shall use the name adjoint matrix and its notation both for real and complex spaces.

Theorem 8.4.2. *In an orthonormal basis, the matrix of $\mathcal{A}^*$ is obtained by reflecting the matrix of $\mathcal{A}$ across the main diagonal and taking its conjugate (in the complex case): $[\mathcal{A}^*] = [\mathcal{A}]^*$.* ♣

The adjoint transformations obey the same laws concerning the operations as matrices (see Exercise 8.4.1).

The characteristic and minimal polynomials and eigenvalues of $\mathcal{A}$ and $\mathcal{A}^*$ are the same in a real Euclidean space and are conjugates in the complex case (see Exercise 8.4.7).

The simple connection between the invariant subspaces below will play an important role in the applications.

Theorem 8.4.3. *U is an invariant subspace of $\mathcal{A}$ if and only if $U^\perp$ is an invariant subspace of $\mathcal{A}^*$.* ♣

Proof. Assume that U is an invariant subspace of $\mathcal{A}$. To show that $U^\perp$ is an invariant subspace of $\mathcal{A}^*$, we have to verify $\mathbf{z} \in U^\perp \Rightarrow \mathcal{A}^*\mathbf{z} \in U^\perp$. In other words, if $\mathbf{z}$ is orthogonal to every $\mathbf{u} \in U$, then the same also holds for $\mathcal{A}^*\mathbf{z}$. Indeed, $\mathbf{u} \cdot (\mathcal{A}^*\mathbf{z}) = (\mathcal{A}\mathbf{u}) \cdot \mathbf{z} = 0$ since $\mathcal{A}\mathbf{u} \in U$ and $\mathbf{z} \in U^\perp$. We obtain the converse by the same method or using $(\mathcal{A}^*)^* = \mathcal{A}$ and $(U^\perp)^\perp = U$. ∎

Exercises 8.4

8.4.1. Prove the properties of the adjoint transformation.

$$(\mathcal{A}+\mathcal{B})^* = \mathcal{A}^* + \mathcal{B}^*, \quad (\lambda\mathcal{A})^* = \overline{\lambda}\mathcal{A}^*, \quad (\mathcal{A}\mathcal{B})^* = \mathcal{B}^*\mathcal{A}^*, \quad (\mathcal{A}^*)^* = \mathcal{A}.$$

8.4.2. Consider the plane with the usual inner product and determine the adjoints of the transformations.
- (a) Reflection across the x-axis.
- (b) Reflection across an arbitrary line containing the origin.
- (c) Rotation around the origin by $+90$.
- (d) Rotation around the origin by an arbitrary angle.
- (e) Perpendicular projection onto the x-axis.
- (f) Perpendicular projection to an arbitrary line containing the origin.
- (g) Projection parallel to the y-axis onto the line $y = x$.
- (h) The linear transformation mapping the points of the x-axis to themselves and rotating the points of the y-axis by -90 degrees.

8.4.3. Consider the usual inner product in the space and let $\mathbf{c}$ be a fixed vector. Let $\mathcal{A}$ assign the cross product $\mathbf{u} \times \mathbf{c}$ to $\mathbf{u}$. Determine $\mathcal{A}^*$.

8.4.4. Consider the Euclidean spaces defined in Exercise 8.1.4 and determine the adjoint of the second derivation $f \mapsto f''$.

8.4.5. Show that if $\mathcal{A}^2 = \mathcal{O}$, then $\mathcal{A}\mathbf{x} \perp \mathcal{A}^*\mathbf{x}$ for every $\mathbf{x}$. Is the converse true?

8.4.6. Consider a pair of eigenvectors of $\mathcal{A}$ and $\mathcal{A}^*$. Prove that either the corresponding eigenvalues are conjugates, or the two eigenvectors are perpendicular.

8.4.7. Verify that the characteristic and minimal polynomials and eigenvalues of $\mathcal{A}$ and $\mathcal{A}^*$ are the same in a real Euclidean space and are conjugates in the complex case. (The conjugate of a polynomial is obtained by conjugating the coefficients.)

8.4.8. Show $\operatorname{Ker}\mathcal{A}^* = (\operatorname{Im}\mathcal{A})^\perp$ and $\operatorname{Im}\mathcal{A}^* = (\operatorname{Ker}\mathcal{A})^\perp$.

8.4.9. Verify that the images and kernels of $\mathcal{A}$ and $\mathcal{A}^*$ have the same dimensions.

8.4.10. Let $A \in \mathbf{C}^{k\times n}$, $\mathbf{b} \in \mathbf{C}^k$, and consider the system of linear equations $A\mathbf{x} = \mathbf{b}$ with n unknowns and k equations. Show that this is solvable if and only if $\mathbf{b}$ is orthogonal to every solution of the homogeneous system $A^*\mathbf{z} = \mathbf{0}$. (Orthogonality refers to the usual Euclidean space $\mathbf{C}^k$.)

8.4.11. Prove.
- (a) $\mathcal{A}^*\mathcal{A} = \mathcal{O} \Rightarrow \mathcal{A} = \mathcal{O}$.
- (b) $\operatorname{Ker}(\mathcal{A}^*\mathcal{A}) = \operatorname{Ker}\mathcal{A}$ and $\operatorname{Im}(\mathcal{A}^*\mathcal{A}) = \operatorname{Im}\mathcal{A}^*$.

8.4.12. Prove that $\mathcal{A}^*\mathcal{B} = \mathcal{O}$ implies
(a) $\operatorname{Im}\mathcal{A}$ and $\operatorname{Im}\mathcal{B}$ are orthogonal subspaces; and
(b) $\operatorname{Ker}(\mathcal{A} + \mathcal{B}) = \operatorname{Ker}\mathcal{A} \cap \operatorname{Ker}\mathcal{B}$.
Decide whether the converses are true or false.

*8.4.13. Show that $\mathcal{A}^*\mathcal{B} = \mathcal{B}\mathcal{A}^* = \mathcal{O}$ implies $\operatorname{Im}(\mathcal{A} + \mathcal{B}) = \operatorname{Im}\mathcal{A} \oplus \operatorname{Im}\mathcal{B}$.

***8.4.14.** Let V be a finite-dimensional vector space over the real or complex field and we consider arbitrary inner products on it.
(a) For which transformations $\mathcal{A} \in \operatorname{Hom} V$ is $\mathcal{A}^*$ the same for every inner product?
(b) Let S_1 and S_2 be two inner products. Prove that every $\mathcal{A} \in \operatorname{Hom} V$ has the same adjoint with respect to S_1 and S_2 if and only if $S_1 = \lambda S_2$ with some $\lambda \neq 0$.

8.5. Normal, Self-adjoint, and Unitary Transformations

In this section we study (finite-dimensional) *complex* Euclidean spaces. With the help of the adjoint, we can characterize which transformations have an *orthonormal* eigenbasis. In other words: which transformations have a diagonal matrix in some *orthonormal* basis. We shall discuss two important special cases in more detail. The situation is somewhat different for real Euclidean spaces; we shall investigate it in the next section.

Definition 8.5.1. A transformation $\mathcal{A}$ is *normal* if it commutes with its adjoint: $\mathcal{A}\mathcal{A}^* = \mathcal{A}^*\mathcal{A}$. ♣

Theorem 8.5.2. *In a finite-dimensional complex Euclidean space, a transformation $\mathcal{A}$ has an orthonormal eigenbasis if and only if $\mathcal{A}$ is normal, i.e., $\mathcal{A}\mathcal{A}^* = \mathcal{A}^*\mathcal{A}$.* ♣

Proof. We prove sufficiency first, i.e., a normal transformation has an orthonormal eigenbasis.

We claim that if two transformations commute, then they have a common eigenvector. Consider an eigenspace of one of the transformations. This is an invariant subspace of the other transformation by Exercise 6.4.6. Restricting this transformation onto this subspace, it has an eigenvector. This is a common eigenvector of the two transformations.

So, if $\mathcal{A}$ and $\mathcal{A}^*$ commute, they have a common eigenvector $\mathbf{e}$. Multiplying by a suitable scalar, we can achieve $\|\mathbf{e}\| = 1$. This vector will be the first element in the orthonormal eigenbasis.

Let $U = \langle \mathbf{e} \rangle^\perp$ be the orthogonal complement of $\langle \mathbf{e} \rangle$. As $\langle \mathbf{e} \rangle$ is an invariant subspace of $\mathcal{A}$ and $\mathcal{A}^*$, thus U is an invariant subspace of $\mathcal{A}^*$ and $\mathcal{A}$, by Theorem 8.4.3. Repeating the process on U, etc., finally we get an orthonormal eigenbasis.

Turning to the proof of necessity, we have to show that if $\mathcal{A}$ has an orthonormal eigenbasis, then $\mathcal{A}\mathcal{A}^* = \mathcal{A}^*\mathcal{A}$. In an eigenbasis, the matrix $[\mathcal{A}]$ is diagonal. As the basis is orthonormal, we have $[\mathcal{A}^*] = [\mathcal{A}]^*$, thus also $[\mathcal{A}^*]$ is a diagonal matrix. Two diagonal matrices always commute, therefore the corresponding transformations $\mathcal{A}$ and $\mathcal{A}^*$ commute, as well. ∎

We describe another interesting characterization of normal transformations. Further equivalent conditions are given in Exercise 8.5.10.

Theorem 8.5.3. *A transformation $\mathcal{A}$ is normal if and only if $\mathcal{A}^*$ is a polynomial of $\mathcal{A}$, i.e., $\mathcal{A}^* = f(\mathcal{A})$ for some $f \in \mathbf{C}[x]$.* ♣

Proof. If $\mathcal{A}^* = f(\mathcal{A})$, then $\mathcal{A}\mathcal{A}^* = \mathcal{A}f(\mathcal{A}) = f(\mathcal{A})\mathcal{A} = \mathcal{A}^*\mathcal{A}$ since $\mathcal{A}$ commutes with its powers and $\mathcal{I}$.

To prove the converse, assume $\mathcal{A}\mathcal{A}^* = \mathcal{A}^*\mathcal{A}$. By the previous theorem, $[\mathcal{A}]$ is diagonal in some orthonormal basis and $[\mathcal{A}^*] = [\mathcal{A}]^*$. Let $\lambda_1, \dots, \lambda_k$ be the elements in the main diagonal of $[\mathcal{A}]$. Then $\overline{\lambda_1}, \dots, \overline{\lambda_k}$ are the elements in the main diagonal of $[\mathcal{A}]^*$.

Let n denote the number of distinct λ_j's. In the following we construct a polynomial $f = \alpha_0 + \alpha_1 x + \dots + \alpha_{n-1}x^{n-1} \in \mathbf{C}[x]$ satisfying $\mathcal{A}^* = f(\mathcal{A})$. Switching from transformations to diagonal matrices, we get the system of equations

$$\alpha_0 + \alpha_1\lambda_j + \dots + \alpha_{n-1}\lambda_j^{n-1} = \overline{\lambda_j}, \quad j = 1, 2, \dots, k$$

with α_i's as unknowns. Keeping only one copy of the identical equations belonging to equal λ's, we get an $n \times n$ system of equations. Its determinant is a Vandermonde determinant generated by the distinct λ_j's, so it is not 0. Thus, the system of equations has a (unique) solution providing an appropriate polynomial f. (Instead, we could also have referred to Theorem 3.2.4 about the interpolation polynomials.) ∎

The two most important classes of normal transformations are the self-adjoint and unitary transformations satisfying $\mathcal{A}^* = \mathcal{A}$ and $\mathcal{A}^* = \mathcal{A}^{-1}$, respectively.

Definition 8.5.4. A transformation $\mathcal{A}$ is *self-adjoint* if $\mathcal{A}^* = \mathcal{A}$. ♣

It is clear that a self-adjoint transformation is normal, so it has an orthonormal eigenbasis. Exactly those normal transformations are self-adjoint where all eigenvalues are real (Exercise 8.5.1).

Definition 8.5.5. A transformation $\mathcal{A}$ is *unitary* if $\mathcal{A}^* = \mathcal{A}^{-1}$. ♣

It is clear that a unitary transformation is normal, so it has an orthonormal eigenbasis. Exactly those normal transformations are unitary where all eigenvalues are of absolute value 1 (Exercise 8.5.3).

Another useful characterization of unitary transformations is that they preserve the inner product, norm, and distance.

Theorem 8.5.6. *$\mathcal{A}$ is unitary if and and only if any of the following conditions holds:*

I. *$\mathcal{A}$ preserves the inner product, i.e., $\mathbf{x}, \mathbf{z} \in V \Rightarrow (\mathcal{A}\mathbf{x}) \cdot (\mathcal{A}\mathbf{z}) = \mathbf{x} \cdot \mathbf{z}$.*

II. *$\mathcal{A}$ preserves the norm, i.e., $\mathbf{x} \in V \Rightarrow \|\mathcal{A}\mathbf{x}\| = \|\mathbf{x}\|$.*

III. *$\mathcal{A}$ preserves the distance, i.e., $\mathbf{x}, \mathbf{z} \in V \Rightarrow \delta(\mathcal{A}\mathbf{x}, \mathcal{A}\mathbf{z}) = \delta(\mathbf{x}, \mathbf{z})$.*

Proof.

I. $(\mathcal{A}\mathbf{x}) \cdot (\mathcal{A}\mathbf{z}) = \mathbf{x} \cdot (\mathcal{A}^*\mathcal{A}\mathbf{z}) = \mathbf{x} \cdot \mathbf{z}$ holds for every $\mathbf{x}$ and $\mathbf{z}$ if and only if $\mathcal{A}^*\mathcal{A} = \mathcal{I}$.

II. The norm is a special inner product, so preserving the inner product implies preserving the norm. For the converse, we apply that the inner product can uniquely be expressed from the norm.

III. Distance is defined with the norm, so preserving the norm implies preserving the distance. The converse follows from $\|\mathbf{u}\| = \delta(\mathbf{u}, \mathbf{0})$. ∎

Exercises 8.5

8.5.1. (a) Prove that a normal transformation is self-adjoint if and only if all its eigenvalues are real numbers.
(b) Is it true that if all eigenvalues of a transformation are real numbers, then it is self-adjoint?

8.5.2. Show that if $\mathcal{A}$ is self-adjoint, then $\mathcal{A}, \mathcal{A}^2, \dots, \mathcal{A}^m, \dots$ are either all distinct, or there are at most two distinct transformations among them.

8.5.3. (a) Prove that a normal transformation is unitary if and only if all its eigenvalues are of absolute value 1.
(b) Is it true that if all eigenvalues of a transformation are of absolute value 1, then it is unitary?

8.5.4. True or false? Every linear transformation of a finite-dimensional complex vector space is normal for some inner product.

8.5.5. Consider the usual Euclidean space $\mathbf{C}^4$. Which transformations are normal, self-adjoint, and unitary?

$$\mathcal{A}\begin{pmatrix} u_1 \\ u_2 \\ u_3 \\ u_4 \end{pmatrix} = \begin{pmatrix} 0 \\ u_1 \\ u_2 \\ u_3 \end{pmatrix}; \quad \mathcal{B}\begin{pmatrix} u_1 \\ u_2 \\ u_3 \\ u_4 \end{pmatrix} = \begin{pmatrix} u_4 \\ u_1 \\ u_2 \\ u_3 \end{pmatrix}; \quad \mathcal{C}\begin{pmatrix} u_1 \\ u_2 \\ u_3 \\ u_4 \end{pmatrix} = \begin{pmatrix} u_1 \\ u_1 \\ u_1 \\ u_1 \end{pmatrix};$$

$$\mathcal{D}\begin{pmatrix} u_1 \\ u_2 \\ u_3 \\ u_4 \end{pmatrix} = \begin{pmatrix} u_1 + u_2 + u_3 + u_4 \\ u_1 + u_2 + u_3 + u_4 \\ u_1 + u_2 + u_3 + u_4 \\ u_1 + u_2 + u_3 + u_4 \end{pmatrix}; \quad \mathcal{E}\begin{pmatrix} u_1 \\ u_2 \\ u_3 \\ u_4 \end{pmatrix} = \begin{pmatrix} u_1 - u_2 \\ u_2 - u_3 \\ u_3 - u_4 \\ u_4 - u_1 \end{pmatrix}.$$

8.5.6. Let $\mathcal{A}$ and $\mathcal{B}$ be self-adjoint. Does this imply that $\mathcal{A} + \mathcal{B}, \lambda\mathcal{A}, \mathcal{A}^2$, and $\mathcal{A}\mathcal{B}$ are self-adjoint? Solve the problem also for unitary and normal transformations instead of self-adjoint ones.

8.5.7. (a) Comment on the argument. “Claim: The product of two self-adjoint transformations is self-adjoint. Proof: (i) a transformation is self-adjoint if and only if it has a diagonal matrix with real entries in an orthonormal basis; (ii) the product of two such matrices is again a matrix of this type; (iii) so multiplying the suitable matrices of two self-adjoint transformations, the product will be a matrix of this type, hence the product of the transformations is self-adjoint.”

(b) Verify that the product $\mathcal{AB}$ of self-adjoint transformations $\mathcal{A}$ and $\mathcal{B}$ is self-adjoint if and only if $\mathcal{AB} = \mathcal{BA}$.

8.5.8. Show that the eigenspaces of a normal transformation are pairwise orthogonal. Is the converse true?

8.5.9. Let A be the matrix of a normal transformation $\mathcal{A}$ in some basis $\mathbf{b}_1, \dots, \mathbf{b}_n$. True or false?

(a) If the basis $\mathbf{b}_1, \dots, \mathbf{b}_n$ is orthonormal, then $AA^* = A^*A$.

(b) If $AA^* = A^*A$, then the basis $\mathbf{b}_1, \dots, \mathbf{b}_n$ is orthonormal.

8.5.10. Prove that the normality of $\mathcal{A}$ is equivalent to any of the following conditions.

(a) $\mathbf{x} \in V \Rightarrow \|\mathcal{A}\mathbf{x}\| = \|\mathcal{A}^*\mathbf{x}\|$.

(b) $\mathcal{A}$ and $\mathcal{A}^*$ have the same eigenvectors.

(c) $\operatorname{Ker}(\mathcal{A} - \lambda\mathcal{E}) = \operatorname{Ker}(\mathcal{A}^* - \overline{\lambda}\mathcal{E})$ for every λ.

(d) $\operatorname{Im}(\mathcal{A} - \lambda\mathcal{E}) = \operatorname{Im}(\mathcal{A}^* - \overline{\lambda}\mathcal{E})$ for every λ.

(e) $\operatorname{Ker}(\mathcal{A} - \lambda\mathcal{E}) \perp \operatorname{Im}(\mathcal{A} - \lambda\mathcal{E})$ for every λ.

8.5.11. Verify that if the normal transformations $\mathcal{A}$ and $\mathcal{B}$ satisfy $\mathcal{AB} = \mathcal{O}$, then also $\mathcal{BA} = \mathcal{O}$ holds.

8.5.12. Show that the normal transformations $\mathcal{A}$ and $\mathcal{B}$ share a common orthonormal basis if and only if $\mathcal{AB} = \mathcal{BA}$.

8.5.13. Prove that if the normal transformations $\mathcal{A}$ and $\mathcal{B}$ commute, i.e., $\mathcal{AB} = \mathcal{BA}$, then also $\mathcal{AB}$ is normal. Is the converse true?

8.5.14. Show that a transformation is normal if and only if it is the product of a self-adjoint and a unitary transformation that commute.

8.5.15. Verify that, in a complex Euclidean space, every transformation has an upper triangle matrix in some orthonormal basis.

8.5.16. Let V be an n-dimensional complex Euclidean space and let $\lambda_1, \dots, \lambda_n$ and $\mu_1, \dots, \mu_n$ be the roots of the characteristic polynomials of $\mathcal{A}$ and $\mathcal{A}^*\mathcal{A}$ (counted with multiplicity). Prove.

(a) $\mu_1, \dots, \mu_n$ are non-negative real numbers.

(b) $\sum_{j=1}^n |\lambda_j|^2 \le \sum_{j=1}^n \mu_j$.

(c) In (b), equality holds if and only if $\mathcal{A}$ is normal.

8.5.17. Verify that a transformation preserves orthogonality if and only if it is a scalar multiple of a unitary transformation.

8.5.18. Prove the properties of any matrix of a unitary transformation in an orthonormal basis.
   (a) The usual inner product of two distinct column vectors in $\mathbf{C}^n$ is 0, and the inner product of a column vector with itself is 1.
   (b) The same holds for rows instead of columns.
   (c) The determinant of the matrix has absolute value 1.
   (d) Any entry in the matrix has the same absolute value as its cofactor.

8.6. Symmetric and Orthogonal Transformations

In this section, we discuss a few types of transformations of (finite-dimensional) *real* Euclidean spaces. There are much fewer transformations with orthonormal eigenbases, i.e., with diagonal matrices in orthonormal bases: only the symmetric transformations, the analogues of self-adjoint ones have this property. This is the Principal Axis Theorem playing an important role in geometry in describing quadratic curves and surfaces. Orthogonal transformations are the analogues of unitary ones, and their nicest matrices have a slightly more complicated form in general.

Definition 8.6.1. $\mathcal{A}$ is *symmetric* if $\mathcal{A}^* = \mathcal{A}$. ♣

Theorem 8.6.2 (Principal Axis Theorem). *A transformation $\mathcal{A}$ has an orthonormal eigenbasis if and only if $\mathcal{A}$ is symmetric.* ♣

Proof. If there exists an orthonormal eigenbasis, then the matrix of $\mathcal{A}$ in this basis is diagonal. A diagonal matrix is symmetric, so $\mathcal{A}$ is symmetric.

For the converse, we assume that $\mathcal{A}$ is symmetric. We show first that $\mathcal{A}$ has an eigenvector. As the minimal polynomial over the real field is the product of irreducible polynomials of degree 1 or 2, $\mathcal{A}$ has an invariant subspace W of dimension 1 or 2, by Theorem 6.5.5. If $\dim W = 1$, then any non-zero element in W is an eigenvector. So, assume $\dim W = 2$ and consider the matrix of $\mathcal{A}$ (restricted to W) in an orthonormal basis. $\mathcal{A}$ is symmetric, thus so is this matrix:

$$A = \begin{pmatrix} \alpha & \beta \\ \beta & \delta \end{pmatrix}.$$

Its characteristic polynomial is $k_{\mathcal{A}} = x^2 - (\alpha + \delta)x + (\alpha\delta - \beta^2)$. The discriminant is $(\alpha + \delta)^2 - 4(\alpha\delta - \beta^2) = (\alpha - \delta)^2 + 4\beta^2 \geq 0$, so $k_{\mathcal{A}}$ has a (real) root. This root is an eigenvalue of $\mathcal{A}$, so $\mathcal{A}$ has an eigenvector.

Let $\mathbf{e}_1$ be an eigenvector of $\mathcal{A}$ of norm 1. Then the orthogonal complement $U = \langle \mathbf{e}_1 \rangle^{\perp}$ is an invariant subspace of $\mathcal{A}^* = \mathcal{A}$. We repeat the process on U, etc. This finally leads to an orthonormal eigenbasis. ■

From a geometric aspect, we can consider a symmetric matrix as the matrix of a symmetric bilinear function. Then the Principal Axis Theorem states that a symmetric bilinear function can be orthogonalized so that the basis is orthonormal with respect to the given inner product in the Euclidean space. This means that there exist *perpendicular* eigendirections to the quadratic curves and surfaces in the plane and space.

Therefore the quadratic form corresponding to the given curve or surface will be a sum of squares with the eigenvalues as coefficients.

Another interpretation of this form of the Principal Axis Theorem is that two symmetric bilinear functions can be orthogonalized simultaneously if at least one of them is an inner product, i.e., it is positive definite.

The equivalence of the two forms of the Principal Axis Theorem related to transformations and bilinear functions is not quite straightforward as the matrix obeys different rules when switching bases. But there is no problem in this case because the bases are *orthonormal* and then the change in the matrix is the same for transformations and bilinear functions.

Now we discuss the analogue of unitary transformations.

Definition 8.6.3. $\mathcal{A}$ is *orthogonal* if $\mathcal{A}^* = \mathcal{A}^{-1}$. ♣

Theorem 8.6.4. *A transformation $\mathcal{A}$ is orthogonal if and only if its matrix in some orthonormal basis consists of 2×2 and 1×1 blocks on the main diagonal where the 2×2 blocks are of the form*

$$\begin{pmatrix} \cos\theta & -\sin\theta \\ \sin\theta & \cos\theta \end{pmatrix}$$

and the 1×1 blocks are ± 1 (all other entries are 0 in the matrix). ♣

In other words, V is the direct sum of pairwise orthogonal $\mathcal{A}$-invariant "planes" and "lines" and $\mathcal{A}$ acts as a rotation on each of them.

In general, there exists no orthonormal eigenbasis, e.g., a rotation of the plane (around the origin) is an orthogonal transformation but it has no eigenvector if the angle is not a multiple of π.

Proof. If there exists such a basis, then a simple calculation shows that multiplying the matrix with its transpose we get the identity matrix. So, the transpose of the matrix equals the inverse. Thus, the same holds for the transformation (as the basis is orthonormal).

To prove the converse, assume that $\mathcal{A}$ is orthogonal: $\mathcal{A}^* = \mathcal{A}^{-1}$. Similar to the proof of the previous theorem, $\mathcal{A}$ has an invariant subspace W of dimension 1 or 2. If $\dim W = 1$, then any non-zero element in W is an eigenvector and the corresponding eigenvalue is ± 1 as $\mathcal{A}$ is orthogonal.

If $\dim W = 2$, consider the matrix of $\mathcal{A}$ (restricted to W) in an orthonormal basis. $\mathcal{A}^* = \mathcal{A}^{-1}$ implies that the same holds for the matrices, too. Therefore $D = \det A = \det A^* = \det A^{-1} = 1/D$, so $D = \pm 1$.

If $A = \begin{pmatrix} \alpha & \beta \\ \gamma & \delta \end{pmatrix}$, then $A^* = \begin{pmatrix} \alpha & \gamma \\ \beta & \delta \end{pmatrix}$ and $A^{-1} = \begin{pmatrix} \delta/D & -\beta/D \\ -\gamma/D & \alpha/D \end{pmatrix}$ (by the formula using the cofactors). Hence,

$$A^* = A^{-1} \iff \alpha = \delta/D, \gamma = -\beta/D, \beta = -\gamma/D, \delta = \alpha/D.$$

If $D = -1$, then $\beta = \gamma$ (and $\alpha = -\delta$). As A is symmetric, it has an orthonormal eigenbasis by Theorem 8.6.2. Further, $k_A = x^2 - 1$ implies that the two eigenvalues are 1 and -1. (No other value could occur as $\mathcal{A}$ is orthogonal.)

If $D = 1$, then

$$A = \begin{pmatrix} \alpha & \beta \\ -\beta & \alpha \end{pmatrix}$$

where $\alpha^2 + \beta^2 = 1$. So,

$$A = \begin{pmatrix} \cos\theta & -\sin\theta \\ \sin\theta & \cos\theta \end{pmatrix}$$

for some θ.

Consider now $U = W^{\perp}$. Then U is an invariant subspace of $\mathcal{A}^* = \mathcal{A}^{-1}$. It is easy to check that $\mathcal{A}$ and $\mathcal{A}^{-1}$ have the same invariant subspaces, so U is an invariant subspace of $\mathcal{A}$. We repeat the process on U, etc. This finally leads to an orthonormal basis satisfying the requirements. ∎

Exercises 8.6

8.6.1. Prove that if a transformation $\mathcal{A}$ is both symmetric and orthogonal, then $\mathcal{A}^2 = \mathcal{I}$. Is the converse true?

8.6.2. Verify that the eigenspaces of a (a) symmetric; (b) orthogonal transformation are pairwise orthogonal.

8.6.3. Consider the usual Euclidean space $\mathbf{R}^4$. Which transformations are symmetric or orthogonal? Find an orthonormal eigenbasis for the symmetric ones and an orthonormal basis satisfying the conditions of Theorem 8.6.4 for the orthogonal ones. In both cases, write the matrix belonging to this basis.

$$\mathcal{A}\begin{pmatrix} u_1 \\ u_2 \\ u_3 \\ u_4 \end{pmatrix} = \begin{pmatrix} u_4 \\ u_3 \\ u_2 \\ u_1 \end{pmatrix}; \qquad \mathcal{B}\begin{pmatrix} u_1 \\ u_2 \\ u_3 \\ u_4 \end{pmatrix} = \begin{pmatrix} u_1 + u_2 \\ u_1 - u_2 \\ u_3 + u_4 \\ u_3 - u_4 \end{pmatrix};$$

$$\mathcal{C}\begin{pmatrix} u_1 \\ u_2 \\ u_3 \\ u_4 \end{pmatrix} = \begin{pmatrix} u_4 \\ u_1 \\ u_2 \\ u_3 \end{pmatrix}; \qquad \mathcal{D}\begin{pmatrix} u_1 \\ u_2 \\ u_3 \\ u_4 \end{pmatrix} = \begin{pmatrix} u_1 + u_2 + u_3 + u_4 \\ u_1 + u_2 + u_3 + u_4 \\ u_1 + u_2 + u_3 + u_4 \\ u_1 + u_2 + u_3 + u_4 \end{pmatrix};$$

$$\mathcal{E}\begin{pmatrix} u_1 \\ u_2 \\ u_3 \\ u_4 \end{pmatrix} = \begin{pmatrix} (u_1 - u_2 + u_3 - u_4)/2 \\ (u_1 - u_2 - u_3 + u_4)/2 \\ (u_1 + u_2 + u_3 + u_4)/2 \\ (u_1 + u_2 - u_3 - u_4)/2 \end{pmatrix}.$$

8.6.4. Is $\mathcal{A}^* = -\mathcal{A}^{-1}$ true for some $\mathcal{A}$?

8.6.5. True or false?

(a) If $\mathcal{A}$ is symmetric, then also $\mathcal{A}^2$ is symmetric.
(b) If $\mathcal{A}^2$ is symmetric, then also $\mathcal{A}$ is symmetric.
(c) If $\mathcal{A}^2$ is symmetric and $\mathcal{A}$ has a diagonal matrix (in a not necessarily orthonormal basis), then also $\mathcal{A}$ is symmetric.
(d) If $\mathcal{A}$ is orthogonal, then also $\mathcal{A}^2$ is orthogonal.
(e) If $\mathcal{A}^2$ is orthogonal, then also $\mathcal{A}$ is orthogonal.

8.6.6. Assume that $\mathcal{A}\mathcal{A}^* = \mathcal{A}^*\mathcal{A}$ and $\mathcal{A}^k$ is orthogonal for some $k > 1$. Show that $\mathcal{A}$ is orthogonal also. Is a similar statement true if orthogonal is replaced by symmetric?

8.6.7. Assume that $\operatorname{Ker}\mathcal{A} = \mathbf{0}$ and $\mathcal{A}^* = \mathcal{A}^m$ for some $m > 1$. Prove that $\mathcal{A}$ is orthogonal.

8.6.8. Let $(k, t) = 1$ and assume that $\mathcal{A}$ is invertible. Verify that both $\mathcal{A}^k$ and $\mathcal{A}^t$ are symmetric if and only if $\mathcal{A}$ is symmetric. Prove a similar statement for orthogonal transformations, too.

8.6.9. Prove the analogues of Exercises 8.5.17–8.5.18 for orthogonal transformations.

8.6.10. Give a geometric characterization of the symmetric and orthogonal transformations in the plane and space.

8.6.11. Show that if $\mathcal{A}^*$ is a polynomial of $\mathcal{A}$, then V is the direct sum of pairwise orthogonal $\mathcal{A}$-invariant subspaces of dimensions 1 or 2. Is the converse true?

*8.6.12. Prove that $\mathcal{A}^*$ is a polynomial of $\mathcal{A}$ if and only if $\mathcal{A}\mathcal{A}^* = \mathcal{A}^*\mathcal{A}$.

Chapter 9

Combinatorial Applications

Linear algebra has a very wide range of applications. One might think, however, that substantive achievements need many prerequisites both from linear algebra and from the other relevant areas. In this chapter, we show that a great deal of combinatorial applications require hardly any previous knowledge, still allowing us to prove significant and surprising results. As we shall see, quite often we have to use just some basic facts about systems of equations (Gaussian elimination and consequences, Section 3.1). There will be, of course, also parts when we rely on deeper material (including, e.g., the structure of finite fields, too).

We selected the applications not just strictly from combinatorics and graph theory, but we discuss several questions from number theory and geometry, as well. The word "combinatorial" in the title of the chapter refers, thus, rather to the combinatorial *character* of the problems. (Important applications of different character are related mostly to various parts of analysis, such as differential and difference equations, but we do not touch on these fields in this book.) The selection of the topic was arbitrary and subjective; we strived to show many types of problems and methods. We favored questions which lead to other interesting relations not necessarily connected to linear algebra.

9.1. Nice Polynomials

Our introductory problem is about polynomials and seems to lack any relation to linear algebra.

Theorem 9.1.1. *Every non-zero polynomial f has a non-zero multiple $g = fh$ such that g contains only terms with prime exponents.* ♣

The theorem holds for polynomials over any field, but also for polynomials, e.g., with integer coefficients. This latter follows from solving the problem for rational coefficients and then multiplying h by the least common multiple of the denominators of its coefficients.

We present three different proofs using linear algebra relying on systems of equations, dimensions of subspaces, and the Dimension Theorem for linear maps.

It will be clear from the proofs that the theorem remains valid if we replace the primes by any infinite sequence of integers in the requirement for the exponents. (The assertion is astounding if we take some interesting sequence such as primes, Fibonacci numbers, or powers of two; we just have to check that the assertion is not trivial for that particular sequence.)

First proof. Let $f = \alpha_0 + \alpha_1 x + \cdots + \alpha_t x^t$, $\alpha_t \neq 0$, and we search h in the form $h = \beta_0 + \beta_1 x + \cdots + \beta_n x^n$ where the suitable value of n will be chosen later. Performing the multiplication fh, the coefficients of the terms with non-prime exponents must be 0; this means a homogeneous system of linear equations with variables β_j and coefficients coming from the values α_i. (The first few equations are $\alpha_0\beta_0 = 0$, $\alpha_0\beta_1 + \alpha_1\beta_0 = 0$, $\alpha_0\beta_4 + \cdots + \alpha_4\beta_0 = 0$, etc.) There are $n+1$ variables. Each equation corresponds to a non-prime number from the set $\{0, 1, 2, \ldots, n+t\}$; hence there are $n+t+1-\pi(n+t)$ equations where $\pi(s)$ stands for the number of positive primes not greater than s. If there are more variables than equations, then the homogeneous system of linear equations must have a non-trivial solution yielding a suitable polynomial h. The required inequality $n+1 > n+t+1-\pi(n+t)$ holds if and only if $\pi(n+t) > t = \deg f$. Thus, choosing n accordingly, this implies the existence of g satisfying the condition. ∎

Second proof. Let s be a positive integer to be specified later and let V be the usual vector space of polynomials having degree at most s (including the 0 polynomial). We consider the following two subsets of V: W_1 contains the polynomials (of degree at most s) where only terms with prime exponents occur and W_2 consists of the multiples of f. Clearly, W_1 and W_2 are subspaces in V; further $\dim V = s+1$, $\dim W_1 = \pi(s)$, and $\dim W_2 = s-t+1$. If $\dim W_1 + \dim W_2 > \dim V$, then $W_1 \cap W_2 \neq \mathbf{0}$ (see Exercise 4.6.6), and any non-zero element in $W_1 \cap W_2$ works as g. The condition $\dim W_1 + \dim W_2 > \dim V$ holds if and only if $\pi(s) > t = \deg f$ (which is the same as the bound obtained in the first proof). ∎

Third proof. Let W_1 and V be the same as in the second proof and consider the linear map $\mathcal{A} : W_1 \to V$ which assigns to each polynomial its remainder when divided by f. Then $\operatorname{Ker} \mathcal{A}$ is the set of polynomials (of degree at most s) divisible by f and containing only terms with prime exponents. If $\dim \operatorname{Im} \mathcal{A} < \dim W_1$, then $\operatorname{Ker} \mathcal{A} \neq \mathbf{0}$ by the Dimension Theorem 5.4.1; thus any non-zero element of $\operatorname{Ker} \mathcal{A}$ can be chosen as g. Clearly, $\operatorname{Im} \mathcal{A}$ is a subspace of the polynomials of degree at most $t-1$, so $\dim \operatorname{Im} \mathcal{A} \leq t$; further $\dim W_1 = \pi(s)$. Hence $\dim \operatorname{Im} \mathcal{A} < \dim W_1$ follows from $\pi(s) > t = \deg f$ (obtained also in the previous proofs). ∎

Exercises 9.1

9.1.1. *A guessing game.*

(a) Alice thought of 20 integers; let these be $x_1, x_2, \dots, x_{20}$. Bob can ask the value of any expression formed from these 20 numbers using addition and subtraction (e.g., $x_1 + 8x_2 - 7x_3$). The next question can depend on the answer to the previous one. Find the minimal number of questions allowing Bob to guess the 20 integers.

*(b) What happens if Bob knows that each integer is positive?

*(c) Solve the previous two problems if multiplication is also permitted, i.e., Bob can ask the value of any multivariate polynomial in variables x_i and having integer coefficients (e.g., $x_1 + 8x_2^3x_5$).

9.1.2. *Weights.* We have 13 weights. Disregarding any of them, we can form two groups of six from the other 12 so that the sum of weights is equal in the two groups. Prove that all 13 weights are equal.

9.1.3. *Weights again.* Is the assertion of the previous exercise still valid if we do not require that the two groups of equal total weight should contain the same number of weights?

***9.1.4.** *Boring vectors.*

(a) Let us call a vector in $\mathbf{R}^m$ boring if at most two distinct values occur among its coordinates (e.g., each coordinate is -1 or $\sqrt{2}$). What is the minimal number of terms needed to represent

(i) $\begin{pmatrix} 1 \\ 2 \\ \vdots \\ m \end{pmatrix}$;

(ii) any element in $\mathbf{R}^m$ as the sum of boring vectors?

(b) What is the answer if we modify the definition of boring vectors so that at most k distinct values occur among their coordinates?

(c) Does it make any difference if we replace real numbers by the modulo p field F_p?

Remark. The reason for the cautious formulation in part (c) is due to our lack of knowledge for the case of general vectors, i.e., we cannot determine the exact minimum for all values of m and p concerning the field F_p.

9.1.5. *A weird contest.*

(a) A class of 32 students organizes a contest of several rounds. In each round, two teams of arbitrary sizes compete. Even one student can form a team and no student is obliged to participate in every round. The only requirement is that any two students should be opponents sometime (i.e., they should be members of different teams in at least one round). What is the minimal possible number of rounds?

*(b) What is the minimal number of rounds if any two students have to be opponents in *exactly* one round?

***9.1.6.** *Small and large intersections.*

(a) Do there exist (a1) countably infinite; (a2) continuum many sets of positive integers such that the intersection of any two sets is infinite but the intersection of any three is finite?

(b) Solve the problem if the second condition is modified so that the intersection of any three sets is empty.

9.1.7. *Odd distances.* Are there four points in the (a) (ordinary three-dimensional) space; *(b) plane where the distance of any two points is an odd integer?

*9.1.8 *Cauchy's functional equation.* Consider the functions $f : \mathbf{R} \to \mathbf{R}$ satisfying $f(a+b) = f(a) + f(b)$ for every $a, b \in \mathbf{R}$.

(a) Show that $f(x) = cx$ for every rational x (for some constant c).

(b) If f is continuous in some point, then $f(x) = cx$ for every real x.

(c) If f is bounded in an (arbitrarily small) interval, then $f(x) = cx$ for every real x.

(d) There exist functions $f(x) \neq cx$ satisfying Cauchy's functional equation.

9.1.9. *Halving.* Can the set of the positive (a) rational; *(b) real numbers be partitioned into two disjoint, non-empty subsets each closed under addition?

*9.1.10 *Polynomials and periodic functions.*

(a) Is the identity function $f(x) = x$ defined on (a1) $\mathbf{Q}$; (a2) $\mathbf{R}$ representable as the sum of (finitely many) periodic functions?

(b) Is a polynomial function with real coefficients of degree $n \geq 1$ the sum of (b1) n; (b2) $n+1$ periodic $\mathbf{R} \to \mathbf{R}$ functions?

9.2. Fibonacci Numbers

In this section, we determine the formula for the *Fibonacci numbers* discussed already in Exercise 4.6.8. Recall that the sequence of Fibonacci numbers is defined by the recursion $\varphi_0 = 0$, $\varphi_1 = 1$, $\varphi_{j+1} = \varphi_j + \varphi_{j-1}$, $j = 1, 2, \ldots$. The first few terms are 0, 1, 1, 2, 3, 5, 8, 13, 21, 34, ...

Theorem 9.2.1.

$$\varphi_n = \frac{1}{\sqrt{5}}\left(\left(\frac{1+\sqrt{5}}{2}\right)^n - \left(\frac{1-\sqrt{5}}{2}\right)^n\right). ♣$$

Remark. If we know the formula, then we can easily verify it by induction. This, however, gives no explanation how one can find this formula; it is absolutely improbable to guess it even after studying (say) a thousand values. The proofs below also provide the route leading to the formula.

The first two proofs use linear algebra and the third one relies on power series from analysis (though the algebraic theory of formal power series would work, as well). Similar methods apply also in general to *recursive sequences* playing an important role in several branches of mathematics.

First proof. As a generalization of the Fibonacci sequence, let us call a sequence $\alpha_0, \alpha_1, \ldots$ of real numbers a Φ-sequence if it satisfies the condition $\alpha_{j+1} = \alpha_j + \alpha_{j-1}$, $j = 1, 2, \ldots$ where the first two terms α_0 and α_1 are arbitrary real numbers. A scalar multiple of a Φ-sequence and the sum of two Φ-sequences are again Φ-sequences (the sum of two sequences and a scalar multiple of a sequence mean the usual operations adding the corresponding terms of the sequences and multiplying every term with the scalar).

We try to express the Fibonacci sequence $F = (\varphi_0, \varphi_1, \dots)$ with Φ-sequences that have a simple formula. It is easy to see that the only arithmetic progression among the Φ-sequences is the constant zero sequence, so this does not help in solving our problem. We are, however, more lucky with geometric sequences: $(\alpha, \alpha\rho, \alpha\rho^2, \dots)$ (where $\alpha \neq 0$) is a Φ-sequence if and only if $\rho^2 = \rho + 1$ yielding $\rho_1 = (1+\sqrt{5})/2$ and $\rho_2 = (1-\sqrt{5})/2$.

Let $S_m = (1, \rho_m, \rho_m^2, \dots)$, $m = 1, 2$, and we want to represent the Fibonacci sequence as $F = \gamma_1 S_1 + \gamma_2 S_2$. There are Φ-sequences on both sides of the equality, so it holds exactly if it holds for the first two terms of the sequences as the equality gets inherited for all later terms by the recursion. The equalities for the first two terms are $\varphi_0 = 0 = \gamma_1 + \gamma_2$ and $\varphi_1 = 1 = \gamma_1\rho_1 + \gamma_2\rho_2$. Solving this system of linear equations, we get $\gamma_1 = 1/\sqrt{5}, \gamma_2 = -1/\sqrt{5}$. So $\varphi_n = \gamma_1\rho_1^n + \gamma_2\rho_2^n$ yielding the formula in the theorem (by the concrete values ρ_m and γ_m). ∎

Remark. One can ask where linear algebra was used, since the proof in its above form works with high-school methods. Linear algebra is what provides the perspective and the background. We are looking for a suitable spanning set in the vector space of the Φ-sequences to represent the Fibonacci sequence F as a linear combination. The dimension of this vector space is two as the first two terms can be chosen freely. (Precisely: A natural basis is given by the two Φ-sequences starting with $1, 0$ and $0, 1$: $(1, 0, 1, 1, 2, 3, 5, \dots)$ and $(0, 1, 1, 2, 3, 5, 8, \dots)$ that are the Fibonacci sequence and its translate.) In this two-dimensional vector space, we want to find a spanning set of sequences that can be easily handled. The geometric sequences S_1 and S_2 are okay as two linearly independent sequences form a basis.

This shows that it was not accidental that we found the representation above. The quotients of the geometric sequences were obtained as the roots of a quadratic equation and we only would have been in trouble if the equation had a multiple root. Complex roots would have caused no difficulty; then we just have to operate with complex sequences. The case of multiple roots can be handled, too, even for recursions of more terms; see Exercise 9.2.3.

Second proof. Let $\mathcal{F} : \mathbf{R}^2 \to \mathbf{R}^2$ be the linear transformation

$$\mathcal{F}\begin{pmatrix} a \\ b \end{pmatrix} = \begin{pmatrix} b \\ a+b \end{pmatrix}.$$

Then

$$\mathcal{F}\begin{pmatrix} \varphi_{j-1} \\ \varphi_j \end{pmatrix} = \begin{pmatrix} \varphi_j \\ \varphi_{j-1} + \varphi_j \end{pmatrix} = \begin{pmatrix} \varphi_j \\ \varphi_{j+1} \end{pmatrix}, \text{ so } \mathcal{F}^n\begin{pmatrix} 0 \\ 1 \end{pmatrix} = \begin{pmatrix} \varphi_n \\ \varphi_{n+1} \end{pmatrix}.$$

Assume that $\mathbf{b}_1, \mathbf{b}_2$ is an eigenbasis of $\mathcal{F}$ in $\mathbf{R}^2$. Let λ_1 and λ_2 be the eigenvalues; then $\mathcal{F}^n\mathbf{b}_m = \lambda_m^n\mathbf{b}_m$, $m = 1, 2$. If

$$\begin{pmatrix} 0 \\ 1 \end{pmatrix} = \delta_1\mathbf{b}_1 + \delta_2\mathbf{b}_2, \tag{9.2.1}$$

then

$$\begin{pmatrix}\varphi_n\\ \varphi_{n+1}\end{pmatrix} = \mathcal{F}^n\begin{pmatrix}0\\1\end{pmatrix} = \delta_1\lambda_1^n\mathbf{b}_1 + \delta_2\lambda_2^n\mathbf{b}_2. \tag{9.2.2}$$

So, φ_n is equal to the first coordinate on the right-hand side of (9.2.2).

To compute the eigenvectors, we write the matrix of $\mathcal{F}$ in the basis

$$\begin{pmatrix}1\\0\end{pmatrix}, \begin{pmatrix}0\\1\end{pmatrix} : [\mathcal{F}] = \begin{pmatrix}0 & 1\\ 1 & 1\end{pmatrix}.$$

Then the characteristic polynomial is $k_{\mathcal{F}} = x^2 - x - 1$, its roots $\lambda_{1,2} = (1 \pm \sqrt{5})/2$ are the eigenvalues and an eigenbasis is

$$\mathbf{b}_{1,2} = \begin{pmatrix}1\\ (1 \pm \sqrt{5})/2\end{pmatrix}.$$

Substituting into (9.2.1), we get $\delta_1 = -\delta_2 = 1/\sqrt{5}$ and (9.2.2) yields the formula in the theorem. ∎

Third proof. We consider the power series $F(z) = \sum_{n=0}^{\infty} \varphi_n z^n$. The definition of the Fibonacci numbers implies $\varphi_n < 2^n$ by induction, so $F(z)$ is absolutely convergent for $|z| < 1/2$. This is useful, as absolutely convergent series obey similar laws of operations as finite sums.

We write the power series $F(z)$, $zF(z)$, and $z^2F(z)$ one below the other:

$$\begin{aligned}
F(z) &= \varphi_0 + \varphi_1 z + \varphi_2 z^2 + \ldots + \quad \varphi_n z^n + \ldots,\\
zF(z) &= \qquad\quad \varphi_0 z + \varphi_1 z^2 + \ldots + \varphi_{n-1} z^n + \ldots,\\
z^2F(z) &= \qquad\qquad\quad + \varphi_0 z^2 + \ldots + \varphi_{n-2} z^n + \ldots.
\end{aligned}$$

Subtracting the sum of the last two rows from the first one, nearly all terms get canceled on the right-hand side and we obtain $(1 - z - z^2)F(z) = z$ or $F(z) = z/(1 - z - z^2)$.

We decompose this algebraic fraction for $F(z)$ into partial fractions and expand these into power series. Let $\mu_{1,2} = (-1 \pm \sqrt{5})/2$ be the roots of $z^2 + z - 1$. Then

$$F(z) = \frac{-z}{z^2 + z - 1} = \frac{\beta_1}{1 - \frac{z}{\mu_1}} + \frac{\beta_2}{1 - \frac{z}{\mu_2}} \tag{9.2.3}$$

with suitable constants β_1 and β_2. Performing the multiplications (for $z \neq \mu_1, \mu_2$) we get the equivalent condition $-z = \mu_1\beta_1(\mu_2 - z) + \mu_2\beta_2(\mu_1 - z)$. So $0 = \mu_1\mu_2(\beta_1 + \beta_2)$ and $\mu_1\beta_1 + \mu_2\beta_2 = 1$ yielding $\beta_1 = -\beta_2 = 1/\sqrt{5}$. Substituting back into the right-hand side of (9.2.3) and expanding the functions $(1 - z/\mu_m)^{-1}$ into infinite geometric series, we obtain

$$F(z) = \beta_1 \sum_{n=0}^{\infty}\left(\frac{z}{\mu_1}\right)^n + \beta_2 \sum_{n=0}^{\infty}\left(\frac{z}{\mu_2}\right)^n = \sum_{n=0}^{\infty}\left(\frac{\beta_1}{\mu_1^n} + \frac{\beta_2}{\mu_2^n}\right)z^n.$$

The coefficient of z^n is φ_n on the left-hand side and is the formula of the theorem on the right-hand side. (Comparing the values in the second and third proofs, we have $\mu_m = 1/\lambda_m$ and $\beta_m = \delta_m$, $m = 1, 2$.) ∎

Exercises 9.2

9.2.1. Compute α_{1000} if $\alpha_1 = \alpha_2 = 1$ and
(a) $\alpha_k = \alpha_{k-1} + 2\alpha_{k-2}$ for $k \geq 3$;
(b) $\alpha_k = 2\alpha_{k-1} + \alpha_{k-2}$ for $k \geq 3$.

9.2.2. Compute α_{1111} if $\alpha_1 = 3, \alpha_2 = 7$ and
(a) $\alpha_k = 2\alpha_{k-1} - \alpha_{k-2}$ for $k \geq 3$;
(b) $\alpha_k = \alpha_{k-1} - \alpha_{k-2}$ for $k \geq 3$;
(c) $\alpha_k = \alpha_{k-1} - \alpha_{k-2} + \alpha_{k-3} - \alpha_{k-4}$ for $k \geq 5$.

*9.2.3. Consider the complex sequences $\alpha_0, \alpha_1, \dots$ satisfying $\alpha_k = \mu_1\alpha_{k-1} + \dots + \mu_t\alpha_{k-t}$ for every $k \geq t$ where t is a fixed positive integer, $\mu_1, \dots, \mu_t$ are fixed complex numbers and $\mu_t \neq 0$. Let the distinct complex roots of the polynomial $f = x^t - \mu_1 x^{t-1} - \dots - \mu_t$ be $\lambda_1, \dots, \lambda_r$ with multiplicities $s_1, \dots, s_r$. Prove that $\alpha_n = \sum_{j=1}^{r} g_j(n)\lambda_j^n$ where g_j is a polynomial of degree not greater than $s_j - 1$, $j = 1, 2, \dots, r$ and the coefficients of g_j only depend on the initial values $\alpha_0, \dots, \alpha_{t-1}$.

9.2.4. We use the notation of the previous exercise. Verify that a complex sequence $\alpha_0, \alpha_1, \dots$ is periodic for *every* system $\alpha_0, \dots, \alpha_{t-1}$ of initial values if and only if every root of f is a root of unity and f has no multiple roots.

9.2.5. Determine β_n if $\beta_0 = 0$, $\beta_1 = 1$, and $\beta_k = \beta_{k-1} + \beta_{k-2} + 2$ for $k \geq 2$.

9.2.6. Prove.
(a) The closest integer to $\tau_n = (1/\sqrt{5})\left((1 + \sqrt{5})/2\right)^n$ is the nth Fibonacci number φ_n.
(b) The difference of φ_n and τ_n tends to 0 if $n \to \infty$ (i.e., φ_n is nearly a geometric sequence).

9.2.7. (a) Compute the number of tilings of a $2 \times n$ rectangle with 2×1 dominoes.
*(b) Compute the number of tilings of a $3 \times n$ rectangle with 2×1 dominoes.
*(c) Let ψ_n denote the number of tilings of a $3 \times n$ rectangle with 3×1 dominoes. Prove $1.46^n < \psi_n < 1.47^n$ for n large enough.

9.2.8. How many subsets of $\{1, 2, \dots, n\}$ do not contain two consecutive integers?

9.2.9. Show that every positive integer is the sum of distinct Fibonacci numbers.

9.2.10. Verify the equalities for Fibonacci numbers.
(a) $\varphi_{m+n} = \varphi_{m-1}\varphi_n + \varphi_m\varphi_{n+1}$;
(b) $\varphi_{2n-1} = \varphi_{n-1}^2 + \varphi_n^2$;
(c) $\varphi_1 + \varphi_2 + \dots + \varphi_n = \varphi_{n+2} - 1$;
(d) $\varphi_1 + 2\varphi_2 + \dots + n\varphi_n = (n+1)\varphi_{n+2} - \varphi_{n+4} + 2$;
(e) $\varphi_1^2 + \varphi_2^2 + \dots + \varphi_n^2 = \varphi_n\varphi_{n+1}$;
(f) $\varphi_n^2 = \varphi_{n-1}\varphi_{n+1} + (-1)^{n+1}$;
(g) $\varphi_{3n} = 5\varphi_n^3 + 3(-1)^n\varphi_n$;
(h) $\varphi_n = \sum_{j=0}^{\lfloor (n-1)/2 \rfloor} \binom{n-j-1}{j}$;
(i) $\varphi_{2n} = \sum_{j=1}^{n} \binom{n}{j}\varphi_j$.

9.2.11. Prove that two consecutive Fibonacci numbers are always coprime. What can we assert for the second neighbors and the third neighbors?

9.2.12. Show that every integer $m > 0$ has infinitely many multiples among the Fibonacci numbers.

*9.2.13. Verify $k \mid n \iff \varphi_k \mid \varphi_n$, moreover $\varphi_{(k,n)} = (\varphi_k, \varphi_n)$.

9.2.14. (a) Let $a > b > 0$ be integers such that the Euclidean algorithm consists of *exactly* n steps. Determine the smallest possible b.

*(b) Solve the problem if, instead of minimal non-negative remainders, we perform the Euclidean algorithm with remainders of minimal absolute value.

*9.2.15. Compute the sum of the infinite series formed of the reciprocals of Fibonacci numbers with indices powers of two.

***9.2.16.** We do not need parentheses for products of more than two factors as multiplication is associative. Consider now a non-associative operation. How many values can assume a product of n factors (i.e., how many ways can we place the parentheses)?

**9.2.17. How many ways can we partition a convex n-gon into triangles using diagonals with no intersection inside the polygon?

9.3. Search for Squares

In this section, we prove that if none of sufficiently many distinct positive integers has greater prime divisors than a given threshold, then the product of some (one, more, or all) of the integers is a square. (A concrete instance of the problem occurred in Exercise 3.3.12.) Surprisingly, the linear algebraic solution of this question plays an important role in the factorization of large numbers. As no really efficient factorization algorithm is known at present, any improvement has a practical significance, too, due to the connection of factorization and public-key cryptosystems widely used for military, diplomatic, and business purposes. Another interesting feature of the search for squares problem is that the field of modulo 2 residue classes arises naturally in the solution. The generalizations for cubes and higher powers lead to further interesting and partly unsolved problems (see Exercises 9.3.1–9.3.4).

Theorem 9.3.1. *Let $P = \{p_1, \dots, p_k\}$ be a set of k distinct positive prime numbers and let $C = \{c_1, \dots, c_{k+1}\}$ a set of $k+1$ be distinct positive integers without prime factors outside P. Then some non-empty product of distinct elements of C is a square.* ♣

Note that the theorem is false for $|C| = k$ instead of $k+1$; take, e.g., $c_j = p_j$.

First proof. We can form $2^{k+1} - 1$ products from the elements of the non-empty subsets D of C. Considering the prime factorization of such a product, there are 2^k options for the parities of the exponents of the primes p_i. By the pigeonhole principle, there must be two products where the exponent of each prime has the same parity in both products. Deleting the common factors from the two products and multiplying the remaining ones, the resulting product is a square. ∎

Second proof. Let $F = F_2$ be the modulo 2 field and assign the vectors $\mathbf{c}_j \in F^k$ to the integers $c_1, \dots, c_{k+1}$ by

$$\mathbf{c}_j = \begin{pmatrix} \gamma_{1j} \\ \vdots \\ \gamma_{kj} \end{pmatrix}, \qquad \gamma_{ij} = \begin{cases} 1, & \text{if the exponent of } p_i \text{ in } c_j \text{ is odd;} \\ 0, & \text{if the exponent of } p_i \text{ in } c_j \text{ is even.} \end{cases}$$

Thus, such a vector shows the parity of the exponents of the primes in the corresponding integer and the modulo 2 addition of the vectors $\mathbf{c}_j$ corresponds to the multiplication of the numbers c_j. These $k+1$ vectors in F^k are linearly dependent, so we can write $\mathbf{0}$ as their non-trivial linear combination. Consider the product of the integers c_j corresponding to the vectors $\mathbf{c}_j$ that occur in this linear combination with a coefficient non-zero, i.e., 1. As the sum of these vectors is $\mathbf{0}$, the product of the integers c_j is a square. ∎

One could think that the second proof using linear algebra is too big a gun for the problem. This is not the case, as it provides an *efficient, fast algorithm* to create a square product (in contrast with the first proof based on the pigeonhole principle): we can find a non-trivial linear combination by solving a homogeneous system of linear equations, e.g., by Gaussian elimination.

This fact plays an important role in the *factorization* of large integers. No really efficient algorithm is known for factorization but the quickest algorithms make use of such a square product, too.

We sketch a factorization algorithm. Let N be a large odd composite number and the task is to find a non-trivial factor. It is easy to recognize and factor the complete powers (see Exercise 9.3.7). So it is enough to consider when N has at least two distinct prime divisors.

Assume that

$(*)$ $u^2 \equiv v^2 \pmod N$

$(**)$ $u \not\equiv \pm v \pmod N$

i.e., $N \mid u^2 - v^2$ but $N \nmid u \pm v$. Then the gcd of N and $u+v$ is a non-trivial divisor of N. We can compute this gcd quickly by the Euclidean algorithm.

How can we find u and v satisfying $(*)$ and $(**)$? We consider first $(*)$. We try to construct integers b such that the remainder c of b^2 under division by N is only divisible by small primes. We can hope to find such a c if we choose b a slightly greater integer than, e.g., $\sqrt{N}$, $\sqrt{2N}$, etc. We fix a bound R depending on N, and we keep b if no prime divisor of c exceeds R, otherwise we discard it. Then we try a new value for b, etc. We collect $k+1$ pairs b_j, c_j where $k = \pi(R)$ is the number of primes up to R. Then $b_j^2 \equiv c_j \pmod N$, $j = 1, 2, \dots, k+1$. It is worth computing the gcd of b and N by the Euclidean algorithm since this may yield a non-trivial factor of N. It is, however, very improbable that we are so lucky. So, we can assume that the integers b_j (hence also c_j) are coprime to N.

The second proof of Theorem 9.3.1 provides a *quick algorithm* to select some of the integers c_j whose product is a square v^2. Let u be the product of the corresponding numbers b_j. Then multiplying these congruences $b_j^2 \equiv c_j \pmod{N}$, we obtain $u^2 \equiv v^2 \pmod{N}$.

We now have to check $(**)$. We claim that it holds with at least probability $1/2$. Let $N = q_1^{t_1} \dots q_s^{t_s}$ (where $s \geq 2$) be the (unknown, but existent) standard form of N. Then the congruence $u^2 \equiv v^2 \pmod{N}$ is equivalent to the system of congruences $u^2 \equiv v^2 \pmod{q_j^{t_j}}$, $j = 1, 2, \dots, s$. As q^t is a power of an odd prime, $u^2 \equiv v^2 \pmod{q^t}$ holds if and only if $u \equiv \pm v \pmod{q^t}$. The only exception could be if both $u + v$ and $u - v$ are divisible by q, but then $q \mid (u + v) + (u - v) = 2u$. Since $(q, 2) = 1$, this implies $q \mid u$ contradicting $(u, N) = 1$. Therefore the system of congruences $u^2 \equiv v^2 \pmod{q_j^{t_j}}$, $j = 1, 2, \dots, s$ is satisfied in 2^s ways: in each congruence $u \equiv v$ or $u \equiv -v \pmod{q_j^{t_j}}$. There are only two instances among these 2^s cases when $u \equiv \pm v \pmod{N}$: in each congruence $u \equiv v$ or in each congruence $u \equiv -v \pmod{q_j^{t_j}}$. The selection of u and v from the solutions of $u^2 \equiv v^2 \pmod{N}$ can be considered essentially random, so we assume that each of the 2^s cases has the same probability. Then $u \not\equiv \pm v \pmod{N}$ holds with probability $1 - 2/2^s \geq 1/2$. Therefore it is nearly certain that sooner or later we find a pair u, v satisfying $(**)$, and this leads to a non-trivial divisor of N.

Why is this algorithm not really efficient? The devil is in the overlooked details: how should we search the promising integers b and how should we set the bound R. Deep number-theoretic results can somewhat increase the chance to find suitable candidates b. In defining R we face the dilemma that if R is small, then we only have to collect a few values of b_j but we rarely find a suitable one, and if R is big, then though we relatively often find good integers b_j, we need a lot of them. Again, deep number-theoretic considerations help us find the optimal R.

In Exercises 9.3.1–9.3.4, we discuss generalizations of Theorem 9.3.1 for cubes and higher powers instead of squares.

Exercises 9.3

9.3.1. *Cubes.* Let $P = \{p_1, \dots, p_k\}$ be a set of k distinct positive prime numbers and let $C = \{c_1, \dots, c_t\}$ be a set of t distinct positive integers without prime factors outside P. Prove:

(a) If $t \leq 2k$, then it can occur that no non-empty product of distinct elements of C is a cube.

(b) If $t \geq 2 \cdot 3^k + 1$, then some non-empty product of distinct elements of C is a cube.

(c) If $t \geq k + 1$, then some non-empty product of elements of C is a cube so that every c_j occurs at most twice among the factors.

***9.3.2.** *Chevalley's theorem.* Let $p > 0$ be a prime and let $f_i(x_1, x_2, \dots, x_t)$, $i = 1, 2, \dots, k$ be polynomials in t variables with integer coefficients and constant term 0 satisfying $\sum_{i=1}^{k} \deg f_i < t$. Prove that the system of congruences $f_i(x_1, x_2, \dots, x_t) \equiv 0 \pmod{p}$, $i = 1, 2, \dots, k$ has a non-trivial solution. Which well-known theorem is the special case when every polynomial has degree one?

*9.3.3. *Cubes again.* Let $P = \{p_1, \dots, p_k\}$ be a set of k distinct positive primes and let $C = \{c_1, \dots, c_t\}$ be a set of t distinct positive integers without prime factors outside P. Prove that if $t \geq 2k + 1$, then some non-empty product of distinct elements of C is a cube.

*9.3.4. *Higher powers.* Generalize the previous exercise to qth powers instead of cubes where q is an arbitrary prime.

Remark. The corresponding statement can also be verified by different methods if q is a prime power but the question is unsolved for a general integer q.

9.3.5. *Divisibility of sums.* Let C be a set of t positive integers. Prove:

(a) If $t \geq n$, then some non-empty sum of distinct elements of C is divisible by n.

*(b) If $t \geq 2n - 1$, then some non-empty sum of n distinct elements of C is divisible by n.

9.3.6. (a) Verify that for every integer $n > 1$, there exist three integers such that the sum s of their squares is divisible by n but not by n^2.

(b) Show that $(n, s/n) = 1$ is achievable.

9.3.7. Design a fast algorithm for recognizing and factoring complete powers.

9.3.8. *Factorization.* The scheme of another factoring algorithm is that we try to write a large odd number N in the form $N = x^2 - y^2$. Therefore we check the integers $x \geq \sqrt{N}$ whether $x^2 - N$ is a square. Using that a square must have special residues mod 3, 5, 7, 8, etc., we can significantly reduce the search for x.

(a) If we look at the residues of $x^2 - N$ mod s distinct primes, then roughly what is the ratio of the numbers x for which this method reveals that $x^2 - N$ cannot be a square?

(b) When does this method find (relatively) quickly a decomposition $N = de$: if d is small and e is large, or if d and e are close to each other?

(c) Apply this method to factor 86519 and 584189. Use just a calculator and sieve only mod 3 and mod 8.

9.4. Oddtown and Eventown

We impose various conditions on the sizes of subsets and their pairwise intersections in a k-element set and ask for the maximal number of such subsets. Surprisingly, a minimal change in the conditions implies a dramatic change in the answer. We illustrate this by a little tale.

Once upon a time, in the remote land of Combinatorics, there was a small town of 32 inhabitants. These people loved to create clubs. The only requirement was that two clubs cannot have exactly the same membership since then these are the same club under two different names. Even the club with no members was announced and this proved to become a very peaceful association with no hard disputes.

As there were more and more clubs, the houses were getting crowded with the thousands of club-cards and the printers were unable to produce the new ones. So, the citizens decided to introduce stricter measures. There were two motions before the prestigious senate (consisting of all citizens, of course). Both motions prescribed that any two clubs may share only an even number of members. The only slight difference between them was that the Oddtown Party suggested that each club should have an odd number of members, whereas the Eventown Party insisted that each club should have an even number of members. As both parties had 16–16 representatives in the senate, voting resulted in a tie. Therefore they asked the opinion of the highly respected Old Linear Algebra from the next village. Here is what he said.

Theorem 9.4.1 (Oddtown). *Let* $|X| = k$ *and* $H_1, \dots, H_n$ *be distinct subsets in* X *such that every* $|H_j|$ *is odd and* $|H_t \cap H_j|$ *is even if* $t \neq j$. *Then* $\max n = k$. ♣

Theorem 9.4.2 (Eventown). *Let* $|X| = k$ *and* $H_1, \dots, H_n$ *be distinct subsets in* X *such that every* $|H_j|$ *is even and* $|H_t \cap H_j|$ *is even if* $t \neq j$. *Then* $\max n = 2^{\lfloor k/2 \rfloor}$. ♣

So, in the tale, Eventown allows the formation of $2^{16} = 65536$ clubs, whereas Oddtown reduces this number to 32.

The proofs of the two theorems are based on a common principle. We extend the notion of inner product and orthogonality to vector spaces over the modulo 2 field F_2 and this will help to characterize the sizes of the subsets and their intersections. Another approach is presented in Exercise 9.4.4.

Proof. The stated maximums can be achieved: In Oddtown we choose the one-element subsets, and in Eventown we form $\lfloor k/2 \rfloor$ (disjoint) pairs of elements, and take all subsets consisting of such pairs. (Using the terminology of the tale, there are 16 couples in Eventown and the two members of a couple can join a club only together.)

To prove that these are the maximums, we shall use the inner product: If F is a field, then for vectors $\mathbf{u} = (u_1, \dots, u_k)$ and $\mathbf{v} = (v_1, \dots, v_k) \in F^k$ we define $\mathbf{u} \cdot \mathbf{v} = u_1 v_1 + \cdots + u_k v_k (\in F)$ (exactly as over the reals). This is a symmetric bilinear function, but the positive definite property makes no sense in general and also a non-zero vector can be orthogonal to itself (see Exercises 9.4.13–9.4.14).

If $X = \{x_1, \dots, x_k\}$, we can characterize any $H \subseteq X$ by a vector $\mathbf{h} = (h_1, \dots, h_k) \in F^k$ where $h_s = 1$ if $x_s \in H$, and $h_s = 0$ if $x_s \notin H$. Then $\mathbf{h} \cdot \mathbf{h}' = |H \cap H'|$: in the inner product, we add 1 for every common element of H and H'. Since the assumptions of our theorems are related to the parity of the sizes of sets, we choose F as the modulo 2 field F_2. Then $\mathbf{h} \cdot \mathbf{h}'$ is 0 or 1 depending on the parity of $|H \cap H'|$. This also holds for $H = H'$, i.e., $\mathbf{h} \cdot \mathbf{h}$ is 0, if $|H|$ is even, and 1, if $|H|$ is odd.

Using the above formulation, the Oddtown theorem states that if the vectors $\mathbf{h}_1, \dots, \mathbf{h}_n \in F^k$ satisfy $\mathbf{h}_t \cdot \mathbf{h}_j = 1$ for $t = j$, and 0 for $t \neq j$, then $n \leq k$. We shall prove this by showing that our vectors are linearly independent. In F^k there are at most k linearly independent vectors, so $n \leq k$.

Take a linear combination $\delta_1 \mathbf{h}_1 + \cdots + \delta_n \mathbf{h}_n = \mathbf{0}$. The inner product of the two sides and $\mathbf{h}_j$ gives $\delta_1(\mathbf{h}_1 \cdot \mathbf{h}_j) + \cdots + \delta_j(\mathbf{h}_j \cdot \mathbf{h}_j) + \cdots + \delta_n(\mathbf{h}_n \cdot \mathbf{h}_j) = 0$. Using $\mathbf{h}_j \cdot \mathbf{h}_j = 1$

and $\mathbf{h}_t \cdot \mathbf{h}_j = 0$ for $t \neq j$, this reduces to $\delta_j = 0$. So, $\delta_j = 0$ for every j, i.e., the vectors $\mathbf{h}_j$ are linearly independent, indeed.

To prove the Eventown theorem, we need some further preparations. We define orthogonality and orthogonal complement over $F = F_2$ as over $\mathbf{R}$: Two vectors, $\mathbf{a}$ and $\mathbf{b}$ are orthogonal in $V = F^k$, if their inner product $\mathbf{a} \cdot \mathbf{b} = 0$ and $U^\perp$ is the set of vectors orthogonal to every element of a subspace U: $U^\perp = \{\mathbf{x} \in V \mid (\mathbf{u} \in U \Rightarrow \mathbf{u} \cdot \mathbf{x} = 0)\}$. It is easy to see that $U^\perp$ is a subspace but Theorem 8.1.7 is no longer true: $U \cap U^\perp \neq \mathbf{0}$ and $\langle U, U^\perp \rangle \neq V$ may occur. The equality $\dim U + \dim U^\perp = \dim V$, however, remains valid. We discuss all this in Exercise 9.4.15.

Returning to Eventown, let $H_1, \dots, H_n$ be subsets of X such that all $|H_j|$ and $|H_t \cap H_j|$ are even. This means that the vectors $\mathbf{h}_j$ corresponding to the subsets H_j are orthogonal to each other and to themselves. As $(\mathbf{a}+\mathbf{b}) \cdot (\mathbf{a}+\mathbf{b}) = \mathbf{a} \cdot \mathbf{a} + 2(\mathbf{a} \cdot \mathbf{b}) + \mathbf{b} \cdot \mathbf{b} = \mathbf{a} \cdot \mathbf{a} + \mathbf{b} \cdot \mathbf{b}$ in F_2, any two vectors are orthogonal in the subspace U spanned by the vectors $\mathbf{h}_j$. This implies $U \subseteq U^\perp$, thus $\dim U \leq \dim U^\perp$. Using $\dim U + \dim U^\perp = \dim V$, we obtain $\dim U \leq \lfloor \dim V/2 \rfloor = \lfloor k/2 \rfloor$. Then $n \leq |U| = 2^{\dim U} \leq 2^{\lfloor k/2 \rfloor}$ as stated. ∎

Exercises 9.4

9.4.1. In a k-element set, determine the number of subsets of size divisible by (a) 2; (b) 3.

9.4.2. Consider the function $H \mapsto \mathbf{h}$ used in the proof of Theorems 9.4.1–9.4.2.
(a) What is the relation between the subsets H and H' if the vectors $\mathbf{h}$ and $\mathbf{h}'$ differ in each component?
(b) Which operation between H and H' corresponds to the addition (over F_2) of the vectors $\mathbf{h}$ and $\mathbf{h}'$?

9.4.3. Give a new proof to the Oddtown theorem based on the linear independence over $\mathbf{Q}$ instead of F_2.

9.4.4. Let $|X| = k$, $H_1, \dots, H_n$ (distinct) subsets in X, and assign to every H_j the vector $\mathbf{h}_j$ as in the proof of Theorems 9.4.1–9.4.2. The $k \times n$ matrix A whose columns are the vectors $\mathbf{h}_1, \dots, \mathbf{h}_n$ is called the *incidence matrix* of the set system $H_1, \dots, H_n$.
(a) What are the elements of the product matrix $B = A^T A$?
(b) Based on (a), give new proofs to the Oddtown and Eventown theorems.

9.4.5. Let ξ_k denote the number of possibilities of forming k clubs in Oddtown of k inhabitants according to the rules of Theorem 9.4.1. Prove.
(a) $\xi_k > 1$ if $k > 3$;
(b) $\xi_k \to \infty$ if $k \to \infty$;
*(c) $2^{k^2/8}/k! < \xi_k < 2^{k^2}/k!$.

9.4.6. *Reverse Oddtown.* What is the maximal possible number of clubs in a town of k inhabitants if every club has an even number of members and any two clubs share an odd number of common members?

9.4.7. *Exploring Numberlands.*

(a) Triple City has k inhabitants, the membership of no clubs is divisible by 3, but the number of common members of any two clubs is a multiple of 3. At most how many clubs can exist in Triple City?

(b) And in Fort Four?

(c) Show that there cannot be more than $2k$ clubs in Sixvillage.

Remark. It is unsolved whether there can be more than k clubs at all.

9.4.8. (a) Let $|X| = k$ where $3 \mid k$. At most how many subsets H_j of X can satisfy $|H_j| \equiv 2 \pmod 3$ and $|H_t \cap H_j| \equiv 1 \pmod 3$ if $t \neq j$?

(b) Solve the problem if $3 \nmid k$.

9.4.9. *Colorful Oddtown.*

(a) In red-and-blue Oddtown with k inhabitants, n red clubs R_j and n blue clubs B_j are formed so that $|R_j \cap B_j|$ is odd for every j and $|R_t \cap B_j|$ is even if $t \neq j$. What is the maximum of n?

(b) Show that we get the same result if we require $|R_t \cap B_j|$ to be even only for $t < j$ instead of $t \neq j$.

***9.4.10.** *Uniform intersections.* In a k-element set, at most how many subsets can have pairwise exactly one common element?

9.4.11. *Strict rules.* In Pseudolibertyville the laws for forming clubs are: (i) There must be fewer clubs than inhabitants; (ii) any two inhabitants are the common members of exactly the same positive number of clubs; (iii) every club has at least two members and no two clubs can have exactly the same members. How many clubs are in Pseudolibertyville?

****9.4.12.** *Restricted intersections.* Let $m \leq k$. At most how many subsets of a k-element set can satisfy that their pairwise intersections have at most m sizes (i.e., the numbers $|H_t \cap H_j|$ where $t \neq j$ assume at most m distinct values)?

In Exercises 9.4.13–9.4.16, p denotes a positive prime, $F = F_p$, $V = F^k$, and U is a subspace in V. We define the inner product, orthogonality, and $U^\perp$ in V as in the case $p = 2$.

9.4.13. *Small subspaces.*

(a) Let $F = F_2$ and let U be a subspace in V with no non-zero vector orthogonal to itself. Prove $|U| \leq 2$.

*(b) Let p be an odd prime and $F = F_p$. What is the maximal size of a subspace U in V with no non-zero vector orthogonal to itself?

9.4.14. *Isotropic vectors.*

(a) For which p and k does $V = (F_p)^k$ contain a non-zero vector orthogonal to itself?

(b) For which p and k do the vectors orthogonal to themselves form a subspace in V? What is the dimension of this subspace?

9.4.15. *U and* $U^\perp$.
(a) Show an example for $U \cap U^\perp \neq \mathbf{0}$ and $\langle U, U^\perp \rangle \neq V$.
(b) Verify $\dim U + \dim U^\perp = \dim V$.
(c) Prove $U \cap U^\perp = \mathbf{0} \iff \langle U, U^\perp \rangle = V$.

9.4.16. *Rampant orthogonality*. Determine whether there exists a subspace U in V satisfying $U = U^\perp$, if
(a) $p = 2$, $k = 10$;
(b) $p = 5$, $k = 11$;
(c) $p = 13$, $k = 30$;
(d) $p = 23$, $k = 2$;
(e) $p = 43$, $k = 20$.

9.4.17. *New clubs.*
(a) In Oddtown of k inhabitants there are n clubs satisfying the conditions of Theorem 9.4.1. Can we form a new club if the number of clubs is not maximal (i.e., $n < k$)?
(b) Investigate the analogous question in Eventown.

***9.4.18.** *Liberal Eventown*. We cancel the condition in Theorem 9.4.2 that every $|H_j|$ is even. Prove that the maximal number of clubs remains the same if k is even and increases just by one if k is odd.

9.4.19. *Threefold*. At most how many subsets H_j of a nine-element set can satisfy that every $|H_j|$ and every $|H_t \cap H_j|$ is a multiple of 3?

9.5. Nice Graphs

We prove a surprising theorem about graphs relying on some deeper linear algebra. Or, in a more funny interpretation, the key is that 1, 3, 5, and 15 are all positive divisors of 15 since this is the final step in the proof.

Consider a finite undirected graph without loops, multiple edges, triangles, and 4-cycles, and assume that exactly $d > 0$ edges start from every vertex (i.e., the girth is greater than four and the degree of every vertex is d, or, in other words, the graph is d-regular). What can be asserted about the number of vertices?

Pick an arbitrary vertex. The other endpoints of the d edges from here give d other vertices. Each of the new vertices has $d - 1$ new edges. Neither two edges, nor their endpoints can coincide since then we would get a triangle or a 4-cycle. So, the other endpoints of these edges yield $d(d - 1)$ new vertices. Therefore, the graph must have at least $1 + d + d(d - 1) = d^2 + 1$ vertices.

For which values of d can equality hold? For $d = 1$ a single edge, for $d = 2$ a 5-cycle satisfy the requirements. For $d = 3$, the Petersen graph works (see Exercise 9.5.1). More and more complicated computations show that no such graph exists for $d = 4$, 5, and 6. But for $d = 7$, A. J. Hoffman and R. R. Singleton made a brilliant construction by gluing several Petersen graphs together. And what can be said for larger values of d?

Theorem 9.5.1 (Hoffman–Singleton). *Assume that a graph contains no shorter cycle than 5, the degree of every vertex is $d > 0$, and the graph has $d^2 + 1$ vertices. Then d can only be* 1, 2, 3, 7, *or* 57. ♣

Remark. The case when $d = 57$ was a long-standing unsolved problem. Finally, it was proved in 2020 that no such graph exists.

Proof. The *adjacency matrix* of a graph of n vertices is an $n \times n$ matrix $A = (\alpha_{ij})$, $i, j = 1, 2, \dots, n$, where

$$\alpha_{ij} = \begin{cases} 1, & \text{if there is an edge between vertices } i \text{ and } j\text{; and} \\ 0 & \text{otherwise.} \end{cases}$$

We immediately obtain that the element β_{ij} in the matrix $B = A^2$ is the number of common neighbors of vertices i and j.

Let G be a graph with $n = d^2 + 1$ vertices satisfying the conditions of the theorem. The discussion before stating the theorem yields that any two vertices in G are either neighbors, or have exactly one common neighbor. Hence, the $n \times n$ adjacency matrix A of G satisfies

$$A^2 + A - (d-1)I = J \tag{9.5.1}$$

where I is the $n \times n$ identity matrix and every entry in J is 1.

A is a symmetric matrix, so, by the Principal Axis Theorem 8.6.2, it has $n = d^2 + 1$ linearly independent eigenvectors in the Euclidean space $\mathbf{R}^n$ (moreover, these are pairwise orthogonal). Equation (9.5.1) easily implies that if $\mathbf{v}$ is an eigenvector of A with eigenvalue λ, then $\mathbf{v}$ is also an eigenvector of J with eigenvalue

$$\mu = \lambda^2 + \lambda - (d-1). \tag{9.5.2}$$

The eigenvalues of J are n with multiplicity 1 and 0 with multiplicity $n - 1$. Then, by (9.5.2), d is an eigenvalue of A with multiplicity 1 (this could have been checked directly, too), and all other eigenvalues λ of A satisfy $\lambda^2 + \lambda - (d-1) = 0$. Solving this quadratic equation, we obtain that the other eigenvalues of A are

$$\lambda_1 = \frac{-1 + \sqrt{4d-3}}{2} \quad \text{and} \quad \lambda_2 = \frac{-1 - \sqrt{4d-3}}{2}$$

with multiplicity m_1 and m_2 where

$$m_1 + m_2 = n - 1 = d^2. \tag{9.5.3}$$

The trace of A is 0 as it is the sum of the elements in the main diagonal. On the other hand, the trace is the sum of the eigenvalues counted with multiplicity. Therefore

$$0 = m_1\lambda_1 + m_2\lambda_2 + d = \frac{m_1 - m_2}{2}\sqrt{4d-3} - \frac{m_1 + m_2}{2} + d. \tag{9.5.4}$$

This can only hold if $m_1 = m_2$ or $s = \sqrt{4d-3}$ is an integer.

In the first case, (9.5.3) and (9.5.4) imply $d^2 = 2d$, i.e., $d = 2$. In the second case, plugging in $d = (s^2 + 3)/4$ into (9.5.4), we get

$$-s^4 + 2s^2 + 16s(m_1 - m_2) + 15 = 0.$$

Then s divides 15, so $s = 1, 3, 5$, or 15. This gives $d = 1, 3, 7$, or 57. ∎

Exercises 9.5

We only consider finite undirected graphs without loops and multiple edges. The *degree* of a vertex is the number of edges starting from this vertex. A graph is *regular* if all vertices have the same degree.

The *spectrum* of a graph is the set of the eigenvalues of its adjacency matrix.

Assume that a graph G has n vertices and m edges. The *incidence matrix* of G is an $n \times m$ matrix C with elements $\gamma_{ij} = 1$ if vertex i is an endpoint of edge j, and 0 otherwise.

9.5.1. Construction of the Petersen graph: Let A_1, A_2, A_3, A_4, and A_5 be the vertices of a regular pentagon. Applying a shrinking similarity from the center to them, we obtain the vertices B_1, B_2, B_3, B_4, and B_5 of a smaller regular pentagon. These 10 points will be the vertices of the Petersen graph. To obtain the edges, we connect the adjacent vertices of the original pentagon (i.e., we draw the five sides of the pentagon); connect every vertex A_i of the original pentagon with the corresponding vertex B_i of the small pentagon, and finally we connect every second vertex of the small pentagon (i.e., we draw the five sides of a star pentagon). We obtain thus 15 edges.

   (a) Check that the Petersen graph contains no triangles and 4-cycles, and every vertex has degree 3.
   (b) Compute the spectrum of the Petersen graph.

9.5.2. We assign the real (or complex) numbers $x_1, \ldots, x_n$ to the n vertices $v_1, \ldots, v_n$ of a graph G. Verify that the non-zero vector

$$\mathbf{x} = \begin{pmatrix} x_1 \\ \vdots \\ x_n \end{pmatrix}$$

is an eigenvector of the adjacency matrix of G if and only if, for every vertex v_i, we have to multiply x_i by the same number to obtain the sum of the numbers x_j assigned to the neighbors of v_i.

9.5.3. Determine the spectra of the graphs with n vertices.

   (a) A complete graph (or clique): there is an edge between any two vertices.
   (b) A 1-factor: $n = 2k$ and every vertex has exactly one neighbor (the graph is the union of disjoint edges).

(c) A complete bipartite graph: $n = 2k$ and every vertex $a_1, \dots, a_k$ is connected with every vertex $b_1, \dots, b_k$ and there are no other edges.

(d) A star graph: the center is connected with every other vertex and there are no other edges.

*(e) An n-cycle.

9.5.4. What is the relation between the spectra of a regular graph and its complement?

***9.5.5.** Let A be the adjacency graph of a graph G. Prove that there exists a polynomial f satisfying $f(A) = J$ if and only if G is connected and regular. (J is a matrix with all elements 1.)

9.5.6. Write the incidence matrices of the graphs in Exercises 9.5.1 and 9.5.3.

9.5.7. Let C be an incidence matrix of a graph G. What is the combinatorial meaning of the elements in the matrices C^TC and CC^T?

9.5.8. For which graphs do the (suitably chosen) adjacency and incidence matrices coincide?

9.5.9. Let Λ be the maximal eigenvalue of a graph G and let δ and Δ be the minimal and maximal degrees in G. Verify $\delta \leq \Lambda \leq \Delta$.

***9.5.10.** Let Λ be the maximal eigenvalue of a graph G. Prove that the vertices of G can be colored by (at most) $\Lambda + 1$ colors so that no edge has endpoints of the same color.

9.5.11. *Finite projective planes*: Let X be a finite set; its elements are called points and certain subsets are called lines. For a given $n \geq 2$, we say that X is a (non-degenerate) *finite projective plane* of order n if

(i) any two (distinct) lines contain exactly one common point and any two (distinct) points are contained in exactly one line;

(ii) every line contains exactly $n + 1$ points and every point is contained in exactly $n + 1$ lines.

It can be shown that both the number of points and the number of lines are $|X| = n^2 + n + 1$. Exercise A.11.13 discusses a construction of projective planes of any (prime or) prime power order. It is unsolved whether there exist projective planes of other orders, as well.

Finite projective planes provide the non-trivial extremal solutions in Exercise 9.4.10 (see the remark after the second proof of the exercise), and the proof of Theorem 9.6.2 uses projective planes in incognito.

(a) We write a real number onto each point of a finite projective plane so that the sum of the numbers on each line is 0. Show that every number is 0.

*(b) We color the points of a projective plane of order n arbitrarily red and blue. Prove that, on some line, there must be a difference not less than $\sqrt{n}$ between the two colors.

9.6. Sidon Sets

In this section, we deal with a favorite problem of Erdős. Sidon sets are finite or infinite sequences $a_1 < a_2 < \ldots$ of positive integers where the sums $a_i + a_j$, $i \leq j$ (or equivalently, the differences $a_i - a_j$, $i \neq j$) are all distinct. These occurred first in Simon Sidon's investigations of Fourier series around 1930.

At most how many elements can a Sidon set have in the interval $[1, n]$? We shall prove that the maximum is about $\sqrt{n}$. This contains two statements: On the one hand, there exists a Sidon set between 1 and n which has about $\sqrt{n}$ elements (lower estimate for the maximum), and on the other hand, no Sidon set can have substantially more elements within these limits (upper estimate for the maximum).

Erdős and Turán showed in 1941 that the maximum in question is at most $n^{1/2} + 2n^{1/4}$. Later Lindström improved this bound to $n^{1/2} + n^{1/4} + 1$ by a different method, but the same follows also from a more precise execution of the Erdős–Turán proof (Theorem 9.6.4). Relying on a result of J. Singer, Erdős and S. Chowla proved independently in 1944 that there does exist a Sidon set of size $n^{1/2} - n^{\varrho}$ with a suitable positive constant $\varrho < 1/2$ (Theorem 9.6.3). These two theorems together mean that the maximal size of a Sidon set in the interval $[1, n]$ is asymptotically $\sqrt{n}$ with very good (upper and lower) error terms. It is a much stronger conjecture that the difference of the maximum and $\sqrt{n}$ is bounded (independently of n). For a proof or disproof Erdős offered 1000 US dollars, but there has been no improvement in the last 80 years.

Let $s = s(n)$ denote the maximal possible number of a Sidon set up to n. We first establish some simple upper bound for s. There are $\binom{s+1}{2}$ sums $a_i + a_j$ which are distinct and fall between 2 and $2n$, thus $\binom{s+1}{2} < 2n$, and so $s < 2\sqrt{n}$. We get a better upper bound by considering the differences $a_i - a_j > 0$; also these $\binom{s}{2}$ numbers are distinct, they are less than n, hence $\binom{s}{2} < n$, and so $s < \sqrt{2n} + 1$. Thus we attain immediately the order of magnitude $\sqrt{n}$; just the coefficient $\sqrt{2}$ of $\sqrt{n}$ has to be reduced to 1.

From the opposite direction, it is much less clear, how we can achieve the order of magnitude $\sqrt{n}$. The example of powers of two yields $\log_2 n$, and the greedy algorithm guarantees only $\sqrt[3]{n}$ (see Exercise 9.6.1). But a nice elementary construction of Erdős provides already $\sqrt{n/2}$ elements (see Exercise 9.6.2), and as mentioned, we can lift the coefficient of $\sqrt{n}$ to 1.

Let us start constructing really big Sidon sets. We do this for special types of n first, and then use the result to handle a general n.

Theorem 9.6.1. *Let p be an arbitrary positive prime and let $n = p^2 + p + 1$. There exists a Sidon set in the interval $[1, n]$ which has $\lceil \sqrt{n} \rceil = p + 1$ elements.* ♣

Theorem 9.6.1 is a consequence of a surprising and much sharper statement of independent interest.

Theorem 9.6.2. *Let p be any positive prime. Then there exist $p + 1$ integers a_i such that the differences $a_i - a_j$, $i \neq j$ are (not just distinct, but are) pairwise incongruent modulo $p^2 + p + 1$.* ♣

Remark. The number of differences in Theorem 9.6.2 is $p^2 + p$, and there are just that many non-zero residues modulo $p^2 + p + 1$. This means that each non-zero residue has exactly one representation as a difference $a_i - a_j$.

It is clear that the integers a_i in Theorem 9.6.2 must be pairwise incongruent themselves, so they can be chosen between 1 and $n = p^2+p+1$. Thus Theorem 9.6.1 follows, indeed.

Proof of Theorem 9.6.2. We use some basic facts about finite fields (see Section A.11).

Consider the finite field M of p^3 elements and its subfield F_p of p elements. Let Δ be a generator of the cyclic multiplicative group of M, i.e.,

$$M = \{0, \Delta, \Delta^2, \dots, \Delta^{p^3-1} = 1\}. \tag{9.6.1}$$

The non-zero elements of F_p form a subgroup in the multiplicative group of M. This cyclic subgroup is generated by Δ^n, where $n = (p^3 - 1)/(p - 1) = p^2 + p + 1$. Thus

$$F_p = \{0, \Delta^n, \Delta^{2n}, \dots, \Delta^{(p-1)n} = \Delta^{p^3-1} = 1\}.$$

Consider M as a vector space over F_p. By the above, elements Δ^i and Δ^j in M are linearly dependent over F_p if and only if

$$i \equiv j \pmod{n}. \tag{9.6.2}$$

We now construct the integers a_i. We fix an arbitrary $\Theta \in M \setminus F_p$ and form the elements $\Theta + \gamma_i$, where $\gamma_1, \dots, \gamma_p$ are the elements of F_p. By (9.6.1), we can write

$$\Theta + \gamma_i = \Delta^{a_i} \tag{9.6.3}$$

and so we obtain p integers a_i, $1 \le i \le p$. Finally, let $a_{p+1} = 0$.

We verify that these numbers meet the requirements, i.e., the differences $a_i - a_j$, or equivalently, the sums $a_i + a_j$ are pairwise incongruent modulo $p^2 + p + 1$.

Assume $a_i + a_j \equiv a_k + a_m \pmod{p^2 + p + 1}$. If none of the four numbers is $a_{p+1} = 0$, then, by (9.6.2) and (9.6.3), this means

$$\Delta^{a_i}\Delta^{a_j} = \Delta^{a_i+a_j} = \Delta^{a_k+a_m} = \Delta^{a_k}\Delta^{a_m},$$

i.e.,

$$(\Theta + \gamma_i)(\Theta + \gamma_j) - \gamma(\Theta + \gamma_k)(\Theta + \gamma_m) = 0$$

for some $\gamma \in F_p$. Since the degree of Θ is 3 over F_p, it cannot be a root of a polynomial of degree at most 2. Therefore only $\gamma = 1$ and $\{\gamma_i, \gamma_j\} = \{\gamma_k, \gamma_m\}$ are possible which means that the corresponding pairs of integers a_i are the same, too, as stated.

The proof runs the same way if $a_{p+1} = 0$ occurs among the four integers a_i. ∎

Remark. Theorem 9.6.2 and its proof remain valid if p is a prime power. All this is closely related to finite projective planes (see Exercise 9.5.11).

Theorem 9.6.3. *If n is large enough, then the interval $[1, n]$ contains a Sidon set having at least $n^{1/2} - n^{0.27}$ elements.* ♣

Proof. Consider the largest prime p satisfying $p^2 + p + 1 \leq n$ and perform the previous construction of $p + 1$ elements for $p^2 + p + 1$. By a very deep result in number theory, there is a prime between $n^{1/2} - n^{0.27}$ and $n^{1/2}$ if n is large enough. Therefore $p > n^{1/2} - n^{0.27}$, thus verifying the theorem for a general n. ∎

Remark. In the transition to an arbitrary n, we used that the primes occur densely. If we know that there is a prime between N and $N + N^c$ for N large enough, then the error term in our theorem can be reduced to $n^{c/2}$. But the question of gaps between consecutive primes is a very hard problem.

For other proofs of Theorem 9.6.3, see Exercises 9.6.3 and 9.6.4.

Now we turn to the sharp upper bound of the size of Sidon sets.

Theorem 9.6.4. *A Sidon set in interval* $[1, n]$ *has at most* $n^{1/2} + n^{1/4} + 1$ *elements.* ♣

First proof. Let t be a fixed integer to be specified later. We push a segment of length $t - 1$ through the interval $[0, n]$, i.e., we consider the intervals $[-t + 1, 0]$, $[-t + 2, 1]$, $\dots$, $[n, n + t - 1]$. We take a Sidon set of size s, and denote the number of its elements in these intervals by $A_1, A_2, \dots, A_{n+t}$. Then clearly

$$\sum_{i=1}^{n+t} A_i = ts. \tag{9.6.4}$$

We now count the pairs $\{a_i, a_j\}$, $i > j$ that fall into such an interval. We count them with the suitable multiplicity, i.e., each pair is counted as many times as many intervals contain it. Let D be the total number of such pairs. Then clearly

$$D = \sum_{i=1}^{n+t} \binom{A_i}{2} = \sum_{i=1}^{n+t} \frac{A_i^2}{2} - \sum_{i=1}^{n+t} \frac{A_i}{2}. \tag{9.6.5}$$

On the other hand, if the difference $a_i - a_j$ in a pair is d, then this pair falls into exactly $t - d$ intervals. By the Sidon property, every such d can occur at most once, thus

$$D \leq \sum_{d=1}^{t-1} (t - d) = \frac{t(t-1)}{2}. \tag{9.6.6}$$

Combining (9.6.5) and (9.6.6), we obtain

$$\sum_{i=1}^{n+t} A_i^2 - \sum_{i=1}^{n+t} A_i \leq t(t-1). \tag{9.6.7}$$

Using (9.6.4) and the inequality for arithmetic and geometric means, we can estimate the left-hand side of (9.6.7) from below:

$$\sum_{i=1}^{n+t} A_i^2 - \sum_{i=1}^{n+t} A_i \geq \frac{\left(\sum_{i=1}^{n+t} A_i\right)^2}{n+t} - ts = \frac{t^2 s^2}{n+t} - ts. \tag{9.6.8}$$

From (9.6.7) and (9.6.8), we infer

$$s^2 - s\left(\frac{n}{t} + 1\right) - \left(\frac{n}{t} + 1\right)(t - 1) \leq 0.$$

Solving this quadratic inequality, we get

$$s \le \frac{n}{2t} + \frac{1}{2} + \sqrt{n + t + \frac{n^2}{4t^2} - \frac{n}{2t} - \frac{3}{4}}.$$

Choosing $t = \lfloor n^{3/4} \rfloor + 1$, we arrive at the statement of the theorem. ∎

Second proof. We shall estimate the sum of certain differences $a_i - a_j$ from both sides. Let

$$K = \sum_{0<i-j\le r} (a_i - a_j), \tag{9.6.9}$$

where r will be specified later. The sum in (9.6.9) contains

$$(s-1) + (s-2) + \cdots + (s-r) = rs - \frac{r(r+1)}{2} = rw$$

terms, where

$$w = s - \frac{r+1}{2}. \tag{9.6.10}$$

Each term is a difference $a_i - a_j$, and these are all distinct by the Sidon property. Thus K is not less than the sum of the first rw positive integers, i.e.

$$K \ge \frac{rw(rw+1)}{2} > \frac{r^2w^2}{2}. \tag{9.6.11}$$

On the other hand, the sum in (9.6.9) contains, e.g.,

$$(a_s - a_{s-1}) + (a_{s-1} - a_{s-2}) + \cdots + (a_2 - a_1) < a_s \le n,$$

and many other telescoping sums, which can be estimated from above similarly. Their general form is

$$(a_{s-\nu} - a_{s-\nu-\mu}) + (a_{s-\nu-\mu} - a_{s-\nu-2\mu}) + \cdots < a_{s-\nu} \le n, \quad 0 \le \nu < \mu \le r.$$

Moreover, the entire K consists of such telescoping sums; the indices cover all maximal arithmetic progressions between 1 and s where the difference is at most r. There are μ arithmetic progressions with difference μ, so there are $1 + 2 + \cdots + r = r(r+1)/2$ telescoping sums. Each such telescoping sum is not greater than n, hence

$$K \le \frac{nr(r+1)}{2}. \tag{9.6.12}$$

Combining (9.6.11) and (9.6.12) and multiplying by $2/r^2$, we obtain the inequality $w^2 < n + n/r$. Taking a square root and substituting (9.6.10), we get

$$s < \frac{r+1}{2} + \sqrt{n + \frac{n}{r}}.$$

Choosing $r = \lfloor n^{1/4} \rfloor + 1$, we arrive at the statement of the theorem. ∎

Exercises 9.6

9.6.1. Show that the greedy algorithm yields a Sidon set between 1 and n of at least $\sqrt[3]{n}$ elements.

Remark. Moreover, the greedy algorithm provides an *infinite* Sidon set having at least $\sqrt[3]{n}$ elements up to *every* n. It is surprising that it took a long time to surpass this order of magnitude: Ajtai, Komlós, and Szemerédi proved in 1981 the existence of an infinite Sidon set that has at least $c\sqrt[3]{n \log n}$ elements up to every (sufficiently large) n with a suitable positive constant c. Even this was just "slightly" better than $\sqrt[3]{n}$ obtained by the greedy algorithm. In 1997 Ruzsa improved the bound significantly to $cn^{\sqrt{2}-1-\varepsilon}$, though even this is very far from the order of magnitude $n^{1/2-\varepsilon}$ conjectured by Erdős (where ε is an arbitrarily small positive number).

9.6.2. Let $p > 0$ be a prime and $a_i = 1 + 2ip + \langle i^2 \bmod p\rangle$, $i = 0, 1, \ldots, p-1$, where $\langle i^2 \bmod p\rangle$ denotes the least non-negative residue of i^2 modulo p. Verify that this is a Sidon set in $[1, n]$ of size $\sqrt{n/2}$ for $n = 2p^2$.

***9.6.3.** Let $p > 0$ be a prime. There exist p integers a_i such that the sums $a_i + a_j$, $i \le j$, are (not just distinct, but are) pairwise incongruent modulo $p^2 - 1$.

Remark. An equivalent formulation is that the differences $a_i - a_j$, $i \ne j$, are (not just distinct, but are) pairwise incongruent modulo $p^2 - 1$. There are $p^2 - p$ such differences and $p^2 - 2$ non-zero residues modulo $p^2 - 1$. This means that nearly all residues can be represented as a difference $a_i - a_j$. We can see from the proof that the missing residues are the multiples of $p + 1$. We can deduce Theorem 9.6.3 from this exercise just as we did from Theorem 9.6.2. (The same holds also for the next exercise.)

***9.6.4.** Let $p > 0$ be a prime. There exist $p - 1$ integers a_i such that the differences $a_i - a_j$, $i \ne j$, are (not just distinct, but are) pairwise incongruent modulo $p^2 - p$.

9.6.5. *Infinite Sidon sets.* Prove that for any $\varepsilon > 0$ there exists an infinite Sidon set that has at least $(1/\sqrt{2} - \varepsilon)\sqrt{n}$ elements for *infinitely many* n (cf. the remark after Exercise 9.6.1).

Remark. It is unsolved whether the same holds with 1 instead of $1/\sqrt{2}$.

9.6.6. *Sums of more terms.* Let $h \ge 2$ be a fixed integer, and consider sequences in the interval $[1, n]$ such that the h-fold sums are all distinct. (The Sidon sets are the special case $h = 2$.)

*(a) Prove the existence of a sequence having about $n^{1/h}$ terms.

(b) Show that there is a constant $c = c(h)$ depending only on h such that every sequence has at most $c(h)n^{1/h}$ terms.

Remark. It is an unsolved problem whether, similar to the Sidon sets, we can reduce $c(h)$ to $1 + \varepsilon$, i.e., the maximal size is asymptotically $n^{1/h}$ for $h > 2$. The proof of Theorem 9.6.4 does not work since we cannot switch sums to differences if $h \ne 2$.

9.6.7. *Another Erdős favorite: All subset sums are distinct.* Let us call a set of integers $1 \le a_1 < \cdots < a_s \le n$ pretty if all sums formed of distinct elements a_i are distinct.

(a) Find a pretty sequence of $s = 1 + \lfloor \log_2 n \rfloor$ terms.

(b) Verify $s \le \log_2 n + \log_2 \log_2 n + 1$ for any pretty sequence.

*(c) Improve the upper bound in (b) to $s \le \log_2 n + (\log_2 \log_2 n)/2 + 2$.

Remark. Surprisingly, the lower bound can be improved: we can replace the term 1 by 2 if n is large enough. Erdős offered 500 US dollars for deciding whether the difference $\max s - \log_2 n$ remains bounded independently of n. Note that to show $\max s \ge \lfloor \log_2 n \rfloor + c$ (where c is a constant), it is sufficient to find a set H of that many elements for a *single* $n = 2^j$: Let $a_1, \dots, a_s$ be a suitable set for $n = 2^j$. Then the set $1, 2, \dots, 2^{t-1}, 2^t a_1, \dots, 2^t a_s$ will work for $2^{j+t} \le n < 2^{j+t+1}$. It can happen, however that the lower bound mentioned above is sharp, i.e., the value of c cannot be increased even from 2 to 3.

**9.6.8. *Products.* Consider integers in the interval $[1, n]$ such that the products of arbitrarily many distinct elements are distinct. Clearly, the primes form such a set. Prove that any such set can have at most $\pi(n) + 2n^{2/3}$ elements where $\pi(n)$ is the number of primes up to n.

Remark. It can be shown that the order of magnitude of the maximum is $\pi(n)$ plus a constant times $\sqrt{n}/\log n$.

****9.6.9.** *Arithmetic progressions.* Show that if n is large enough, then there exist $n/e^{c\sqrt{\log n}}$ integers between 1 and n not containing a three-term arithmetic progression. ($c > 0$ is a suitable constant independent of n.)

***9.6.10.** *Avoiding monochromatic arithmetic progressions.* Prove that the integers $1, 2, \dots, 2000$ can be colored red and blue so that no 18-term monochromatic arithmetic progression arises.

9.7. Hilbert's Third Problem

Perhaps the most influential talk ever in mathematics was David Hilbert's lecture titled Mathematical problems at the second International Congress of Mathematicians in Paris in 1900. It was a grandiose vision of many important research questions. The 23 packages of problems had an essential influence on the main trends in 20th century mathematics; some problems are still open. The third problem about the equidissectibility of polyhedra proved to be the easiest; it was solved by M. Dehn a few months after the congress.

The origin of the problem is a theorem by F. Bolyai and P. Gerwien about the equidissectibility of polygons. They proved independently that any two polygons having the same area are equidissectible, i.e., we can cut one of them with (segments or) lines into finitely many polygons and reassemble the other polygon of these parts (see Exercise 9.7.1). In other words, we can partition each polygon into the same number of smaller polygons so that the corresponding small polygons are congruent (or isometric, i.e., a distance-preserving transformation maps one of them into the other).

Bolyai asked whether a similar result holds for polyhedra having the same volume. Hilbert's third problem was to decide this question. Confirming the expectation, the answer is negative.

Theorem 9.7.1 (Hilbert's third problem). *A cube and a regular tetrahedron of the same volume cannot be cut into finitely many polyhedra so that the corresponding parts are congruent.* ♣

Proof. To obtain a contradiction, assume that the cube and the tetrahedron are equidissectible and let $\beta_1, \ldots, \beta_m$ be all angles between two faces of the polyhedra arising in the decomposition. These values also include the angles $\pi/2$ between two faces of the cube and α between two faces of the tetrahedron. A simple computation shows $\cos\alpha = 1/3$.

Let V be the usual vector space of the real numbers over $\mathbf{Q}$ and let W be the subspace spanned by the numbers β_i. (Clearly, $\dim W \leq m$.) The quotient of α and $\pi/2$ is irrational (see Exercise 9.7.2), so $\pi/2$ and α are linearly independent vectors in W. Therefore, we can extend these two vectors to a basis of W. Hence, there exists a linear map $f : W \to \mathbf{Q}$ satisfying $f(\pi/2) = 0$ and $f(\alpha) = 1$. Linearity implies $f(\xi + \psi) = f(\xi) + f(\psi)$ and $f(r\xi) = rf(\xi)$ for every $\xi, \psi \in W$ and $r \in \mathbf{Q}$. In particular $f(r\pi) = 2rf(\pi/2) = 0$ holds, too.

We introduce the Dehn-invariant. This is a function D assigning a real number to every polyhedron P by the rule

$$D(P) = \sum_e |e| \cdot f(\beta),$$

where the summation is for all edges e of P, $|e|$ is the length of e, β is the angle between the two faces containing e, and f is the function defined above.

We show that D is additive: If P is the union of polyhedra $P_1, P_2, \ldots, P_k$ where P_i and P_j have no common inner point for $i \neq j$, then

$$D(P) = \sum_{i=1}^{k} D(P_i). \tag{9.7.1}$$

Consider any term $|e| \cdot f(\beta)$ in $D(P)$ on the left-hand side. If the edge e and the angle β remain unchanged after the decomposition, then the same term occurs in exactly one P_i on the right-hand side. If the edge e was cut into segments e_j with no change in the angle β, then $\sum_j |e_j| \cdot f(\beta) = |e| \cdot f(\beta)$ occurs on the right-hand side. If the angle β was cut into parts β_s, then $\sum_s |e| \cdot f(\beta_s)$ appears on the right-hand side. As f is linear, this sum equals $|e| \cdot f(\beta)$. Similarly, (9.7.1) remains valid under any combination of these cases. Finally, we show that if a new edge e occurs inside P or on one of its faces, then the sum of terms induced by e is 0 on the right-hand side of (9.7.1). The sum of the angles β_t on the edge e between the faces of the polyhedra P_t is 2π or π. Thus e contributes to the sum $\sum_{i=1}^{k} D(P_i)$ on the right-hand side of (9.7.1) with

$$\sum_t |e| \cdot f(\beta_t) = |e| \cdot f(\sum_t \beta_t) = |e| f(r\pi) = 0 \text{ (where } r = 1 \text{ or } 2).$$

This completes the verification of (9.7.1).

(9.7.1) implies that the Dehn-invariant of equidissectible polyhedra must be equal. For a cube C with edge length c, we have $D(C) = 12cf(\pi/2) = 0$. For a regular tetrahedron T with edge length t, we have $D(T) = 6tf(\alpha) = 6t \neq 0$. This contradiction proves that C and T are not equidissectible. ■

Remarks.

1. It could be a natural restriction to allow only congruences in the space that preserve orientation and thus to exclude reflections across planes: If the mirror image of a polyhedron P is P', then, in general, P cannot be effectively carried into P' in the (three-dimensional) space (the left and right hands cannot be moved into each other). It can be proved, however, that if two polyhedra are equidissectible, this can also be done using only congruences that preserve orientation. Therefore, it is irrelevant whether we allow any congruences or just the ones preserving orientation.

2. The difference between the situations in the plane and the space clearly indicate that equidissectibility is an effective tool to deal with the notion of the area of polygons (an idea used already by Euclid), but in handling the volume of polyhedra one cannot avoid some sort of limits.

3. If instead of "normal" geometric equidissectibility we allow decompositions into any subsets, this causes a dramatic change. It can be shown e.g., that, in this paradoxical sense, a ball is equidecomposable into two(!) balls congruent to the original ball (Banach–Tarski paradox). It was a long-standing unsolved problem whether a square and a circle of the same area are equidecomposable; the affirmative answer was given by M. Laczkovich (moreover, it can be done only using translations).

Exercises 9.7

9.7.1. *Bolyai–Gerwien theorem.* Verify that any two polygons of the same area are equidissectible.

9.7.2. *Angles and cosines.*
- (a) Show that α/π is an irrational number if $\cos\alpha = 1/3$.
- (b) Prove that if both γ/π and $\cos\gamma$ are rational numbers, then $\cos\gamma = 0$, $\pm 1/2$, or ± 1.

*9.7.3. *Rectangles and squares.* Prove that a rectangle can be cut into finitely many (not necessarily congruent) squares if and only if the ratio of its sides is a rational number.

9.7.4. *Prisms and tetrahedra.*
- (a) Are any two prisms having the same volume equidissectible? (The bases can be arbitrary polygons.)
- *(b) Let A, B, C, and D be the vertices on the base of a cube in this order, and let A', B', C', and D' be the corresponding vertices on the top. Are the tetrahedra $ABCB'$ and $ABCC'$ (having the same base and height) equidissectible?

*9.7.5. *Squares and triangles.* Demonstrate that a square and a triangle of equal areas are not equidissectible using *only translations.*

9.7.6. *Squares and cubes.*
(a) Show that a square can be dissected into n squares if and only if $n \neq 2$, 3, or 5.
(b) Prove that if n is large enough, then a cube can be dissected into n cubes.
*(c) Verify the previous statement for every $n \geq 48$.

***9.7.7.** *Triangles.* Prove.
(a) Every triangle can be dissected into n *similar* triangles if and only if $n \neq 2, 3$, or 5.
(b) Every triangle can be dissected into n *congruent* triangles if and only if n is a square.
(c) If n is a square, the sum of two squares, or the triple of a square, then there exists a triangle dissectible into n congruent triangles similar to the original triangle.

Remark. The converse of (c) is also true. If, however, the small (congruent) triangles do not have to be similar to the big one, then we get many other values of n, e.g., for the regular triangle.

9.8. Volume and Determinant

In this section, we show that the geometric meaning of the determinant is the signed volume. Our argument is valid for vector spaces over an arbitrary field but we shall confine the discussion to $\mathbf{R}$ to emphasize the direct geometric connection.

In the plane, any two vectors span a (possibly degenerate) parallelogram with the origin as a vertex. Similarly, in the space, any three vectors span a parallelepiped (a prism with a parallelogram base). Generalizing these notions, we say that in an n-dimensional vector space V over $\mathbf{R}$, any n vectors span a(n n-dimensional) *parallelepiped.* Its edges are the given vectors and their translated copies and its vertices are the endpoints of the sums formed of the given vectors. For $n = 2$ and 3, the vertices are $\mathbf{0}$, $\mathbf{a}_1$, $\mathbf{a}_2$, and $\mathbf{a}_1 + \mathbf{a}_2$ in the plane, and $\mathbf{0}$, $\mathbf{a}_1$, $\mathbf{a}_2$, $\mathbf{a}_3$, $\mathbf{a}_1 + \mathbf{a}_2$, $\mathbf{a}_1 + \mathbf{a}_3$, $\mathbf{a}_2 + \mathbf{a}_3$, and $\mathbf{a}_1 + \mathbf{a}_2 + \mathbf{a}_3$ in the space. An n-dimensional parallelepiped has 2^n vertices.

We want to define the signed volume of a parallelepiped. This is a function $D\colon V^n \to \mathbf{R}$ assigning a real number to every system of n vectors. D has to satisfy certain conditions induced by the properties expected from the volume.

If we multiply an edge of a parallelepiped by λ keeping the other edges unaltered, then the volume gets multiplied by λ. Similarly, if an edge $\mathbf{a}$ is decomposed into a sum $\mathbf{a}' + \mathbf{a}''$ keeping the other edges unchanged, then the sum of the volumes of the two arising parallelepipeds equals the volume of the original one. These two conditions mean that D is an *n-linear* function, i.e., it preserves sums and scalar multiples in each variable.

Another requirement is that if the n vectors in V span a subspace of smaller dimension, i.e., the parallelepiped is degenerate, then the volume is 0. In other words, if the vectors $\mathbf{a}_1, \dots, \mathbf{a}_n$ are linearly dependent, then $D(\mathbf{a}_1, \dots, \mathbf{a}_n) = 0$.

Finally, we prescribe that the volume of the unit cube is 1. Therefore, we fix a basis $\mathbf{e}_1, \dots, \mathbf{e}_n$ in V and set $D(\mathbf{e}_1, \dots, \mathbf{e}_n) = 1$.

Theorem 9.8.1. *Let V be an n-dimensional vector space over* $\mathbf{R}$ *and let* $\mathbf{e}_1, \dots, \mathbf{e}_n$ *be a fixed basis in* V. *Then there exists exactly one function* $D: V^n \to \mathbf{R}$ *such that*

(i) *D is linear in each variable;*

(ii) *D assigns 0 to every system of linearly dependent vectors; and*

(iii) $D(\mathbf{e}_1, \dots, \mathbf{e}_n) = 1$.

If α_{ij} is the ith coordinate of the vector $\mathbf{a}_j$ in the basis $\mathbf{e}_1, \dots, \mathbf{e}_n$, then

$$D(\mathbf{a}_1, \dots, \mathbf{a}_n) = \begin{vmatrix} \alpha_{11} & \alpha_{12} & \dots & \alpha_{1n} \\ \alpha_{21} & \alpha_{22} & \dots & \alpha_{2n} \\ \vdots & \vdots & \ddots & \vdots \\ \alpha_{n1} & \alpha_{n2} & \dots & \alpha_{nn} \end{vmatrix}. \tag{9.8.1}$$

So $D(\mathbf{a}_1, \dots, \mathbf{a}_n)$ is the determinant of the matrix whose columns are the (coordinate) vectors $\mathbf{a}_j$. ♣

By the theorem, the signed volume of a parallelepiped can be defined as the determinant belonging to its edges.

We note that the theorem can serve as an alternative definition of the determinant.

Proof. First, assume that a function $D: V^n \to \mathbf{R}$ satisfies conditions (i)–(iii). A special case of (ii) yields that

(iv) if two vectors $\mathbf{a}_j$ are equal, then $D(\mathbf{a}_1, \dots, \mathbf{a}_n) = 0$.

We show that

(v) swapping two vectors, D switches to its negative.

Let us swap (say) $\mathbf{a}_1$ and $\mathbf{a}_2$. Then, by (i) and (iv),

$$\begin{aligned} 0 &= D(\mathbf{a}_1 + \mathbf{a}_2, \mathbf{a}_1 + \mathbf{a}_2, \mathbf{a}_3, \dots, \mathbf{a}_n) \\ &= D(\mathbf{a}_1, \mathbf{a}_1, \mathbf{a}_3, \dots, \mathbf{a}_n) \\ &\quad + D(\mathbf{a}_1, \mathbf{a}_2, \mathbf{a}_3, \dots, \mathbf{a}_n) + D(\mathbf{a}_2, \mathbf{a}_1, \mathbf{a}_3, \dots, \mathbf{a}_n) + D(\mathbf{a}_2, \mathbf{a}_2, \mathbf{a}_3, \dots, \mathbf{a}_n) \\ &= D(\mathbf{a}_1, \mathbf{a}_2, \mathbf{a}_3, \dots, \mathbf{a}_n) + D(\mathbf{a}_2, \mathbf{a}_1, \mathbf{a}_3, \dots, \mathbf{a}_n), \end{aligned}$$

so $D(\mathbf{a}_2, \mathbf{a}_1, \mathbf{a}_3, \dots, \mathbf{a}_n) = -D(\mathbf{a}_1, \mathbf{a}_2, \mathbf{a}_3, \dots, \mathbf{a}_n)$.

Let us express $\mathbf{a}_j$ as a linear combination of the basis vectors $\mathbf{e}_i$: $\mathbf{a}_j = \sum_{i=1}^n \alpha_{ij}\mathbf{e}_i$. Then, using (i), $D(\mathbf{a}_1, \dots, \mathbf{a}_n)$ is the sum of n^n terms of type

$$\alpha_{m_1 1}\alpha_{m_2 2} \cdots \alpha_{m_n n} D(\mathbf{e}_{m_1}, \mathbf{e}_{m_2}, \dots, \mathbf{e}_{m_n}). \tag{9.8.2}$$

By conditions (iii), (iv), and (v), every $D(\mathbf{e}_{m_1}, \dots, \mathbf{e}_{m_n})$ is uniquely determined (it is 0 or ± 1), thus also $D(\mathbf{a}_1, \dots, \mathbf{a}_n)$ is uniquely determined. This proves that there is at most one function D that satisfies the conditions.

Also (9.8.1) follows. By (iv), $D(\mathbf{e}_{m_1}, \dots, \mathbf{e}_{m_n}) = 0$ if the same vector $\mathbf{e}$ appears twice. If the vectors $\mathbf{e}$ are distinct, then $m_1, \dots, m_n$ is a permutation of the numbers $1, \dots, n$. Then, by (iii) and (v), $D(\mathbf{e}_{m_1}, \dots, \mathbf{e}_{m_n})$ is 1 or -1 depending on whether an even or an odd number of swaps created the permutation $m_1, \dots, m_n$ from $1, \dots, n$, i.e., whether the permutation $m_1, \dots, m_n$ is even or odd. Therefore $D(\mathbf{a}_1, \dots, \mathbf{a}_n)$ is the sum of $n!$ terms $(-1)^I \alpha_{m_1 1}\alpha_{m_2 2}\dots\alpha_{m_n n}$ where $m_1,\dots,m_n$ assume all permutations of $1,\dots,n$ and I is the number of inversions in $m_1,\dots,m_n$. This equals the determinant of the numbers α_{ij} as it is just the definition of the determinant of the transpose matrix. This proves (9.8.1).

It remains to show that there exists such a function D. By the previous argument, it can only be the determinant. So we have to show that the determinant satisfies conditions (i)–(iii). This immediately follows from the elementary properties of the determinant. ∎

Remarks.

1. We could have finished the proof without referring to the properties of the determinant. Previously we established that the only candidate is the function where $D(\mathbf{a}_1,\dots,\mathbf{a}_n)$ is the sum of the n^n terms in (9.8.2) and every $D(\mathbf{e}_{m_1},\dots,\mathbf{e}_{m_n})$ is uniquely determined (it is 0 or ± 1). Then (iii) is evident, (i) and (iv) follow by a simple computation and they imply (ii) (see Exercise 9.8.1). Of course, this argument was just reproving the corresponding properties of the determinant.

Another option with somewhat less amount of calculation is to show that the functions satisfying (i) and (ii) form a one-dimensional vector space under the naturally arising operations and (iii) selects a unique suitable function D in this vector space (cf. Exercise 9.8.2).

2. As, over a given field, any two n-dimensional vector spaces are isomorphic, we could have worked in $V = \mathbf{R}^n$. Then the usual unit vectors are the natural choice for $\mathbf{e}_1,\dots,\mathbf{e}_n$.

Exercises 9.8

9.8.1. Derive property (ii) from (i) and (iv).

9.8.2. Let F_1 and $F_2 \neq 0$ be $V^n \to \mathbf{R}$ functions satisfying (i) and (ii). Verify $F_1 = \lambda F_2$ for some $\lambda \in \mathbf{R}$.

9.8.3. Let A be the matrix of a linear transformation $\mathcal{A} \in \operatorname{Hom} V$ in the basis $\mathbf{e}_1,\dots,\mathbf{e}_n$. Prove that the functions $F(\mathbf{v}_1,\dots,\mathbf{v}_n) = D(\mathcal{A}\mathbf{v}_1,\dots,\mathcal{A}\mathbf{v}_n)$ and $G(\mathbf{v}_1,\dots,\mathbf{v}_n) = (\det A)\cdot D(\mathbf{v}_1,\dots,\mathbf{v}_n)$ are equal.

Remark. This shows that the transformation multiplies the volume by the determinant.

9.8.4. Prove the Product Theorem 2.2.4 for determinants.

9.8.5. (a) Let P_1, P_2, and P_3 be three points in the plane. We denote the usual cartesian coordinates of P_j by γ_{1j} and γ_{2j} ($j = 1,2,3$). Prove that the points P_j are collinear if and only if

$$\begin{vmatrix} \gamma_{11} & \gamma_{12} & \gamma_{13} \\ \gamma_{21} & \gamma_{22} & \gamma_{23} \\ 1 & 1 & 1 \end{vmatrix} = 0.$$

(b) Formulate and verify the analogous statement in the space for four points to be coplanar.

(c) Are the above results true in a skew coordinate system?

9.8.6. Let $\mathbf{v}_1, \dots, \mathbf{v}_{n-1}$ be any vectors in the Euclidean space $\mathbf{R}^n$ with the usual basis of unit vectors $\mathbf{u}_1, \dots, \mathbf{u}_n$. We define a vector $\mathbf{d}$ by an $n \times n$ determinant where the elements of the first row are $\mathbf{u}_1, \dots, \mathbf{u}_n$ and the ith row is formed of the coordinates of $\mathbf{v}_{i-1}$ for $2 \leq i \leq n$. (So the elements of the first row are vectors and the elements of the other rows are scalars.) Prove that $\mathbf{d}$ is orthogonal to every $\mathbf{v}_i$ and its length is equal to (the absolute value of) the volume of the $(n-1)$-dimensional parallelepiped spanned by these vectors.

Remark. In the special case $n = 3$, we get the well-known formula for the cross product of two vectors in the space: For $\mathbf{a} = (\alpha_1, \alpha_2, \alpha_3)$, $\mathbf{b} = (\beta_1, \beta_2, \beta_3) \in \mathbf{R}^3$, their cross product

$$\mathbf{a} \times \mathbf{b} = \begin{vmatrix} \mathbf{u}_1 & \mathbf{u}_2 & \mathbf{u}_3 \\ \alpha_1 & \alpha_2 & \alpha_3 \\ \beta_1 & \beta_2 & \beta_3 \end{vmatrix}$$

is perpendicular to $\mathbf{a}$ and $\mathbf{b}$ and its length is the area of the parallelogram spanned by them.

Chapter 10

Codes

Coding theory designs and investigates algorithms to detect and correct errors occurring in electronic transmissions with small probability. Instead of sequences of digits corresponding to the original message, we send slightly longer codewords containing some check digits, too. An important aspect is to solve this task with possibly few check digits. Another essential requirement is that both the theoretical basis and the electronic implementation of encoding and decoding should be safe and effective for bulk transfers of data.

We present the basic notions of algebraic coding theory and discuss some efficient codes. Note that there are many other, non-algebraic types of codes and the ciphers in cryptography are called codes but we do not touch these areas.

10.1. Error Detection and Correction

In general, data are stored as 0–1 sequences in the electronic devices. To give an introductory illustration, we restrict ourselves to the special case when altogether at most one bit is transmitted erroneously. We show two simple procedures to detect whether there was an error and how to find the bad bit to regain the correct message.

Let us send the complete message twice. If an error can occur in at most one bit, then the two halves of this double message are the same apart from at most one digit. Thus, if the two parts are identical, then there was no error, and if there is a difference somewhere, then an error has occurred. We cannot correct this error since we do not know which copy of the message is wrong. We can solve the problem by sending the message a third time. Then, by our assumption, at least two of the three copies must be identical and they represent the right message.

Of course, these procedures are very inefficient by wasting too many check digits. The good codes should be able to detect and correct many errors with few check digits possibly by a simple automatic mechanism.

Starting with the systematic mathematical theory, we assume that the data form a 0–1 sequence. We break it into blocks of length n; these are called words. We assign to each word a codeword of length $k > n$ and the codewords will be transmitted. Thus the set of words equals F^n and the codewords form a subset of 2^n elements in F^k where F is the modulo 2 field.

Definition 10.1.1. Let $n < k$ and $F = \mathbf{Z}_2$. A (k, n) *code* is a 2^n-element subset C of F^k. An *encoding function* of this code is an *injective* map $\varphi : F^n \to F^k$ with image $\operatorname{Im} \varphi = C$. The elements of F^n and C are called *words* and *codewords*. To make a clear distinction, we shall mostly use the expression *message word* for the elements of F^n instead of the term word. ♣

Remark. A more general approach is when $F = \mathbf{Z}_p$ or any finite field, i.e., the digits 0 and 1 are replaced, e.g., by $0, 1, \dots, p-1$ for some prime p. We shall, however, restrict ourselves to *binary codes* when $p = 2$. So, the vectors and matrices are over the field $F = \mathbf{Z}_2$ throughout this chapter. From the next section, the codes will be subspaces in the vector space F^k with linear maps $F^n \to F^k$ as encoding functions. Accordingly, we shall switch then the notation of message words and codewords to column vectors in F^n and F^k.

We turn to the precise meaning of error detection. Assume that an error occurred during the transmission of a codeword $\mathbf{c} \in C$ and another $\mathbf{z} \in F^k$ was received by the other party. If $\mathbf{z}$ is a codeword, then the error remained secret, but if $\mathbf{z}$ is not a codeword, then the error was detected.

The error of transmitting a wrong bit is very small (otherwise the entire system would be practically useless), so the vectors $\mathbf{z}$ and $\mathbf{c}$ only differ in a few components in general. Thus to achieve efficient error detecting we have to guarantee that changing only a few digits in a codeword will never result in another codeword.

Definition 10.1.2. A code is *t-error detecting* if changing (at least 1, but) at most t components in any codeword we never get a codeword. ♣

Error detection just reveals the occurrence of an error. Error correction means that we can determine which codeword was deformed.

Definition 10.1.3. A code is *t-error correcting* if changing at most t components in any two (distinct) codewords we never get the same vector. ♣

Clearly, t-error correcting is a much stronger requirement than t-error detecting, and both are closely related to the distance between codewords.

Definition 10.1.4. The *weight* (or *Hamming weight*) $w(\mathbf{v})$ of a vector $\mathbf{v} \in F^k$ is the number of its non-zero components. The *distance* (or *Hamming distance*) $\delta(\mathbf{u}, \mathbf{v})$ of two vectors $\mathbf{u}, \mathbf{v} \in F^k$ is the number of components in which they differ. ♣

The distance is the weight of the difference vector: $\delta(\mathbf{u}, \mathbf{v}) = w(\mathbf{u} - \mathbf{v})$. It satisfies the usual properties required for a distance, so F^k becomes a metric space (see Definition 8.2.6).

Theorem 10.1.5. *A code is t-error detecting if and only if the minimal distance between codewords is at least $t+1$, and it is t-error correcting if and only if this minimal distance is at least $2t+1$.* ♣

Proof. Both assertions follow directly from Definitions 10.1.2–10.1.4. A code is t-error detecting if and only if no vector of distance at most t from a codeword is a codeword. Similarly, a code is t-error correcting if and only if the sets of vectors of distance at most t from codewords are disjoint. ∎

Examples of codes. We start with the repetition codes indicated at the beginning of the section.

E1 *2-repetition code*: The encoding function is to repeat the message word:

$$\varphi\colon \alpha_1\alpha_2\dots\alpha_n \mapsto \alpha_1\alpha_2\dots\alpha_n\alpha_1\alpha_2\dots\alpha_n.$$

Here $k = 2n$, $C = \{\gamma_1\dots\gamma_{2n} \mid \gamma_j = \gamma_{n+j}, 1 \le j \le n\}$. The minimal distance of codewords is 2, the code is 1-error detecting. Of course, it detects several types of other errors, too, but if, e.g., the first and $(n+1)$st digits are wrong, then this 2-error remains undetected since we obtain another codeword.

E2 *3-repetition code*: The encoding function is to write the message word three times:

$$\varphi\colon \alpha_1\alpha_2\dots\alpha_n \mapsto \alpha_1\alpha_2\dots\alpha_n\alpha_1\alpha_2\dots\alpha_n\alpha_1\alpha_2\dots\alpha_n.$$

Here $k = 3n$, $C = \{\gamma_1\dots\gamma_{3n} \mid \gamma_j = \gamma_{n+j} = \gamma_{2n+j}, 1 \le j \le n\}$. The minimal distance of codewords is 3, the code is 2-error detecting and 1-error correcting.

The next code is much more economic than the previous ones.

E3. *Parity-check code*: The encoding function writes a single digit after the message word so that the resulting vector should contain an even number of non-zero digits. Thus, this extra digit is 0 or 1 according to whether the message word contained an even or odd number of non-zero digits and can be uniformly described as the sum of all digits in the message word:

$$\varphi\colon \alpha_1\alpha_2\dots\alpha_n \mapsto \alpha_1\alpha_2\dots\alpha_n\beta, \qquad \text{where} \quad \beta = \sum_{i=1}^{n}\alpha_i.$$

Here $k = n+1$, the codewords are the vectors with an even weight. The minimal distance of codewords is 2, the code is 1-error detecting.

We summarize the usual names of the parameters n, k, and $s = k - n$.

Definition 10.1.6. In a (k, n) code C, k and n are called the *length* and the *dimension* of the code, and $s = k - n$ is the *number of check digits.* ♣

The use of the word dimension is explained by the fact that most codes form a subspace in F^k (see the linear codes in the next section), and this subspace has dimension n.

The efficiency of a code is measured, on the one hand, by the number of errors detected or corrected, i.e., how big the minimal distance between codewords is, and on the other hand, how few check digits are used to achieve this. Both codes in Examples 1 and 3 are 1-error detecting, but the 2-repetition code requires n check digits, while the parity-check code needs only 1, so the latter is much more efficient.

We determine the minimal number of check digits needed to correct all 1-errors. Let us call two vectors neighbors if they only differ in one digit, i.e., their distance is 1. If we consider the $(k+1)$-element subsets in F^k consisting of a codeword and its neighbors, these subsets must be disjoint to guarantee the correction of 1-errors. There are 2^n such subsets, so $2^n(k+1) \leq 2^k$. Substituting $k = n+s$, we obtain $2^s \geq n+s+1$. E.g., for $n = 500$ this means $s \geq 9$. We show in Section 10.3 that the equality $s = 9$ can be attained (cf. $s = 1000$ in the 3-repetition code of Example 3).

Exercises 10.1

10.1.1. Assume that the probability of an error in transmitting a bit is p for each bit independently. What is the probability that transmitting a k-digit codeword
(a) no error occurs;
(b) exactly 1 digit will be wrong;
(c) more than 3 digits will be wrong?
It is interesting and instructive to compare these probabilities for concrete values of k and p, e.g., for $k = 20$ and $p = 10^{-4}$.

10.1.2. Which codes are 1-error detecting and 1-error correcting? The dimension of the codes is $n = 3$. Each codeword starts with the message word $\alpha_1\alpha_2\alpha_3$ and this is followed by the s check digits given below.
(a) $\alpha_1 + \alpha_2 + \alpha_3 + 1$ $(s = 1)$.
(b) $\alpha_1, \alpha_2 + \alpha_3$ $(s = 2)$.
(c) $\alpha_1, \max(\alpha_2, \alpha_3)$ $(s = 2)$.
(d) α_1, γ, where $\gamma = \begin{cases} \alpha_2, & \text{if } \alpha_1 = 1 \\ \alpha_3, & \text{if } \alpha_1 = 0 \end{cases}$ $(s = 2)$.
(e) $\alpha_1 + \alpha_2, \alpha_1 + \alpha_3, \alpha_2 + \alpha_3$ $(s = 3)$.
(f) $\alpha_1, \alpha_1 + \alpha_2, \alpha_1 + \alpha_2 + \alpha_3$ $(s = 3)$.
(g) $\alpha_1, \alpha_1 \cdot \alpha_2, \alpha_1 \cdot \alpha_2 \cdot \alpha_3$ $(s = 3)$.

10.1.3. Verify that the error-detecting and error-correcting capacities of a code remain the same if, in every codeword,
(a) the first digit is changed to its opposite;
(b) every digit is changed to its opposite;
(c) the first two digits are swapped.
How can we generalize these observations?

10.1.4. (a) Prove that the sum of the pairwise distances of any three vectors in F^k is an even number.
(b) Find all m satisfying that the sum of the pairwise distances of any m vectors in F^k is an even number.

10.1.5. Let d denote the distance $\delta(\mathbf{u}, \mathbf{v})$ of the vectors $\mathbf{u}, \mathbf{v} \in F^k$. To given integers q and r, how many vectors $\mathbf{z} \in F^k$ satisfy $\delta(\mathbf{u}, \mathbf{z}) = q$ and $\delta(\mathbf{v}, \mathbf{z}) = r$?

10.1.6. Assume that the pairwise distances of three vectors in F^k are D. Show that D is even and there exists exactly one vector of distance $D/2$ from each of the three vectors.

10.1.7. Let d be the minimal distance of codewords in a (k, n) code C. We construct a new $(k+1, n)$ code C' by writing a single parity-check digit after each codeword of C so that the weight of the new codeword should be even. What is the minimal distance in C'?

10.1.8. Verify $\sum_{i=0}^{t} \binom{k}{i} \leq 2^{k-n}$ for any t-error correcting (k, n) code.

10.1.9. Let $n = 2$. Determine the minimal number of check digits necessary to (a) detect; (b) correct all (A) 1; *(B) 2-errors.

Solve the problem for $n = 3$, too.

10.2. Linear Codes

Definition 10.2.1. A (k, n) code C is *linear* if C is a subspace in F^k. ♣

Note that the zero vector is a codeword in any linear code.

The codes in the previous section in Examples E1–E3 and in Exercise 10.1.2(b), (e), (f) are linear.

For linear codes, it is much easier to see the error-detecting and error-correcting capacities.

Theorem 10.2.2. *A linear code is t-error detecting if and only if the weight of every non-zero codeword is at least $t + 1$, and is t-error correcting if and only if the weight of every non-zero codeword is at least $2t + 1$.* ♣

Proof. The codewords form a subspace, so the difference of two codewords is a codeword. Therefore, the distance of any two distinct codewords is the weight of some non-zero codeword. Conversely, the weight of a non-zero codeword is equal to the distance of this codeword from the zero codeword. Thus we proved that the set of distances between codewords is the same as the set of weights of non-zero codewords. This implies that the minimal distance between codewords is equal to the minimal weight of non-zero codewords. The theorem thus follows from Theorem 10.1.5. ■

E.g., a linear code is 1-error correcting if and only if the weight of every non-zero codeword is at least 3.

If C is a linear code, we use (injective) linear maps $\varphi : F^n \to F^k$ as encoding functions. Accordingly, we switch their notation to uppercase calligraphic Roman letters and characterize them by matrices.

Definition 10.2.3. Let a linear map $\mathcal{A} : F^n \to F^k$ be an encoding function of a (k, n) linear code C and let $G = G_{\mathcal{A}} = [\mathcal{A}]$ be its matrix in the natural bases of the unit vectors in F^n and F^k. This $k \times n$ matrix G is called a *generator matrix* of the code C. ♣

By the definition, the columns of the generator matrix are the codewords belonging to the unit vectors of F^n. The generator matrix of the parity-check code of Example 3 in the previous section is the $(n+1) \times n$ matrix

$$\begin{pmatrix} 1 & 0 & \dots & 0 \\ 0 & 1 & \dots & 0 \\ \vdots & \vdots & \ddots & \vdots \\ 0 & 0 & \dots & 1 \\ 1 & 1 & \dots & 1 \end{pmatrix}.$$

If every codeword starts with the message word, then the generator matrix is of the form

$$G = \begin{pmatrix} I_{n \times n} \\ B_{s \times n} \end{pmatrix}$$

where I is the $n \times n$ identity matrix and B is an $s \times n$ matrix.

Let $\mathbf{v} \in F^n$ be any message word and let $\mathbf{c} = \mathcal{A}\mathbf{v}$ be the corresponding codeword, both considered as column vectors. Then $\mathbf{c} = G_{\mathcal{A}}\mathbf{v}$. So encoding means multiplication by the generator matrix from the left.

The remaining part of this section deals with an implementation of error correction for linear codes. In the next section, we shall learn a much better method.

Error correction means that for every $\mathbf{z} \in F^k$ we determine a codeword $\mathbf{c}$ with minimal distance from $\mathbf{z}$, and if we receive $\mathbf{z}$ in the transmission, we decode it to $\mathbf{c}$. The *error vector* $\mathbf{e} = \mathbf{z} - \mathbf{c}$ has the minimal weight among the differences of $\mathbf{z}$ and the codewords. If $\mathbf{z}$ is a codeword, then $\mathbf{e} = \mathbf{0}$.

Therefore it is worth representing the elements of F^k as translates of codewords by error vectors. This means that we consider the translates or cosets $\mathbf{e} + C$ of the subspace C of the codewords (see Exercise 4.2.16). Thus we partition F^k into 2^{k-n} disjoint classes, each containing $|C| = 2^n$ vectors. One of the classes is C itself (as the translate by $\mathbf{e} = \mathbf{0}$). Note that the name coset can also be understood in its group-theoretic meaning since F^k and F^n can be considered as additive Abelian groups, C is a subgroup in F^k, and $\mathcal{A} : F^n \to F^k$ is a group homomorphism.

We perform the error correction with the help of a *decoding table*. We write C, i.e., the codewords into the first row starting with $\mathbf{0}$. Then we fill in the next rows with the cosets so that the first element is a vector $\mathbf{e}$ of minimal weight in that coset and the other elements are obtained by adding the codewords to $\mathbf{e}$. Thus the elements in the first column are the error vectors and we correct any vector to the codeword in the first row above it. Then we still have to regain the message word from the codeword. If the codeword starts with the message word, then this is straightforward. A method to handle the other cases is discussed in Exercise 10.2.9.

If a code is t-error correcting, then the error vectors containing at most t non-zero coordinates are in distinct cosets, so they are distinct elements of the first column.

Example. Let $\mathcal{A} : \alpha_1\alpha_2 \mapsto \alpha_1\alpha_2\alpha_1\alpha_2\beta$, where $\beta = \alpha_1 + \alpha_2$. This is a linear code; see its decoding table below.

00000	10101	01011	11110
10000	00101	11011	01110
01000	11101	00011	10110
00100	10001	01111	11010
00010	10111	01001	11100
00001	10100	01010	11111
11000	01101	10011	00110
01100	11001	00111	10010

Note that the error vectors with only one coordinate 1 are in distinct rows, so the code is 1-error correcting. E.g., in the case of $\mathbf{z} = 01010$ in the 3rd place of the 6th row, the correct codeword is $\mathbf{c} = 01011$. The error vectors in the last two rows are not unique as in those cosets there are more vectors with a minimal weight. E.g., 00110 could have been the first element in the 7th row. This illustrates well that the last two rows cannot be used for error correction as these vectors contain at least two errors and the code cannot correct them.

This procedure is somewhat inconvenient. In the next section, we shall give a different characterization of linear codes that provides a quick and automatic error correction method.

Exercises 10.2

***10.2.1.** Let $k > n$. True or false?

(a) The number of linear codes divides the number of all codes.

(b) The number of possible linear map encoding functions of linear codes divides the number of possible encoding functions of all codes.

10.2.2. Establish generator matrices of the repetition codes in Examples E1 and E2 in Section 10.1.

10.2.3. Verify that the linear codes in Exercise 10.1.2 are (b), (e), and (f), and construct a generator matrix and a decoding table for each.

10.2.4. Let A be a $k \times n$ 0–1 matrix. Prove that A is a generator matrix of a linear code if and only if its rank $\mathrm{rk}(A)$ is n over $F = \mathbf{Z}_2$.

10.2.5. Verify that the codewords of a linear code are all linear combinations of the columns of its generator matrix.

10.2.6. Prove that the codewords of even weight in a linear (k, n) code form a subspace in F^k and compute its dimension.

10.2.7. We construct new codes from a (k_1, n_1) code C_1 and a (k_2, n_2) code C_2. Let d_i be the minimal distance of codewords and let $\varphi_i : F^{n_i} \to F^{k_i}$ be the encoding functions in C_i, $1 = 1, 2$.

I. We use the same message words and write the codewords of the two codes one after the other. We assume $n_1 = n_2 = n$ and we define a

(k_1+k_2, n) code C_3 by the encoding function $\varphi_3: F^n \to F^{k_1+k_2}$ assigning $\varphi_3(\mathbf{a}) = \varphi_1(\mathbf{a})|\varphi_2(\mathbf{a})$ to every $\mathbf{a} \in F^n$.

II. We write both the message words and the codewords of the two codes one after the other. We define a (k_1+k_2, n_1+n_2) code C_4 by the encoding function $\varphi_4: F^{n_1+n_2} \to F^{k_1+k_2}$ assigning $\varphi_4(\mathbf{a}_1|\mathbf{a}_2) = \varphi_1(\mathbf{a}_1)|\varphi_2(\mathbf{a}_2)$ to every pair $\mathbf{a}_i \in F^{n_i}$, $i = 1, 2$.

III. We write the message words one after the other and form an asymmetric combination of the codewords. We assume $k_1 = k_2 = k$ and define a $(2k, n_1+n_2)$ code C_5 by the encoding function $\varphi_5: F^{n_1+n_2} \to F^{2k}$ assigning $\varphi_5(\mathbf{a}_1|\mathbf{a}_2) = \varphi_1(\mathbf{a}_1)|(\varphi_1(\mathbf{a}_1) + \varphi_2(\mathbf{a}_2))$ to every pair $\mathbf{a}_i \in F^{n_i}$, $i = 1, 2$.

(a) What can be stated about the minimal distances d_3, d_4, and d_5 in the codes C_3, C_4, and C_5?

(b) If C_1 and C_2 are linear codes, then so are C_3, C_4, and C_5. How do we obtain their generator matrices G_3–G_5 from the generator matrices G_1 and G_2 of C_1 and C_2?

10.2.8. Let $F_{(m)}[x]$ denote the usual vector space of polynomials over $F = \mathbf{Z}_2$ of degree at most $m-1$. Then

$$\alpha_0\alpha_1\ldots\alpha_{m-1} \mapsto \alpha_0 + \alpha_1 x + \cdots + \alpha_{m-1}x^{m-1}$$

is an isomorphism between the vector spaces F^m and $F_{(m)}[x]$.

Let $k > n$ and $s = k - n$. We shall identify the vector spaces F^n and F^k with $F_{(n)}[x]$ and $F_{(k)}[x]$ obtained by the above isomorphism.

(a) Let g be a fixed, non-zero polynomial over F of degree s and let the map $\mathcal{A}: F_{(n)}[x] \to F_{(k)}[x]$ be the multiplication by g: $\mathcal{A}f = gf$ for $\deg f \le n-1$. Prove that $\operatorname{Im}\mathcal{A}$ is a linear code and write its generator matrix. A code of this type is a *polynomial code* and g is its *generator polynomial*.

(b) Let $g \ne 0$ and h be polynomials over F where $\deg h = k$. The map $\mathcal{A}: F_{(n)}[x] \to F_{(k)}[x]$ assigns to any polynomial f of degree at most $n-1$ the remainder of the product gf in the long division by h. Show that this defines a linear code if and only if the degree of the gcd of g and h is not greater than $s = k - n$.

(c) We assume $\deg(g, h) = s$ for the code C in (b). Prove that C is a polynomial code in the sense of (a) with a generator polynomial dividing h.

10.2.9. Demonstrate that for any (k, n) linear code C there exists an $n\times k$ 0–1 matrix M such that multiplying the codewords by M from the left we regain the message words (i.e., if the codeword $\mathbf{c} \in C$ corresponds to the message word $\mathbf{v} \in F^n$, then $M\mathbf{c} = \mathbf{v}$). Devise a fast algorithm to find such a matrix M.

10.3. Hamming code

Until now, a linear code was characterized as the *image* of a linear map. In this section, we switch to a characterization as the *kernel* of a linear map.

A (k, n) linear code C is an n-dimensional subspace in F^k. Let $k = n + s$ and let $\mathcal{P} : F^k \to F^s$ be a linear map with $\operatorname{Ker}\mathcal{P} = C$. Then $\operatorname{Im}\mathcal{P} = F^s$ by the dimension theorem.

To construct such a map $\mathcal{P} : F^k \to F^s$, we extend a basis $\mathbf{b}_1, \dots, \mathbf{b}_n$ of C with vectors $\mathbf{d}_1, \dots, \mathbf{d}_s$ to a basis of F^k, prescribe $\mathcal{P}\mathbf{b}_1 = \cdots = \mathcal{P}\mathbf{b}_n = \mathbf{0}$, $1 \le i \le n$, and let $\mathcal{P}\mathbf{d}_1, \dots, \mathcal{P}\mathbf{d}_s$ be an arbitrary basis in F_s. This shows that $\mathcal{P}$ is not unique in general.

Definition 10.3.1. Let C be a (k, n) linear code, $k = n + s$, and let $\mathcal{P} : F^k \to F^s$ be a linear map with kernel C. The $s \times k$ matrix $P = [\mathcal{P}]$ in the natural pair of bases of unit vectors is a *parity-check matrix* of the code C. ♣

To explain the name, we look at the effect of multiplying a row of P with a vector $\mathbf{z} \in F^k$. The result is the sum of those components of $\mathbf{z}$ where a digit 1 stands in the corresponding place of the given row of P. This sum is 0 or 1 depending on whether an even or odd number of 1's occur in these coordinates of $\mathbf{z}$. By the condition, this sum is 0 in every row exactly if $\mathbf{z}$ is a codeword. Thus P checks the parity of the total number of digits 1 in certain components of the vector.

Since $\mathcal{P}$ is not unique in most cases, a code has more parity-check matrices in general (cf. Exercises 10.3.3–10.3.4). By definition, each is characterized by the property that $P\mathbf{z} = \mathbf{0}$ for $\mathbf{z} \in F^k$ if and only if $\mathbf{z}$ is a codeword.

The property $\operatorname{Im}\mathcal{P} = F^s$ is equivalent to $\dim \operatorname{Im}\mathcal{P} = s$. This means that a parity-check matrix has rank s. Conversely, if an $s \times k$ matrix P has rank s and $P\mathbf{z} = \mathbf{0}$ for every codeword $\mathbf{z}$, then P is a parity-check matrix of the code (see Exercise 10.3.3(a)). In general, if a 0–1 matrix has s rows, $k = n + s$ columns, and its rank is s, then it is a parity-check matrix of some linear (k, n) code (see Exercise 10.3.1).

Example. The parity-check matrix of the parity-check code in Example E3 in Section 10.1 is a single row where all the k entries are 1.

We can easily obtain a parity-check matrix P from the generator matrix G (see Exercise 10.3.5). In the special case of *systematic codes* when the codewords start with the message words, the generator matrix is of the form

$$G = \begin{pmatrix} I_{n\times n} \\ B_{s\times n} \end{pmatrix}$$

and a parity-check matrix is $P = \begin{pmatrix} B_{s\times n} & I_{s\times s} \end{pmatrix}$.

How can we use a parity-check matrix for error correction? By definition, $P\mathbf{z} = \mathbf{0}$ if and only if $\mathbf{z}$ is a codeword. An erroneous transmission results in $\mathbf{z} = \mathbf{c} + \mathbf{e}$, where $\mathbf{c}$ is the correct codeword and $\mathbf{e}$ is the error vector. In this case, $P\mathbf{z} = P\mathbf{e}$. The vector $P\mathbf{z} \in F^s$ is called the *syndrome* of $\mathbf{z} \in F^k$.

If only the ith coordinate is 1 in the error vector $\mathbf{e}$, then $P\mathbf{e}$ is the ith column of P. Thus the syndrome of a general vector $\mathbf{z}$ is the sum of the columns of P corresponding to the erroneous coordinates in $\mathbf{z}$. This immediately implies the next theorem.

Theorem 10.3.2. *A linear code is t-error-correcting if and only if adding any $j \leq t$ columns of its parity-check matrix P, the obtained sums are distinct non-zero vectors.* ♣

Thus, the condition of 1-error correction is that no two columns in P are equal and none of them is 0.

We have shown at the end of Section 10.1 that if the number of check digits is $s = k - n$ in a 1-error-correcting (k, n) code, then $n \leq 2^s - s - 1$. We prove that the converse is also true.

Theorem 10.3.3. *Let $s \geq 2$ and $n \leq 2^s - s - 1$. Then there exists a 1-error-correcting linear $(s + n, n)$ code.* ♣

Proof. Let P be an $s{\times}(s{+}n)$ matrix having distinct non-zero columns so that there exist s linearly independent among them. As $2^s - 1 \geq n + s$, F^s has sufficiently many non-zero elements to satisfy the first condition. To meet the independence requirement, we put, e.g., the s unit vectors into the last s columns.

The rank of P is s, so P is a parity-check matrix of some linear code. This code is 1-error-correcting by (the remark after) Theorem 10.3.2. ∎

Returning to the example of $n = 500$ at the end of Section 10.1, Theorem 10.3.3 implies that we can solve 1-error correction with just $s = 9$ check digits (and this is the minimum). It is worth comparing it with the value $s = 1000$ obtained in the 3-repetition code.

We introduce a special name for the above codes in the case $n = 2^s - s - 1$.

Definition 10.3.4. If $n = 2^s - s - 1$ and $k = n + s = 2^s - 1$, then the 1-error correcting linear (k, n) codes are called *Hamming codes.* ♣

By the above, a Hamming code is characterized by having a parity-check matrix where every non-zero element of F^s occurs exactly once among the $2^s - 1$ columns.

Exercises 10.3

In the exercises, $n + s = k$, C is a linear (k, n) code with generator matrix G, and P is an $s \times k$ 0–1 matrix.

10.3.1. Verify that a 0–1 matrix is a parity-check matrix of some linear code if and only if its rows are independent and its columns are dependent.

10.3.2. Construct parity-check matrices of the repetition codes in Examples E1 and E2 of Section 10.1 and of the linear codes in Exercise 10.1.2.

10.3.3. Prove that each of the conditions is equivalent to P being a parity-check matrix of the code C.

(a) $\mathrm{rk}(P) = s$ and $P\mathbf{c} = \mathbf{0}$ for every $\mathbf{c} \in C$.

(b) $\mathrm{rk}(P) = s$ and $PG = 0$.

(c) The rows of P form a basis in $(C)^{\perp}$.

10.3.4. Prove.

(a) The parity-check matrix of C is unique if and only if $s = 1$.

(b) In general, the number of parity-check matrices of C is $\prod_{i=0}^{s-1}(2^s - 2^i)$.

(c) All parity-check matrices of C are obtained as a product MP where P is a fixed parity-check matrix and M is any invertible $s \times s$ matrix.

10.3.5. (a) Verify that if $G = \begin{pmatrix} I_{n\times n} \\ B_{s\times n} \end{pmatrix}$, then $P = \begin{pmatrix} B_{s\times n} & E_{s\times s} \end{pmatrix}$ is a parity-check matrix of C.

(b) Devise a fast algorithm to obtain a parity-check matrix from G in the general case.

10.3.6. Let Q be any 0–1 matrix with k columns satisfying $\operatorname{Ker} Q = C$. (The number of rows is not specified.) Prove that $\operatorname{rk}(Q) = s$ and any s independent rows of Q provide a parity-check matrix of C.

10.3.7. Verify that two vectors in F^k appear in the same row of the decoding table of C if and only if they have the same syndrome.

10.3.8. The minimal distance in C is d if and only if any $d - 1$ columns of a parity-check matrix are independent and there exist d dependent columns.

10.3.9. Show that the minimal distance in C is at most one greater than the number of check digits.

10.3.10. We call $C^\perp$ the *dual code* of C. What is the connection between the generator and parity-check matrices of C and its dual code?

10.3.11. Let $P = \begin{pmatrix} B & I \end{pmatrix}$ be a parity-check matrix of C. Prove that C is self-dual if and only if $B^T = B^{-1}$.

10.3.12. What is the minimal distance in a Hamming code?

10.3.13. Prove that in a Hamming code, the complement of a codeword is a codeword. (The complement differs in each digit from the original vector, cf. Exercise 9.4.2(a)).

10.3.14. Let $s \geq 2$, $n = 2^s - s - 1$, and $k = 2^s - 1$. For $0 \leq j \leq s - 1$, let M_j denote the set of non-negative integers such that the coefficient of 2^j is 1 in their binary representation (i.e., from the right, their $(j + 1)$st binary digit is 1). Clearly $|M_j| = 2^{s-1}$.

We define a (k, n) code C. Instead of using indices $1, 2, \ldots, n$ for the 1st, 2nd, …, nth digits in a message word, we omit the powers of two (including $2^0 = 1$) from them (similar to some hotels where no room has number 13, just that we are superstitious concerning the powers of two). So the indices are $3, 5, 6, 7, 9, \ldots, k = n + s$ as just the powers $1, 2, 2^2, \ldots, 2^{s-1}$ are missing. Let

$$\varphi\colon \alpha_3\alpha_5\alpha_6\ldots\alpha_k \mapsto \alpha_3\alpha_5\alpha_6\ldots\alpha_k\gamma_0\gamma_1\ldots\gamma_{s-1}$$

where $\gamma_j = \sum_{i\in M_j} \alpha_i$, $j = 0, 1, \ldots, s - 1$. Prove that this defines a Hamming code.

10.4. BCH Codes

The Hamming code was born at the dawn of coding theory in the late 1940s. It took more than 10 years to improve 1-error correction to 2-error correction of similar efficiency. The new codes were invented by R. C. Bose and his student D. K. Ray-Chaudhuri and independently by A. Hocquenghem. The name BCH code bears the initials of the inventors.

BCH codes are based on finite fields (see Section A.11). As before, n, k, and $s = k - n$ denote the dimension, length, and the number of check digits of the code. Our goal is to construct an efficient t-error correcting code for a given t. Until now, we fixed n and tried to minimize s. We slightly change our approach by fixing $k = n + s$ and looking for a small s and a large n.

In other words, we want to find a large subspace C in F^k satisfying that its elements, the codewords, are not closer to each other than $2t + 1$. This subspace C is the kernel of a parity-check matrix P. The capacity of t-error correction depends on properties of the columns of P, so we characterize C by constructing a suitable $s \times k$ matrix P. The minimality of s means that P should have as few rows as possible.

Let us disregard temporarily the condition of the independence of the rows of P. If a row is a linear combination of other rows, then deleting this row effects neither the kernel of the matrix, nor properties of its columns related to error correction (cf. Exercises 10.3.7). Thus we can reduce the number of rows until they become independent.

Accordingly, we shall construct $m \times k$ matrices Q satisfying that the sums of at most t columns are distinct non-zero vectors, or equivalently, any $2t$ columns are linearly independent, and m is small. Then the kernel of Q is a t-error correcting linear code. The number of check digits is $s = \text{rk}(Q) \leq m$. We obtain a parity-check matrix P by keeping any s independent rows of Q. Therefore we shall call Q a *quasi-parity-check matrix* of the code C.

Before turning to the general t-error correcting BCH codes, we present the case $t = 2$ separately. Then our first meeting with a complicated construction takes place in a simpler setting and this might help to arrive at a better understanding of the topic. Further, the proof of 2-error correction is much less difficult than for a general t.

Let $q \geq 3$ and $k = 2^q - 1$. By Exercise 10.1.8, the number of check digits in a 2-error correcting code has to satisfy $s \geq 2q - 1$. The 2-error correcting BCH codes are very close to this bound, they guarantee $s \leq 2q$. It can be shown that this always holds with equality (see Exercise 10.4.5). This means that for the same k, a 2-error correcting BCH code only requires twice as many check digits as a Hamming code, whereas there are many more 2-errors than 1-errors ($\binom{k}{2} + k$ versus k).

To characterize a 2-error correcting BCH code, we shall construct its parity-check matrix P. We define a $2q \times k$ quasi-parity-check matrix Q first. The upper q rows of Q should be the same as in the Hamming code: the columns are the non-zero elements in F^q.

To construct the lower q rows of Q, we consider F^q as a field of 2^q elements. The vector space and the field have the same additive structure, so we shall keep the notation F^q for the field. Let Δ be a generator of the multiplicative group of the field. Then the non-zero elements in F^q are the powers of Δ with exponents from 0 to $2^q - 2$. Thus, the columns in the upper half of Q are Δ^j for $j = 0, 1, 2, \dots, 2^q - 2$.

Each column in the lower half of Q should be the cube of the column above it:

$$Q = \begin{pmatrix} 1 & \Delta & \Delta^2 & \dots & \Delta^j & \dots \\ 1 & \Delta^3 & \Delta^6 & \dots & \Delta^{3j} & \dots \end{pmatrix}. \tag{10.4.1}$$

We claim that the kernel of Q is a 2-error correcting code. We have to verify that the sums of at most two columns of Q are distinct non-zero vectors. Equivalently, the system of equations

$$\mathbf{x} + \mathbf{z} = \mathbf{a}, \qquad \mathbf{x}^3 + \mathbf{z}^3 = \mathbf{b}$$

has at most one solution in F^q for every $\mathbf{a} \neq \mathbf{0}$, apart from the symmetry of $\mathbf{x}$ and $\mathbf{z}$.

Subtracting the second equation from the cube of the first one, we obtain $\mathbf{xz}(\mathbf{x} + \mathbf{z}) = \mathbf{a}^3 - \mathbf{b}$, so $\mathbf{xz} = \mathbf{a}^2 - \mathbf{b}/\mathbf{a}$. This implies that both the sum and product of $\mathbf{x}$ and $\mathbf{z}$ are given. By the Vieta formulas, $\mathbf{x}$ and $\mathbf{z}$ can only be the roots of a uniquely determined quadratic equation (if the equation is solvable in the field F^q at all).

As mentioned before, the rank of Q is $2q$, so Q itself is a parity-check matrix of this code. (If this were not the case, then the code would be even more efficient as it would guarantee 2-error correction with less than $2q$ check digits with the same $k = 2^q - 1$.)

We summarize the result in the following theorem.

Theorem 10.4.1 (2-error correcting BCH code). *Let $q \geq 3$, $n = 2^q - 2q - 1$, and $k = 2^q - 1$. The kernel of matrix Q in (10.4.1) is a 2-error correcting linear (k, n) code.* ♣

Example. Let $q = 4$. We need a generator of the multiplicative group of a $2^4 = 16$-element field. There are $\varphi(15) = 8$ such generators. Each has degree 4 over $F = \mathbf{Z}_2$, so it is a root of an irreducible polynomial f of degree 4. Since $f(0) \neq 0$, the constant term is 1, and $f(1) \neq 0$ implies that f consists of an odd number of terms. Thus f can only be one of the four polynomials $x^4 + x + 1$, $x^4 + x^3 + 1$, $x^4 + x^2 + 1$, and $x^4 + x^3 + x^2 + x + 1$. The third one is $(x^2 + x + 1)^2$, so it is not irreducible. The fourth one is irreducible but it divides $x^5 - 1$, so the fifth power of any of its roots is 1. Therefore no root can generate the multiplicative group in the field. The first two polynomials satisfy the requirements, their roots are the eight generators in question. Choosing the first polynomial, let Δ be a root of it. Then $1, \Delta, \Delta^2, \Delta^3$ is a basis in the vector space F^4, these will be the unit vectors. We can compute the coordinates of the other powers by the repeated application of $\Delta^4 = 1 + \Delta$ obtained from $f(\Delta) = 0$. This leads to an 8×15 parity-check matrix.

$$\left(\begin{array}{ccccccccccccccc}
1 & 0 & 0 & 0 & 1 & 0 & 0 & 1 & 1 & 0 & 1 & 0 & 1 & 1 & 1 \\
0 & 1 & 0 & 0 & 1 & 1 & 0 & 1 & 0 & 1 & 1 & 1 & 1 & 0 & 0 \\
0 & 0 & 1 & 0 & 0 & 1 & 1 & 0 & 1 & 0 & 1 & 1 & 1 & 1 & 0 \\
0 & 0 & 0 & 1 & 0 & 0 & 1 & 1 & 0 & 1 & 0 & 1 & 1 & 1 & 1 \\
\hline
1 & 0 & 0 & 0 & 1 & 1 & 0 & 0 & 0 & 1 & 1 & 0 & 0 & 0 & 1 \\
0 & 0 & 0 & 1 & 1 & 0 & 0 & 0 & 1 & 1 & 0 & 0 & 0 & 1 & 1 \\
0 & 0 & 1 & 0 & 1 & 0 & 0 & 1 & 0 & 1 & 0 & 0 & 1 & 0 & 1 \\
0 & 1 & 1 & 1 & 1 & 0 & 1 & 1 & 1 & 1 & 0 & 1 & 1 & 1 & 1
\end{array}\right)$$

The generalization to t-error correcting BCH codes runs similarly to the case $t = 2$. Let $k = 2^q - 1$ and we require $s \le tq$ (in some lucky cases, s is much smaller than this bound). The columns of the (tq, k) quasi-parity-check matrix Q consist of the 1st, 3rd, 5th, ..., $(2t-1)$st powers one below the other of the non-zero elements in the field F^q. If the rows of Q are dependent, then any maximal independent system of them is a parity-check matrix P. We summarize this in the following theorem.

Theorem 10.4.2 (t-error correcting BCH code). *Let q be a fixed positive integer satisfying $k = 2^q - 1 > qt$ and let Δ be a generator of the multiplicative group of the field F^q. Let Q be a $tq \times k$ matrix where the $(j+1)$st column consists of the t vectors Δ^j, Δ^{3j}, Δ^{5j}, ..., $\Delta^{(2t-1)j} \in F^q$ one below the other, $j = 0, 1, \ldots, 2^s - 2$, so*

$$Q = \begin{pmatrix} 1 & \Delta & \Delta^2 & \ldots & \Delta^j & \ldots \\ 1 & \Delta^3 & \Delta^6 & \ldots & \Delta^{3j} & \ldots \\ 1 & \Delta^5 & \Delta^{10} & \ldots & \Delta^{5j} & \ldots \\ \vdots & \vdots & \vdots & & \vdots & \\ 1 & \Delta^{2t-1} & \Delta^{4t-2} & \ldots & \Delta^{(2t-1)j} & \ldots \end{pmatrix}.$$

Then the kernel of Q is a t-error correcting linear code where the number of check digits is $s \le tq$. ♣

We need further preparations to prove that this code is t-error correcting.

We recall the notion of *polynomial codes* introduced in Exercise 10.2.8(a). Let $F_{(m)}[x]$ denote the usual vector space of polynomials over $F = \mathbf{Z}_2$ of degree at most $m - 1$. Then

$$\alpha_0\alpha_1 \ldots \alpha_{m-1} \mapsto \alpha_0 + \alpha_1 x + \cdots + \alpha_{m-1}x^{m-1}$$

is an isomorphism between the vector spaces F^m and $F_{(m)}[x]$. This makes it possible to identify the vector space F^m with $F_{(m)}[x]$.

Definition 10.4.3. Let $k = n + s$ and let $g \ne 0$ be a fixed polynomial over F of degree s. Then the multiples of g in $F_{(k)}$ form a linear (k, n) code called a *polynomial code* with *generator polynomial* g. The encoding function is multiplication by g: $\mathcal{A}f = gf$ for every $g \in F_{(n)}$. ♣

We show that the Hamming code and the BCH codes are polynomial codes and determine their generator polynomials.

The $s \times (2^s - 1)$ matrix $P = \begin{pmatrix} 1 & \Delta & \Delta^2 & \ldots & \Delta^{k-1} \end{pmatrix}$ is a parity-check matrix for a Hamming code. Let $R = \varrho_0 + \varrho_1 x + \cdots + \varrho_{k-1}x^{k-1} \in F_{(k)}[x]$ be the polynomial corresponding to the vector $\mathbf{r} = \varrho_0\varrho_1 \ldots \varrho_{k-1} \in F^k$. Then $\mathbf{r}$ is a codeword if and only if $\mathbf{0} = P\mathbf{r} = \sum_{j=0}^{k-1} \varrho_j \Delta^j$. This means that Δ is a root of R, or equivalently, the minimal polynomial m_Δ of Δ divides R. Therefore a Hamming code is a polynomial code with generator polynomial $g = m_\Delta$.

Turning to the 2-error correcting BCH codes, we get similarly that $\mathbf{r}$ is a codeword if and only if both m_Δ and m_{Δ^3} divide R. So, the generator polynomial is the least common multiple of these two polynomials: $g = [m_\Delta, m_{\Delta^3}]$.

We get the general result similarly.

Theorem 10.4.4. *A general t-error correcting BCH code in Theorem* 10.4.2 *is a polynomial code with generator $g_t = [m_1, m_3, \dots, m_{2t-1}]$ where $m_j = m_{\Delta^j}$.* ♣

Thus the number of check digits is the degree of the generator polynomial g_t. Therefore $s = tq$ holds if and only if each of $\Delta, \Delta^3, \dots, \Delta^{2t-1}$ has degree q over F and no two have the same minimal polynomial (see Exercise 10.4.2).

We are ready to prove Theorem 10.4.2: we verify that the code is t-error correcting.

Proof of Theorem 10.4.2. We verify first that $m_i = m_{2i}$ for every i. Squaring in the field F^q is an isomorphism where the elements $0, 1 \in F$ are fixed points. Thus if $\rho_0 + \rho_1\Theta + \cdots + \rho_v\Theta^v = 0$, then this also holds with Θ^2 instead of Θ. This proves $m_{2i} \mid m_i$. As both minimal polynomials are irreducible, they must be equal. (Also, we can verify the converse divisibility directly by noting that Θ can be obtained from Θ^2 by repeated squarings as $\Theta^{2^q} = \Theta$.)

By the above, $g_t = [m_1, m_2, \dots, m_{2t}]$. To verify t-error correction, we show that the minimal distance of codewords is at least $2t + 1$.

For a proof by contradiction, assume that there exists a non-zero codeword **r** of weight not greater than $2t$. The corresponding polynomial $R \neq 0$ is a multiple of g_t, so Δ^i is a root of R for every $i \leq 2t$. We consider the equalities $R(\Delta^i) = 0$ as a homogeneous system of $2t$ linear equations where the variables are the coefficients of R. Since R has at most $2t$ non-zero coefficients, we can omit some 0 coefficients from R so that exactly $2t$ coefficients remain.

This $2t \times 2t$ homogeneous system of equations has a non-trivial solution, namely the (remaining) coefficients of R. The determinant of the coefficient matrix is a Vandermonde determinant $V = V(\Delta^{i_1}, \Delta^{i_2}, \dots, \Delta^{i_{2t}})$ multiplied by a non-zero constant where $i_1, i_2, \dots, i_{2t}$ are the degrees of the remaining terms in R. The generators of V are distinct, so $V \neq 0$. But then the system only has a trivial solution, a contradiction. ∎

Exercises 10.4

We use the notation of the text: s is the number of check digits, k is the length of the code, Δ is a generator of the multiplicative group of the 2^q-element field, m_i is the minimal polynomial of Δ^i, etc.

10.4.1. Construct a parity-check matrix of a 2-error correcting BCH code for $q = 5$.

10.4.2. Show for a t-error correcting BCH code of parameter q, that $s = tq$ holds if and only if each of $\Delta, \Delta^3, \dots, \Delta^{2t-1}$ has degree q over F and no two have the same minimal polynomial.

***10.4.3.** (a) Verify that $\deg m_i$ is the smallest positive integer v satisfying $v \mid q$ and $(2^q - 1)/(2^v - 1) \mid i$.

(b) Prove that the roots of m_i are the powers of Δ with exponents $i \cdot 2^j$ where $0 \leq j < \deg m_i$.

10.4.4. Consider a 3-error correcting BCH code for $q > 3$.

(a) Determine s for $q = 4$.

*(b) Verify $s = 3q$ for every odd q.

*10.4.5. Prove $s = 2q$ for a 2-error correcting BCH code of parameter q.

***10.4.6.** Show that if $t \leq 2^{q/2-1}$, then $s = tq$ in a t-error correcting BCH code of parameter q.

10.4.7. Let g be the generator polynomial of a polynomial code. Prove:

(a) The weight of every codeword is even if and only if $1 + x \mid g$.

(b) Every vector of even weight is a codeword if and only if $g = 1 + x$.

10.4.8. Verify that the generator polynomial of a BCH code divides $x^k - 1$.

10.4.9. A linear code is *cyclic* if every cyclic permutation of a codeword is a codeword: $\mathbf{z} = \gamma_0\gamma_1\gamma_2 \dots \gamma_{k-1} \in C \Rightarrow \gamma_{k-1}\gamma_0\gamma_1 \dots \gamma_{k-2} \in C$. Prove:

*(a) Cyclic codes are special polynomial codes: C is cyclic if and only if C is a polynomial code and its generator polynomial is a divisor of $x^k - 1$.

(b) Every BCH code is cyclic.

10.4.10. Assume $m < k$ and $\sum_{i=0}^{d-2} \binom{k-1}{i} < 2^m$. Prove the existence of a linear code of length k with $s \leq m$ check digits and minimal distance at least d.

***10.4.11.** *Reed–Muller codes.*

(a) Let q be given, $n = q + 1$, $k = 2^q$, and we define the rows of the generator matrix as the binary representations of integers from 2^q to $2^{q+1} - 1$. (See this matrix after (b) for $q = 3$.) Show that the minimal distance of codewords is 2^{q-1}. This code is called a *first order Reed–Muller code.*

(b) Let $m < q$. The *mth order Reed–Muller code* is a linear (k, n) code with $n = \sum_{i=0}^{m} \binom{q}{i}$ and $k = 2^q$. The first $q + 1$ columns are the same as above and the other columns are the componentwise products of all at least 2-element subsets in the set of the 2nd, 3rd, …, $(q+1)$st columns. Prove that the minimal distance of codewords is 2^{q-m}.

The generator matrices of the first and second order Reed–Muller codes for $q = 3$ are:

$$G(1,3) = \begin{pmatrix} 1 & 0 & 0 & 0 \\ 1 & 0 & 0 & 1 \\ 1 & 0 & 1 & 0 \\ 1 & 0 & 1 & 1 \\ 1 & 1 & 0 & 0 \\ 1 & 1 & 0 & 1 \\ 1 & 1 & 1 & 0 \\ 1 & 1 & 1 & 1 \end{pmatrix}; \quad G(2,3) = \begin{pmatrix} 1 & 0 & 0 & 0 & 0 & 0 & 0 \\ 1 & 0 & 0 & 1 & 0 & 0 & 0 \\ 1 & 0 & 1 & 0 & 0 & 0 & 0 \\ 1 & 0 & 1 & 1 & 0 & 0 & 1 \\ 1 & 1 & 0 & 0 & 0 & 0 & 0 \\ 1 & 1 & 0 & 1 & 0 & 1 & 0 \\ 1 & 1 & 1 & 0 & 1 & 0 & 0 \\ 1 & 1 & 1 & 1 & 1 & 1 & 1 \end{pmatrix}.$$

The last three columns of $G(2,3)$ are the products of the 2nd and 3rd; 2nd and 4th; and 3rd and 4th columns.

Appendix A

Basic Algebra

The Appendix summarizes the most important algebraic notions and theorems *used in the book* to provide an algebraic background *to the other chapters*. Therefore, we just *sketch some* topics mostly without proofs and without building a systematic, general theory of certain areas in algebra.

The chapter consists of an "elementary" and an "advanced" part which differ pretty much in character.

The elementary sections A.1–A.7 treat some very fundamental knowledge with *relatively* detailed explanations. We strongly recommend their thorough study (or review).

The advanced sections A.8–A.11 are used infrequently in the other chapters, most of the book can be understood well without them. The notions appearing here are generally much more difficult than the previous ones and their discussion is more concise.

Sections A.8–A.10 pave the way for Section A.11 about finite fields that play an important role in several applications. In many of these applications, we need only the field of the residue classes modulo a prime (and often just modulo 2). We use the general structure of finite fields primarily in Sections 9.6 and 10.4.

The Appendix, besides facilitating the better understanding of other parts of the book, provides an insight into an important and useful area of independent interest.

A.1. Warm-up

1. Mathematical induction

Induction is a basic tool for proving statements for all positive integers. We check that

(i) the statement holds for $n = 1$ (base step), and

(ii) its truth gets inherited from $n = k$ to $n = k + 1$ for any positive integer k (induction step).

Then the statement is true for every positive integer n.

This is a sort of a domino effect: If infinitely many dominoes stand one behind the other and we push the first one, then it will push the second one, that will push the third one, etc., and all will fall.

The induction step is a conditional one: we do not prove the statement for k (if we could prove it directly, there would be no need for induction), we just assume its validity for $n = k$ and use this to prove that it must hold for $k + 1$, too.

There are several variants of the method; we may start at $n = 0$ or some other initial value, or the induction step may rely on the validity of the theorem for both $n = k$ and $n = k-1$ (in this case we have to check two consecutive initial values), or for all values less than or equal to $n = k$.

Example E1. Prove $5^{n+2} + 6^{2n+1}$ is divisible by 31 for every integer $n \geq 0$.

Solution. Base step: For $n = 0$ we have $31 \mid 5^2 + 6$.

Induction step: We assume $31 \mid 5^{k+2} + 6^{2k+1}$, and show $31 \mid 5^{(k+1)+2} + 6^{2(k+1)+1}$. We have

$$5^{k+3} + 6^{2k+3} = 5(5^{k+2} + 6^{2k+1}) + 31 \cdot 6^{2k+1}.$$

The last term on the right-hand side is a multiple of 31, and by the induction hypothesis, so is the first term, too, thus their sum is divisible by 31.

2. Binomial theorem

The binomial coefficient $\binom{n}{k}$ is the number of subsets of k elements in a set of n elements. We can interpret it as how many ways we can select k elements from n elements so that every element can be selected at most once and the order of the selection is irrelevant. We have

$$\binom{n}{k} = \frac{n!}{k!\,(n-k)!} = \frac{n(n-1)\dots(n-k+1)}{k!} \tag{A.1.1}$$

where $j! = j(j-1)\dots 2 \cdot 1$ for any integer $j > 0$ and $0! = 1$. To verify (A.1.1), we can take first any of the n elements, and whatever this first element is, we can take next any of the remaining $n-1$ elements, etc., and finally we can take any of the remaining $n-k+1$ elements. This gives $n(n-1)\dots(n-k+1)$ options. However, the order of the selection is irrelevant, so we have to divide this by the number of orders of the k elements yielding the expression after the last equality in (A.1.1). Multiplying the numerator and denominator by $(n-k)!$, we obtain the middle expression in (A.1.1).

E.g., $\binom{8}{2} = 8 \cdot 7/2 = 28$, $\binom{n}{1} = n$, $\binom{n}{0} = 1$ (as the empty set is the only 0-element subset, or by $0! = 1$ in the formula).

Theorem A.1.1 (Binomial theorem).

$$(a+b)^n = \binom{n}{0}a^n + \binom{n}{1}a^{n-1}b + \binom{n}{2}a^{n-2}b^2 + \dots + \binom{n}{n}b^n = \sum_{k=0}^{n}\binom{n}{k}a^{n-k}b^k. \clubsuit$$

Proof. Performing the multiplication $(a+b)(a+b)\dots(a+b)$, we take one term, a or b, from each factor, multiply them, and add all these products. E.g., taking a from all the n parentheses, we get the product a^n, taking a from the first two parentheses and b from the other $n-2$ ones, we get a^2b^{n-2}, etc. So we have to add products of type $a^{n-k}b^k$ where $0 \le k \le n$. A product $a^{n-k}b^k$ occurs if we take b from k parentheses, and a from the other $n-k$ parentheses, so we have $\binom{n}{k}$ options to select b. Hence, this term appears $\binom{n}{k}$ times in the sum. ∎

Example E2. Prove $\sum_{k=0}^{n}\binom{n}{k} = 2^n$.

First solution. The left-hand side is the number of all subsets in a set of n elements. The right-hand side counts this directly: for each of the n elements we have two options, whether it belongs to the subset or not.

Second solution. $2^n = (1+1)^n = \sum_{k=0}^{n}\binom{n}{k}1^{n-k}\cdot 1^k = \sum_{k=0}^{n}\binom{n}{k}$.

3. Inclusion-exclusion formula

We have N objects and each may have some of the properties $P_1,\dots,P_k$. We know how many objects have P_1 disregarding whether or not they have some other properties, how many have P_2, how many have (say) P_1, P_4, and P_6 disregarding the others, etc. Let $N_{i_1 i_2 \dots i_j}$ be the number of objects with all properties $P_{i_1}, P_{i_2}, \dots, P_{i_j}$. Then we can find the number N_0 of objects having none of the properties.

Theorem A.1.2 (Inclusion-exclusion formula).

$$N_0 = N - \sum_{i=1}^{k} N_i + \sum_{1\le i_1<i_2\le k} N_{i_1 i_2} - \dots + (-1)^j \sum_{1\le i_1<\dots<i_j\le k} N_{i_1 i_2\dots i_j} + \dots + (-1)^k N_{12\dots k}. \clubsuit$$

Proof. If some object has none of the properties, then it is counted on the right-hand side of the formula exactly once, in N. If it possesses exactly properties $P_{i_1}, P_{i_2}, \dots, P_{i_m}$, then it is counted on the right-hand side $1 - \binom{m}{1} + \binom{m}{2} - \dots + (-1)^m = (1-1)^m = 0$ times. ∎

Example E3. How many results can occur in a fun run with n participants so that nobody's ranking number is the same as the serial number?

Solution. There are $N = n!$ possible outcomes and P_i, $i = 1, 2, \dots, n$ is the property that contestant i has ranking number i. Then $N_i = (n-1)!$, since the other contestants can be ordered arbitrarily. Similarly, $N_{i_1 i_2} = (n-2)!$, etc. So, by the inclusion-exclusion formula

$$N_0 = n! - n(n-1)! + \binom{n}{2}(n-2)! + \dots + (-1)^n = n!\sum_{j=1}^{n}\frac{(-1)^j}{j!}.$$

As $\sum_{j=1}^{n}(-1)^j/j!$ tends to $1/e \approx 0.37$ as $n \to \infty$, we get that the ratio of such outcomes is about 0.37 for large n.

4. Equivalence relation

A *relation* on a set S means selecting some pairs (a, b) (maybe all, maybe none) where $a, b \in S$. E.g., if $S = \{1, 2, 3\}$, then the relation $a < b$ selects the pairs $(1, 2)$, $(1, 3)$, and $(2, 3)$.

If in a relation R the pair (a, b) belongs to the selected ones, then we say that a is in relation R with b, and denote it by aRb.

Relations possessing some important features of equality are called *equivalence relations*.

Definition A.1.3. A relation R on a set S is an *equivalence relation* if it is

(i) reflexive: aRa for every $a \in S$,

(ii) symmetric: $aRb \Rightarrow bRa$, and

(iii) transitive: $aRb, bRc \Rightarrow aRc$. ♣

The importance of equivalence relations lies in their capacity to partition the set into the union of disjoint subsets.

Theorem A.1.4. *If R is an equivalence relation, then the subsets $C_a = \{b \in S \mid aRb\} \subseteq S$ are either disjoint or coincide. So, S is the disjoint union of distinct subsets C_a.* ♣

The subset C_a is called the equivalence class represented by a. So, an equivalence relation creates a partition of S into equivalence classes. Conversely, every partition of S is created by a suitable equivalence relation: aRb exactly if a and b are in the same subset of the partition.

Proof. We have to show that if C_a and C_b are not disjoint, then $C_a = C_b$. We verify $C_a \subseteq C_b$; the converse containment follows similarly or by the symmetric role of a and b.

Let $x \in C_a$, so aRx, we prove bRx. Let $d \in C_a \cap C_b$, thus aRd and bRd. Then, $aRd \Rightarrow dRa$, as R is symmetric. Using dRa and aRx, transitivity yields dRx. Combining bRd and dRx, we get bRx, by transitivity. ∎

Example E4. Let S consist of the non-zero real numbers, and a is in relation with b, if their product is positive: $aRb \iff ab > 0$. We prove that R is an equivalence relation and determine the equivalence classes.

Solution. Reflexivity: $a \cdot a = a^2 > 0$. Symmetry: $ab = ba$, so $ab > 0 \Rightarrow ba > 0$. Transitivity: If $ab > 0$ and $bc > 0$, then multiplying these inequalities $acb^2 > 0$, so dividing by $b^2 > 0$, we get $ac > 0$.

There are two equivalence classes, the positive and the negative real numbers.

Exercises A.1

A.1.1. Verify for every integer $n > 0$:

(a) $\dfrac{1}{1\cdot 2}+\dfrac{1}{2\cdot 3}+\cdots+\dfrac{1}{n(n+1)}=\dfrac{n}{n+1}$;

(b) $1^2+2^2+3^2+\cdots+n^2=n(n+1)(2n+1)/6$.

A.1.2. Prove the inequalities for every integer $n > 0$.

$$2\sqrt{n+1}-2<1+\frac{1}{\sqrt{2}}+\frac{1}{\sqrt{3}}+\cdots+\frac{1}{\sqrt{n}}\le 2\sqrt{n}-1.$$

A.1.3. Show $8 \mid 5^n+4n+7$ for every integer $n \ge 0$.

A.1.4. Consider the sequence $a_0 = 0$, $a_1 = 1$, and $a_{n+1} = 3a_n - 2a_{n-1}$ for $n \ge 1$. Find an explicit formula for a_n.

A.1.5. We "prove" that all real numbers are equal. Taking any n real numbers, we show by induction that they are all equal. This is certainly true for $n = 1$. We assume its truth for $n-1$, and proceed to n. Let $a_1, a_2, \dots, a_n$ be any real numbers. By the induction hypothesis, $a_1 = a_2 = \cdots = a_{n-1}$ and $a_2 = a_3 = \cdots = a_n$. As a_{n-1} occurs in both parties, we have $a_1 = a_2 = \cdots = a_n$. Find the error in the argument.

A.1.6. How many ways can we decompose 2000 into the sum of four positive integers if we count as distinct the sums that differ only in the order of the terms?

A.1.7. Prove:

(a) $\binom{n}{k}=\binom{n}{n-k}$;

(b) $\binom{n}{k}+\binom{n}{k-1}=\binom{n+1}{k}$;

(c) $\binom{k}{k}+\binom{k+1}{k}+\cdots+\binom{n}{k}=\binom{n+1}{k+1}$;

(d) $\binom{n}{0}^2+\binom{n}{1}^2+\cdots+\binom{n}{n}^2=\binom{2n}{n}$.

A.1.8. Give an explicit formula for the sums:

(a) $\binom{n}{0}+\binom{n}{2}+\binom{n}{4}+\binom{n}{6}+\dots$;

(b) $\binom{n}{0}-2\binom{n}{1}+4\binom{n}{2}-8\binom{n}{3}+\dots$.

A.1.9. (a) How many k-digit numbers can be formed from the digits $1, 2, \dots, n$ so that every digit should occur if (a1) $k < n$; (a2) $k = n$; (a3) $k > n$?

(b) Prove:

$$n^k - n(n-1)^k + \binom{n}{2}(n-2)^k - \dots$$

$$+(-1)^{n-2}\binom{n}{n-2}2^k + (-1)^{n-1}\binom{n}{n-1} = \begin{cases} 0, & \text{if } k < n; \\ n!, & \text{if } k = n. \end{cases}$$

A.1.10. Select the equivalence relations from the relations on the set $\mathbf{Z}$ of integers, and determine the number of equivalence classes induced by them.

(a) $aRb \iff 3 \mid a+b$;

(b) $aRb \iff 3 \mid a-b$;

(c) $aRb \iff 3 \mid 4a+5b$;

(d) $aRb \iff a^2 = b^2$;
(e) $aRb \iff ab \geq 0$;
(f) $aRb \iff |a - b| < 1000$.

A.1.11. The following argument claims to prove that if a relation is symmetric and transitive on a set S, then it is reflexive, too. Take any $a \in S$, and pick an arbitrary b satisfying aRb. Then bRa holds, by symmetry, and the two together imply aRa, by transitivity. Is this argument correct?

A.1.12. For a set of n elements, determine the number of
(a) relations;
(b) reflexive relations;
(c) symmetric relations.

A.2. Divisibility and Congruences

In this section, all occurring letters denote integers. First, we summarize some basic facts about divisibility.

Definition A.2.1. a *divides* b if there exists a c satisfying $ac = b$. Notation: $a \mid b$.

The *greatest common divisor* of a and b is the greatest integer dividing both a and b. Notation: (a, b) or $\gcd(a, b)$.

Two integers are *coprime* (or *relatively prime*) if their gdc is 1.

$p \neq \pm 1$ is *irreducible* if ± 1 and $\pm p$ are all its divisors. ♣

Instead of "irreducible", we generally say "prime", see below.

Theorem A.2.2. *Division algorithm: For every $b \neq 0$ and a, there exist unique q and r satisfying $a = bq + r$ and $0 \leq r < |b|$.*

Unique factorization: Every $c \neq 0, \pm 1$ is the product of irreducible elements, and this factorization is unique apart from the order and the signs of the factors.

Standard form: If $n > 1$, then $n = p_1^{k_1} \dots p_r^{k_r}$ where $p_j > 0$ are distinct primes and $k_j > 0$. ♣

The *Euclidean algorithm* is a repeated application of the division algorithm:

$$a = bq_1 + r_1, \quad b = r_1 q_2 + r_2, \quad r_1 = r_2 q_3 + r_3, \quad \dots.$$

The last non-zero remainder is the gcd of a and b. This is an efficient algorithm to compute the gcd of large numbers. It also has several important theoretical consequences: (i) The gcd is not just the largest among the common divisors, but also a multiple of all common divisors. (ii) If $(a, b) = d$, then $d = au + bv$ for suitable integers u and v. (iii) An integer different from 0 and ± 1 is irreducible if and only if it satisfies the *prime* property: If it divides a product, then it must divide (at least) one of the factors. Therefore, "prime" and "irreducible" are equivalent among the integers. This is the key step in proving the unique factorization theorem.

To avoid common mistakes, we call attention to the following facts:

(A) $a \mid bc$ and $a \nmid b$ do **NOT** imply $a \mid c$, e.g., $15 \mid 3 \cdot 20$, but $15 \nmid 3$ and $15 \nmid 20$.
Correct versions:
(i) $a \mid bc, (a,b) = 1 \Rightarrow a \mid c$.
(ii) a is a prime, $a \mid bc, a \nmid b \Rightarrow a \mid c$.

(B) $a \mid c$ and $b \mid c$ do **NOT** imply $ab \mid c$, e.g., $6 \mid 12, 4 \mid 12$, but $24 \nmid 12$.
Correct versions:
(i) $a \mid c, b \mid c$, and $(a,b) = 1 \Rightarrow ab \mid c$.
(ii) $a \mid c$ and $b \mid c \Rightarrow [a,b] \mid c$ (where $[a,b]$ denotes the least common multiple of a and b).

The above properties can be deduced, e.g., from the unique prime factorization theorem.

Now we turn to some basic facts about congruences.

Definition A.2.3. If $m \mid a - b$ where $m > 0$, i.e., a and b give the same remainder upon division by m, then we say that "a is *congruent* to b modulo m" and denote it by $a \equiv b \pmod{m}$ The integer m is called the *modulus*. ♣

The congruence relation has many similar features as equality. It is reflexive, symmetric, and transitive: For every a, b, and c, we have

$$a \equiv a; \qquad a \equiv b \Rightarrow b \equiv a; \qquad (a \equiv b \text{ and } b \equiv c) \Rightarrow a \equiv c \pmod{m}.$$

So, congruence is an equivalence relation (see Definition A.1.3). The induced equivalence classes (see Theorem A.1.4) can be characterized as all integers giving the same remainder upon division by m. Such a class is called a *residue class* (mod m). E.g., $\{\dots, -3, 7, 17, 27, \dots\}$ is a residue class (mod 10).

Also, we can add, subtract, and multiply congruences:

$$a \equiv b \text{ and } c \equiv d \Rightarrow a + c \equiv b + d,\ a - c \equiv b - d,\ ac \equiv bd \pmod{m}.$$

These imply $a \equiv b \Rightarrow a^t \equiv b^t \pmod{m}$ for any integer $t > 0$; moreover, if f is a polynomial with integer coefficients, then $a \equiv b \Rightarrow f(a) \equiv f(b) \pmod{m}$.

We have to be careful about division: we cannot divide congruences even if the quotients are integers. E.g.,

$$24 \equiv 14 \pmod{10} \quad 2 \equiv 2 \pmod{10} \quad \text{but} \quad 24/2 = 12 \not\equiv 14/2 = 7 \pmod{10}.$$

Correct versions:

(i) $ac \equiv bc \pmod{m}$ and $(c,m) = 1 \Rightarrow a \equiv b \pmod{m}$.

(ii) $ac \equiv bc \pmod{m} \Rightarrow a \equiv b \pmod{m/(c,m)}$.

The following function plays a central role in number theory.

Definition A.2.4 (Euler's function $\varphi(n)$). $\varphi(n)$ is the number of integers coprime to n in $\{1, 2, \dots, n\}$. ♣

Theorem A.2.5. *If the standard form of n is $n = p_1^{k_1} \dots p_r^{k_r}$ where $p_j > 0$ are distinct primes and $k_j > 0$, then $\varphi(n) = p_1^{k_1-1}(p_1 - 1) \dots p_r^{k_r-1}(p_r - 1)$.* ♣

Theorem A.2.6 (Euler–Fermat Theorem). *If $(c, m) = 1$, then $c^{\varphi(m)} \equiv 1 \pmod{m}$.* ♣

An important special case is the next theorem.

Theorem A.2.6A (Fermat's Little Theorem). *If p is a prime and $p \nmid c$, then $c^{p-1} \equiv 1 \pmod{p}$.*

An alternative form is: $c^p \equiv c \pmod{p}$ for every c. ♣

Definition A.2.7. Let $(c, m) = 1$. The *order* of c modulo m is the smallest integer $k > 0$ such that $c^k \equiv 1 \pmod{m}$. We denote it by $o_m(c)$, or $o(c)$ if there is no ambiguity about the modulus m. ♣

The order is the smallest period in which the powers of c modulo m get repeated: $c^r \equiv c^s \pmod{m} \iff r \equiv s \pmod{o_c(m)}$. For $s = 0$ we get $c^r \equiv 1 \pmod{m} \iff o_m(c) \mid r$. This implies $o_m(c) \mid \varphi(m)$ by Theorem A.2.6.

The notion of the order for congruences is a special case of the order of an element in a group; see Definition A.8.2.

A *linear Diophantine equation* (in two variables) is $Ax + By = C$ where A, B, C are given integers, A and B are not both zero, and we are looking for integer solutions in x and y.

Theorem A.2.8.

(i) *The linear Diophantine equation $Ax + By = C$ is solvable if and only if $(A, B) \mid C$, and in this case there are infinitely many solutions.*

(ii) *We can obtain a solution (x_0, y_0) from the Euclidean algorithm, and all solutions are*

$$x_t = x_0 + \frac{tB}{(A,B)}, \quad y_t = y_0 - \frac{tA}{(A,B)}, \quad t = 0, \pm 1, \pm 2, \ldots. \quad ♣$$

A *linear congruence* is $ax \equiv b \pmod{m}$, and we are looking for pairwise incongruent solutions in x.

Theorem A.2.9. *The linear congruence $ax \equiv b \pmod{m}$ is solvable if and only if $(a, m) \mid b$, and in this case there are (a, m) pairwise incongruent solutions.* ♣

The equation $Ax + By = C$ can be transformed into the congruence $Ax \equiv C \pmod{|B|}$ or into $By \equiv C \pmod{|A|}$ (if B and A are not zero, resp.). Conversely, the congruence $ax \equiv b \pmod{m}$ can be transformed into the equation $ax - my = b$.

A simultaneous system of congruences means that several congruence conditions with different moduli are imposed on the same variable.

Theorem A.2.10. *A simultaneous system of congruences $x \equiv c_1 \pmod{m_1}$, $x \equiv c_2 \pmod{m_2}$ is solvable if and only if $(m_1, m_2) \mid c_1 - c_2$. If solvable, there is one solution mod $[m_1, m_2]$.* ♣

In the special case when m_1 and m_2 are coprime, the system is solvable for any c_1 and c_2, and there is one solution mod $m_1 m_2$. This can be generalized to the following theorem.

Theorem A.2.10A (Chinese remainder theorem). *Let $m_1, \dots, m_r$ be pairwise coprime. Then the system of congruences $x \equiv c_i \pmod{m_i}$ $i = 1, 2, \dots, r$ is solvable for any integers $c_1, \dots, c_r$ and there is one solution* mod $m_1 m_2 \dots m_r$. ♣

An important corollary of Theorem A.2.10A is that any congruence with a composite modulus can be reduced to congruences of prime power moduli: If the standard form of m is $m = p_1^{\alpha_1} \dots p_r^{\alpha_r}$, then $f(x) \equiv 0 \pmod{m}$ is equivalent to the system $f(x) \equiv 0 \pmod{p_i^{\alpha_1}}$, $i = 1, 2, \dots, r$.

Exercises A.2

A.2.1. Verify that 8 divides the difference of the squares of any two odd numbers.

A.2.2. True or false?
- (a) $(a \mid b \text{ and } a \mid c) \Longrightarrow a \mid 3b - 5c$;
- (b) $a \mid b + c \Longrightarrow (a \mid b \text{ and } a \mid c)$;
- (c) $ab \mid c \Longrightarrow (a \mid c \text{ and } b \mid c)$;
- (d) $(a \mid c \text{ and } b \mid c) \Longrightarrow ab \mid c$;
- (e) $(a \mid b \text{ or } a \mid c) \Longrightarrow a \mid bc$;
- (f) $a \mid bc \Longrightarrow (a \mid b \text{ or } a \mid c)$;
- (g) $(a \mid b + c \text{ and } a \mid b - c) \Longrightarrow (a \mid b \text{ and } a \mid c)$;
- (h) $(a \mid 2b + 3c \text{ and } a \mid 3b + 5c) \Longrightarrow (a \mid b \text{ and } a \mid c)$.

A.2.3. Verify the following divisibilities:
- (i) $a - b \mid a^n - b^n$;
- (ii) $a + b \mid a^{2k+1} + b^{2k+1}$;
- (iii) $a + b \mid a^{2k} - b^{2k}$.

A.2.4. (Digits are understood in decimal system.)
- (a) Prove that 9 divides n if and only if 9 divides the sum of its digits.
- (b) Prove that 11 divides n if and only if 11 divides the alternating sum of its digits.
- (c) We add the digits of 2^{2019}, then add the digits of the number obtained, etc. until we get a one-digit number. Determine this final number.
- (d) Prove that $43 \mid 6^{n+2} + 7^{2n+1}$ holds for every $n \geq 0$.

A.2.5. True or false? ($[k, n]$ denotes the least common multiple of k and n.)
- (a) $k \mid n,\ a \equiv b \pmod{n} \Longrightarrow a \equiv b \pmod{k}$;
- (b) $k \mid n,\ a \equiv b \pmod{k} \Longrightarrow a \equiv b \pmod{n}$;
- (c) $a \equiv b \pmod{n},\ a \equiv b \pmod{k} \iff a \equiv b \pmod{kn}$;
- (d) $a \equiv b \pmod{n},\ a \equiv b \pmod{k} \iff a \equiv b \pmod{[k, n]}$;
- (e) $a \equiv b \pmod{n} \iff ka \equiv kb \pmod{kn}$;
- (f) $a \equiv b \pmod{n},\ c \equiv d \pmod{k} \Longrightarrow ac \equiv bd \pmod{kn}$;
- (g) $a^2 \equiv b^2 \pmod{n} \Longrightarrow a \equiv \pm b \pmod{n}$;
- (h) $a^2 \equiv b^2 \pmod{101} \Longrightarrow a \equiv \pm b \pmod{101}$.

A.2.6. Consider any infinite integer sequence $a_1, a_2, a_3, \dots$. Prove that there are infinitely many a_i such that the difference of any two of them is divisible by 11111.

A.2.7. We take a maximal set of integers where any two give distinct remainders upon division by n. What is the remainder of their sum?

A.2.8. In a fun run with n participants we added the serial and ranking numbers for each runner (e.g., if a contestant wore 7 on her T-shirt and was the 2nd quickest, then we considered the sum $7 + 2 = 9$). We noted that the n sums obtained give distinct remainders upon division by n. Is this possible if n is (a) 99; (b) 100?

A.2.9. (a) There are m trees around a circular clearing with a squirrel on each tree. The squirrels want to get together on one tree, but they are allowed to move only the following way: In every minute, any two squirrels may jump to an adjacent tree. For which values of m can they gather on one tree?

(b) What happens if we modify the admissible step so that the two squirrels must jump to the adjacent trees in opposite directions (i.e., one of them clockwise, and the other counter-clockwise).

A.2.10. (a) Prove the formula for $\varphi(n)$ if n is (a prime or) a power of a prime (Theorem A.2.5).

(b) For which integers n is $\varphi(n)$ (b1) odd; (b2) a power of two?

A.2.11. What are the last two digits of 1357^{8642}?

A.2.12. Prove that a square number plus one cannot have a prime divisor of the form $4k - 1$.

A.2.13. Solve the congruence $8x \equiv 13 \pmod{11}$.

A.2.14. Determine the last three digits of 614^{4302}.

A.2.15. How many 21-digit positive integers have the property that every power of them terminates with the same 20 digits as the original number?

A.2.16. Prove Theorems A.2.8–A.2.10.

A.3. Complex Numbers

The complex numbers were originally introduced to resolve the paradox situation about cubic equations: Cardano's formula failed due to square roots of negative numbers just in the case when the equation had three distinct real roots. Later it turned out that complex numbers constitute a natural extension of the real numbers and proved to be fundamental not just in all parts of mathematics but also, e.g., in modern physics.

Definition A.3.1. The *complex numbers* are of the form $z = a + bi$ where a and b are real numbers and $i^2 = -1$. The two real numbers, $a = \operatorname{Re} z$ and $b = \operatorname{Im} z$ (without the i!) are called the *real part* and the *imaginary part* of z, resp. A complex number is *imaginary* if its real part is 0.

Addition and multiplication are defined in a natural way:

$$(a+bi)+(c+di) = (a+c)+(b+d)i \quad \text{and} \quad (a+bi)(c+di) = (ac-bd)+(ad+bc)i.$$

Thus we perform multiplication by multiplying each term in $a + bi$ with each term in $c + di$ and observe the rule $i^2 = -1$. ♣

The complex numbers can be identified with the points in the plane: $z = a + bi$ corresponds to the point with coordinates a and b. Also, we can represent the complex numbers by vectors in the plane starting from the origin. Addition of complex numbers conforms to the addition of the corresponding vectors.

Definition A.3.2. Consider the complex number $z = a + bi$.

The *conjugate* of z is $\overline{z} = a - bi$.

The *modulus* or *absolute value* of z is $|z| = \sqrt{a^2 + b^2}$.

The *argument* of $z \neq 0$ denoted by $\arg(z)$ is the angle obtained from the rotation of the positive half of the x-axis to the vector. The rotation can be either counter-clockwise or clockwise resulting in a positive or negative argument, resp. Thus, a non-zero complex number has infinitely many arguments which differ in a multiple of 2π if measured in radians. ♣

An example is $\overline{-1+i} = -1-i$, $|-1+i| = \sqrt{2}$, $\arg(-1+i) = 3\pi/4$ (or $-5\pi/4$, etc.).

We get the point $\overline{z}$ by reflecting z over the real axis. A simple calculation shows $\overline{z+w} = \overline{z}+\overline{w}$, and similar laws apply also for multiplication, subtraction, and division.

The geometric meaning of $|z|$ is the length of the vector z. The conjugate and the modulus are connected in the useful equality $|z|^2 = z \cdot \overline{z}$.

Concerning addition and multiplication, the complex numbers behave similarly to the real or rational numbers.

Theorem A.3.3. *The complex numbers form a field: addition and multiplication satisfy the associative, commutative, and distributive laws;* 0 *and* 1 *are the additive and multiplicative identities, resp.; every element has a negative and every non-zero element has a reciprocal.* ♣

The existence of 0, 1, negatives, and reciprocals are equivalent to subtraction and division. Thus, summarizing the above, we can perform the four basic operations among the complex numbers and these obey the usual operational laws.

Division can be executed using the conjugate: If $a + bi \neq 0$, then

$$\frac{c+di}{a+bi} = \frac{(c+di)(a-bi)}{(a+bi)(a-bi)} = \frac{(ac+bd)+(ad-bc)i}{a^2+b^2} = \frac{ac+bd}{a^2+b^2} + \frac{ad-bc}{a^2+b^2}i.$$

A complex number can be characterized also by its modulus and argument.

Theorem A.3.4. *If* $z \neq 0$ *and* $\arg(z) = \varphi$, *then* $z = |z|(\cos\varphi + i\sin\varphi)$.

Multiplying or dividing two complex numbers, the moduli are multiplied or divided, resp., and the arguments are added or subtracted, respectively. ♣

Expressing z as $a+bi$ is the *algebraic*, whereas $|z|(\cos\varphi + i\sin\varphi)$ is the *trigonometric* form of z. Since the arguments behave as exponents, we often use *Euler's form* $z = |z|e^{i\varphi}$ which gives a natural frame for performing multiplication and division. In fact, Euler's form is not just a practical notation, but is a straightforward consequence of deeper considerations in complex analysis.

An example is $1 - i\sqrt{3} = 2(\cos(-\pi/3) + i\sin(-\pi/3)) = 2e^{-i\pi/3}$.

Raising a complex number to an integer power is defined similarly as for real numbers: If n is a positive integer, then

$$z^n = \underbrace{z \cdot \ldots \cdot z}_{n \text{ times}}, \qquad z^0 = 1, \quad \text{and} \qquad z^{-n} = \frac{1}{z^n} \text{ for } z \neq 0.$$

An immediate consequence of the second statement in Theorem A.3.4 is Moivre's formula: If $z \neq 0$, then, for any integer k,

$$z^k = |z|^k(\cos(k\varphi) + i\sin(k\varphi)) = |z|^k e^{ik\varphi}.$$

For an integer $n > 0$, the nth root of z is defined as *all* complex numbers w satisfying $w^n = z$. If $z = 0$, then clearly $w = 0$ is its only nth root, but if $z \neq 0$, then the number of its nth roots is exactly n.

Theorem A.3.5. *If $z \neq 0$ and $\arg(z) = \varphi$, then*

$$\sqrt[n]{z} = \sqrt[n]{|z|}\left(\cos\frac{\varphi + 2k\pi}{n} + i\sin\frac{\varphi + 2k\pi}{n}\right) = \sqrt[n]{|z|}e^{i(\varphi+2k\pi)/n}, \; 0 \leq k \leq n-1.$$

So, the nth roots of z form the vertices of a regular n-gon around the origin. ♣

The nth roots of 1 are called the nth roots of unity.

Definition A.3.6. A complex number w is a *root of unity* if $w^n = 1$ for some integer $n > 0$. In this case, w is called an nth root of unity.

The *order* of w, denoted by $o(w)$, is the smallest positive integer t satisfying $w^t = 1$.

w is a *primitive* nth root of unity if $o(w) = n$. ♣

E.g., $o(i) = 4$, so i is a primitive 4th root of unity and it is also an 8th, 12th, etc. root of unity.

The order is analogous to the order seen at congruences, and both are special cases of the general notion of order of an element in a group; see Definition A.8.2.

Theorem A.3.7. *The nth roots of unity are* $w_k = \cos(2k\pi/n) + i\sin(2k\pi/n) = e^{2\pi ki/n}$, $k = 0, 1, \ldots, n-1$.

w_k is a primitive nth root of unity if and only if $(k, n) = 1$. Hence, there are exactly $\varphi(n)$ primitive nth roots of unity (where φ denotes Euler's function).

A complex number w is a root of unity if and only if its modulus is 1 and its argument is a rational multiple of 2π. If this rational multiplier is r/s, where $(r, s) = 1$ and $s > 0$, then $o(w) = s$.

If w is a root of unity, then $w^t = w^m \iff t \equiv m \pmod{o(w)}$. In particular, $w^t = 1 \iff o(w) \mid t$. ♣

Exercises A.3

A.3.1. Compute:

(a) $\dfrac{7+6i}{2+i} - \dfrac{5+i}{2+3i}$;

(b) i^{9999}

(c) $\sqrt{5-12i}$;

(d) $\sqrt[6]{8i}$;

(e) $\sqrt[5]{-3-i\sqrt{3}}$.

A.3.2. Describe the set of points in the plane corresponding to the complex numbers z satisfying

(a) $\mathrm{Im}(3i-5z)=7$;

(b) $\mathrm{Re}(2iz+5+8i)>9$;

(c) $|z-3+8i|\leq 1$;

(d) $|z|=2\mathrm{Im}\, z$;

(e) $|z-5i|\geq |z+i|$;

(f) $(1+i)\bar{z}=z\sqrt{2}$;

(g) $(z+1)/(z+3)$ is imaginary.

A.3.3. Find the complex roots of the equations

(a) $z^2+8=0$; (b) $z^2+2z+2=0$; (c) $z^2+(1+i)z+5i=0$.

A.3.4. Verify $|z_1+z_2|^2+|z_1-z_2|^2=2|z_1|^2+2|z_2|^2$ for any complex numbers z_1 and z_2. What is the geometric meaning of this equality?

A.3.5. Prove that if each of two positive integers is the sum of two squares, then so is their product. Is the converse true? What about sums of three squares?

A.3.6. Find the image of $(3,4)$ under rotation around the origin by (a) 90; (b) 60 degrees?

A.3.7. Assume $z_1\neq z_2$. Show that $(z_1+z_2)/(z_1-z_2)$ is imaginary if and only if $|z_1|=|z_2|$.

A.3.8. Express $\sin 5x$ using the powers of $\sin x$.

A.3.9. (a) Compute $(1+i)^n$ in two ways; using the trigonometric form and the binomial theorem.

(b) Determine the sum $\binom{n}{0}-\binom{n}{2}+\binom{n}{4}-\binom{n}{6}+\dots$.

A.3.10. Find the algebraic form of all (a) third; (b) sixth roots of unity.

A.3.11. (a) Show that the product and quotient of two roots of unity are roots of unity.

(b) Determine all cases when the sum or the difference of two roots of unity is a root of unity.

A.3.12. Compute the sum and product of all nth roots of unity.

A.3.13. Let $o(z)=k$ and $o(w)=n$.

(a) Compute $o(z^t)$.

(b) Show $o(zw)\mid [k,n]$.

A.3.14. (a) Compute the product of all primitive nth roots of unity.
(b) Determine the sum of all primitive nth roots of unity for $n = 7, 27$, and (**) in general.

A.3.15. We draw squares leaning on the sides of a quadrangle from outside. Prove that the two segments connecting the centers of the opposite squares are perpendicular and of equal length.

A.3.16. Find a simpler form for the sum $\cos x + \cos(2x) + \cdots + \cos(nx)$.

A.4. Operation

Definition A.4.1. A (binary) *operation* on a non-empty set S is a function $S \times S \to S$, so this is a map assigning to every pair $a, b \in S$ a unique element in S. ♣

In general, the operation is called a multiplication and the element assigned to the pair $a, b \in S$ is denoted by ab. Further possible notations are, e.g., $a + b$ (for addition), $a * b$, $a \circ b$, or $f(a, b)$.

Examples of operations

E1 Addition or multiplication among the natural numbers, integers, rational, real, or complex numbers.

E2 Subtraction among the integers, rational, real, or complex numbers. It is not an operation among the natural numbers as, e.g., the difference $3 - 5$ does not exist since there is no *natural* number m satisfying $m + 5 = 3$.

E3 Division among the non-zero rational, real, or complex numbers.

E4 Consider the set $\mathbf{Z}_m = \{0, 1, 2, \ldots, m - 1\}$ of all possible remainders modulo m. We can define addition, subtraction, and multiplication of two remainders by taking the remainder of their sum, difference, and product modulo m, e.g., for $m = 9$, we get $4 + 7 = 2, 4 - 7 = 6$, and $4 \cdot 7 = 1$.

E5 Addition of matrices of the same size and multiplication of square matrices of the same size.

E6 The composition of functions $\mathbf{R} \to \mathbf{R}$, or $X \to X$ in general.

E7 The composition of congruences (distance-preserving transformations) in the plane.

E8 The cross product of the vectors in the space. Note that the inner product of vectors and the multiplication of vectors by scalars do not satisfy Definition A.4.1, since the result of the first one is not a vector, and the two factors are not from the same set in the second one. It is possible to extend the definition of operation to include these important maps.

Remark. It is customary to say that S is closed for the operation. This is a rather improper terminology as the definition of operation includes that performing the operation for any two elements the result belongs to the same set S. Closedness makes sense if an operation is defined on S and we ask whether a given subset $K \subseteq S$ is closed for this operation, i.e., performing the operation for any two elements in K, is it true that the result is also in K. Thus, the closedness of K means that restricting the operation on S to K we obtain an operation on K.

The most important *operational identities* are the associative and commutative laws.

Definition A.4.2. An operation on S is *associative* if $a(bc) = (ab)c$ for every a, b, $c \in S$. ♣

Definition A.4.3. An operation on S is *commutative* if $ab = ba$ for every $a, b \in S$. ♣

The associative law guarantees the unique meaning of a product of more than two factors without using parentheses. If, in addition, the operation is commutative, then the order of the factors can be modified arbitrarily.

Examples. Addition and multiplication of numbers, polynomials, remainders modulo m, addition of matrices are commutative and associative. Multiplication of matrices and composition of functions are associative but not commutative in general. Forming the arithmetic mean of two real numbers is commutative but not associative. Subtraction of integers and cross product of vectors are neither commutative, nor associative.

Remarks.

1. To check the associative law, we do not have to verify that $a(bc)$ and $(ab)c$ are elements of S as this follows from the definition of the operation. We have to determine whether these two elements coincide in *every* case.

2. It makes no sense that an operation is partially commutative or associative. If $ab \neq ba$ for at least one pair $a, b \in S$, then the operation is *not* commutative, otherwise it is commutative. Of course, if an operation is not commutative, there may be many commuting pairs $a, b \in S$ satisfying $ab = ba$.

3. Our experiences about addition and multiplication of numbers could suggest that associative and commutative laws are equally important for operations, or perhaps the commutative law is even more essential. The mathematical truth is just the opposite as associativity surpasses commutativity in significance. First, there are several important associative operations that are not commutative as, e.g., the composition indispensable in most areas of mathematics. Second, several basic operational properties depend on the associative law such as the uniqueness of the inverse of an element (see the discussion after Definition A.4.5) or the connection between the inverses of elements and the inverse operation (Theorem A.4.7).

Definition A.4.4. A *left identity* is an element $e_L \in S$ satisfying $e_L a = a$ for every $a \in S$.

Similarly, a *right identity* $e_R \in S$ satisfies $ae_R = a$ for every $a \in S$.

An *identity* (or *identity element*) e is both a left and right identity, i.e., $ea = ae = a$ for every $a \in S$. ♣

Thus the phrase “identity” in itself always means a two-sided identity element.

The identity of addition is called a *zero* or *zero element* and is denoted by 0. The identity of multiplication is often denoted by 1 (instead of e).

The definition of a left identity must not be interpreted as to every element $a \in S$ there exists an element e_L (depending on a) that satisfies $e_L a = a$, but it requires a *universal* element e_L which fits for every a simultaneously. This does not contradict the eventual existence of more such universal elements; see below.

An operation may have several left identities: e.g., if the product of any two elements is the second factor, then every element is a left identity. If, however, there also exists a right identity, then $e_R = e_B e_R = e_R$ implies $e_B = e_R$, so there can be only one left identity (being a two-sided identity in this case). This implies that an operation can have at most one (two-sided) identity.

Definition A.4.5. Let e denote the (two-sided) identity of an operation. (The previous paragraph showed that this e is unique.)

A *left inverse* of some $a \in S$ is an element $a_L \in S$ satisfying $a_L a = e$.

Similarly, a *right inverse* of a is an element $a_R \in S$ satisfying $aa_R = e$.

An *inverse* of a is an element $a^{-1} \in S$ satisfying $a^{-1}a = aa^{-1} = e$. ♣

Similar to identities, the inverse without the adjective left or right always means a two-sided inverse.

For addition, we call the inverse of a its *negative* and denote it by $-a$. For multiplication of numbers, we can call the inverse of a its *reciprocal* and denote it by $1/a$ (instead of a^{-1}).

Clearly, we can only speak about the left, right, or two-sided inverse of an element if the operation has an identity.

The expressions left inverse, right inverse, or inverse make no sense unless we clarify to which element they belong. Sentences like "an operation has no inverse" are meaningless as generally some elements do, some others do not have an inverse. As an extreme case, it may occur that every element has an inverse, e.g., for the multiplication of positive real numbers. The opposite extreme is not possible as the identity always has an inverse.

Similar to the situation at identities, an element may have several left inverses (see Exercise A.4.6). If, however, the operation is *associative* and a has also a right inverse a_R, then

$$a_L = a_L e = a_L(aa_R) = (a_L a)a_R = ea_R = a_R$$

implies that the left and right inverses of a coincide. So, in this case, a can have only one left inverse (being a two-sided inverse of a). This implies that for associative operations the inverse of an element is *unique*, so an element can have at most one (two-sided) inverse.

Now we discuss the notion of an *inverse operation*. Looking at subtraction of (e.g., real) numbers, $a - b$ means the number c satisfying $c + b = a$. The subtraction of real numbers is an operation because for every a, b there exists exactly one such c. However, subtraction is *not* an operation on the set of natural numbers, as there is no such c for every a, b. Thus, it depends on the actual addition, whether we can execute subtraction; this is a property of the addition. We generalize this in the following definition.

Definition A.4.6. Denote an operation F on S as multiplication. Assume that the equation $xb = a$ has a unique solution for every $a, b \in S$, i.e., there exists exactly one $c \in S$ satisfying $cb = a$. Then the map $L(a, b) = c$ is called the *left inverse operation* of F.

Similarly, if for every $a, b \in S$ there exists exactly one $d \in S$ satisfying $bd = a$, then $R(a, b) = d$ is called the *right inverse operation* of F. ♣

The (left or right) inverse operation of addition is subtraction, and the inverse operation of multiplication of non-zero real numbers is division. If the (original) operation is commutative, then clearly $L = R$.

We know that subtraction of numbers can be composed from addition and taking negatives: $a - b = a + (-b)$. The inverse of an element and the inverse operation have a similar close relationship for any associative operation.

Theorem A.4.7. *Let an associative operation on S be given.*

(i) *If the operation has an identity, and b has an inverse b^{-1}, then the equations $xb = a$ and $by = a$ have a unique solution for every $a \in S$.*

(ii) *If equations $xb = a$ and $by = a$ are solvable for every $a, b \in S$, then the operation has an identity and every element has an inverse.* ♣

Proof.

(i) We claim that if b has an inverse, then multiplying an equality by b^{-1} from either side, the obtained equality is equivalent to the original one. We show this for multiplying by b^{-1} from the right. On the one hand, clearly $h_1 = h_2 \Rightarrow h_1b^{-1} = h_2b^{-1}$. On the other hand, multiplying $h_1b^{-1} = h_2b^{-1}$ by b from the right, we get $(h_1b^{-1})b = (h_2b^{-1})b$, and $(hb^{-1})b = h(b^{-1}b) = he = h$ implies $h_1 = h_2$.

Thus, $xb = a$ is equivalent to $x = ab^{-1}$, so the equation has a unique solution. Similarly, the unique solution of $by = a$ is $y = b^{-1}a$.

(ii) Pick a $b \in S$, and let g be a solution of equation $bx = b$. We show that g does not depend on b. Let $a \in S$ be arbitrary. By the condition, $cb = a$ for some c, so $ag = (cb)g = c(bg) = cb = a$. Thus, g is a right identity. We prove the existence of a left identity h similarly. We saw earlier that necessarily $g = h$, so there exists an identity e.

The left and right inverse of b is obtained as a solution of equations $xb = e$ and $by = e$. As shown earlier, the left and right inverses of an element must be equal, so every element has an inverse. ■

Exercises A.4

A.4.1. Select the operations from the following functions. Which operations are commutative, which are associative? Determine the left and right identities. If the operation has an identity, which elements have an inverse?

(a) Even integers: (a1) addition; (a2) multiplication; (a3) subtraction.

(b) Odd integers: (b1) addition; (b2) multiplication.

(c) Positive integers: (c1) $\max(a, b)$; (c2) $\min(a, b)$; (c3) $[a, b]$ (least common multiple).

(d) Residue classes modulo m, we define the operations with the help of representatives: (d1) maximum, so, e.g., for $m = 10$, the maximum of the class containing 8 and the class containing 13 is defined as the class containing $\max(8, 13) = 13$; (d2) exponentiation; (d3) addition.

(e) Composition for (e1) all translations in the plane; (e2) all rotations in the plane (with an arbitrary angle and around any point); (e3) all translations and rotations in the plane.

(f) All subsets of a set: (f1) union; (f2) symmetric difference(= union minus intersection).

(g) Real numbers: $a \circ b = 2a + 2b$.

(h) Integers: $5 * a = a * 5 = a$ for every a, and $a * b = 5$ if neither of a and b is 5.

(i) Matrix multiplication for real 2×2 matrices where (i1) the second row is 0; (i2) all elements are equal; (i3) the sum of the four elements is 0.

A.4.2. Let X be an arbitrary (finite or infinite) set. Consider the functions $X \to X$ with the operation of composition. Find the identity element. Characterize the functions having a left/right inverse, and determine the number of left/right inverses of these functions.

A.4.3. How many operations can be defined on a set of n elements? How many of them are commutative and how many have an identity?

A.4.4. Consider an associative operation with identity. Prove that if both a and b have an inverse, then ab also has an inverse. Is the converse true?

A.4.5. True or false?

(a) If $ab = ba = b$ for some $a, b \in S$, then a is an identity.

(b) If an operation has an identity and $ab = ba = b$ for some $a, b \in S$, then a is the identity.

(c) If an operation has an identity, $ab = ba = b$ for some $a, b \in S$, and b has an inverse, then a is the identity.

(d) If an associative operation has an identity, $ab = ba = b$ for some $a, b \in S$, and b has an inverse, then a is the identity.

A.4.6. (a) Show an associative operation with identity where some elements have infinitely many left inverses.

(b) Prove that if an associative operation has an identity and every element has a left inverse, then every element has exactly one left inverse which is also a right inverse (and so the two-sided inverse) of that element.

(c) Verify that the identity always has exactly one left inverse and exactly one right inverse (even if the operation is not associative).

(d) Find a non-associative operation with identity where every non-identity element has infinitely many left and infinitely many right inverses.

A.4.7. Modify part (i) in Theorem A.4.7 if we reduce the requirement on b to have just a left inverse (instead of a two-sided inverse) but keep the assumption that the operation is associative and has an identity.

A.4.8. Prove that if an operation is associative and the equations $xb = a$ and $by = a$ are solvable for every $a, b \in S$, then these equations have exactly one solution for every a, b.

A.4.9. Show an example that none of the statements in Theorem A.4.7 remain valid if the operation is not associative.

A.5. Fields

The notion of a commutative field generalizes the properties of addition and multiplication of rational, real, or complex numbers.

Definition A.5.1. A set F of at least two elements is a *commutative field* if

(i) two operations, called addition and multiplication, are defined on F;

(ii) addition is associative and commutative, it has a zero, and every element has a negative;

(iii) multiplication is associative and commutative, it has an identity, and every non-zero element has an inverse;

(iv) addition and multiplication are connected by the *distributive law*: $a(b + c) = ab + ac$ for every $a, b, c \in F$. ♣

The adjective "commutative" refers to the commutativity of multiplication. Omitting this requirement, we speak about a *skew field* but then we have to insert a second distributive law $(b + c)a = ba + ca$ in (iv). E.g., the *quaternions* form a skew field; see Example E5 in Section 5.6. In the present chapter (and in most parts of the book) a field always means a commutative field.

Requirements (i)–(iv) are called the *field axioms*.

According to the previous section, we denote the zero of a field by 0, the (multiplicative) identity by 1, the negative of a by $-a$, and if $a \neq 0$, the (multiplicative) inverse of a by a^{-1} (or $1/a$).

In a field F, the equation $b + x = a$ has a unique solution for every $a, b \in F$, by Theorem A.4.7, so *subtraction* makes sense. Similarly, the equation $bx = a$ has a unique solution for every $b \neq 0$ and $a \in F$, giving rise to *division* (except for dividing by zero). In fact, we can even replace the requirements about zero and negatives, or identity and inverses by subtraction and division, resp., in the definition of a field. So, we can summarize the notion of a field as a set with the four basic operations satisfying the usual operational identities.

Note that in a non-commutative field we have to distinguish left division and right division as multiplication (among the non-zero elements) has two inverse operations. Subtraction is unique here, too, as addition is commutative.

Examples of fields

E1 As mentioned in the introduction, the rational, real, and complex numbers provided the model to the notion of a field. These fields are denoted by $\mathbf{Q}$, $\mathbf{R}$, and $\mathbf{C}$, resp.

E2 The residue classes modulo a prime p form a field with the addition and multiplication defined using the representatives of the classes. One has to show first that the operations make sense by verifying that they do not depend on which representatives were chosen from the classes (cf. with Exercise A.4.1(d)). Except the multiplicative inverse, all other properties follow from the corresponding properties of integers. Rephrasing the requirement about the multiplicative inverse for the representatives, we need that the linear congruence $ax \equiv 1 \pmod{p}$ is solvable for every $a \not\equiv 0 \pmod{p}$. By Theorem A.2.9, this holds, since p is a prime, so $(a, p) = 1$.

We denote the field of the modulo p residue classes by F_p or $\mathbf{Z}_p$. This is a *finite field* of p elements. The general description of finite fields is discussed in Section A.11.

E3 The real numbers $a + b\sqrt{2}$ where a and b are rational numbers form a field. The existence of a multiplicative inverse can be shown by the usual rationalization, and the other field axioms are straightforward. If we want to have a similar construction with $\sqrt[3]{5}$ instead of $\sqrt{2}$, then we have to consider the real numbers $a + b\sqrt[3]{5} + c\sqrt[3]{25}$ where a, b, and c are rational numbers. This is a field but it is hard to prove the existence of multiplicative inverses. See Section A.10 where we discuss fields of this type in detail.

E4 The algebraic fractions(= quotients of polynomials with real coefficients) form a field with the usual addition and multiplication.

Exercises A.5

A.5.1. Which of the following sets form a (commutative) field with the usual addition and multiplication?

(a) Subsets of $\mathbf{R}$: (a1) fractions with an odd denominator (also 1 is an odd number); (a2) the numbers $a + b\sqrt{7}$ with rational a and b; (a3) the non-negative rational numbers.

(b) The even residue classes modulo $2m$ (represented by $0, 2, \dots, 2m - 2$) where (b1) $2m = 10$; (b2) $2m = 20$.

(c) The functions $f : \mathbf{R} \to \mathbf{R}$ satisfying (c1) $f(0)=0$; (c2) $a\neq 0 \Rightarrow f(a)= 0$.

(d) The 2×2 real matrices of the form

(d1) $\begin{pmatrix} 0 & 0 \\ a & 0 \end{pmatrix}$; (d2) $\begin{pmatrix} 0 & 0 \\ 0 & a \end{pmatrix}$; (d3) $\begin{pmatrix} a & 2a \\ 4a & 8a \end{pmatrix}$;

(d4) $\begin{pmatrix} a & b \\ a & b \end{pmatrix}$; (d5) $\begin{pmatrix} a & b \\ -b & a \end{pmatrix}$.

A.5.2. Let $m > 1$ be a fixed integer and consider the m^2 "complex numbers" $a + bi$ where a and b are modulo m residue classes. We define addition and multiplication as seen at the complex numbers, so, e.g., for $m = 5$ we have $(2 + 3i)(1 + 4i) = (2 - 12) + (3 + 8)i = i$. Do we get a field if

(a) $m = 2$; (b) $m = 3$; (c) $m = 5$?

A.5.3. Which of the following sets form a (commutative) field with the addition $\oplus$ and multiplication $\odot$? (The uncircled signs mean the usual operations.)

(a) The real numbers with $a \oplus b = \sqrt[3]{a^3 + b^3}$ and $a \odot b = ab$.
(b) The real numbers with $a \oplus b = 5(a + b)$ and $a \odot b = ab$.
(c) The real numbers with $a \oplus b = a + b$ and $a \odot b = 5ab$.
(d) The real numbers with $a \oplus b = a + b - 1$ and $a \odot b = a + b - ab$.
(e) The positive real numbers with $a \oplus b = ab$ and $a \odot b = a^{\log_{10} b}$.
(f) The complex numbers with the usual addition and $(a + bi) \odot (c + di) = ac + bdi$.
(g) The complex numbers with the usual addition and $(a + bi) \odot (c + di) = (ad + bc) + (bd - ac)i$.

A.5.4. Two fields are *isomorphic* if there exists a *bijection* between the two sets that *preserves the operations*, so there is a bijection $\varphi\colon F_1 \to F_2$ satisfying $\varphi(a + b) = \varphi(a) + \varphi(b)$ and $\varphi(ab) = \varphi(a)\varphi(b)$ for every $a, b \in F_1$. This means that the two fields are essentially the same, just the elements and operations are denoted differently.

(a) Prove that no two of the fields **Q**, **R**, **C**, and F_p are isomorphic.
*(b) Find all fields in Exercises A.5.1 and A.5.3 that are isomorphic to one of the fields **Q**, **R**, **C**, and F_p.

A.5.5. A *subfield* in a field F is a subset $K \subseteq F$ that is a field with (the restrictions of) the operations of F. E.g., **R** is a subfield in **C**, and **Q** is a subfield in **R**.

(a) Verify that **R** and **C** have infinitely many subfields.
(b) Show that **Q** and F_p have no proper subfields, i.e., the only subfield in them are themselves.
(c) Prove that if a field F has no proper subfields, then F is isomorphic to **Q or F_p for some prime p.
(d) Verify that **R** has no subfield isomorphic to some F_p.

*A.5.6. Can we define an addition $\oplus$ and a multiplication $\odot$ on the set of integers so that the integers form a field

(a) with the addition $\oplus$ and the usual multiplication;
(b) with the usual addition and the multiplication $\odot$;
(c) with the addition $\oplus$ and the multiplication $\odot$?

A.6. Rings

A ring is an algebraic structure with two operations with weaker requirements for the multiplication than in fields.

Definition A.6.1. A non-empty set R is a *ring* if

(i) two operations, called addition and multiplication, are defined on R;

(ii) addition is associative and commutative, it has a zero, and every element has a negative;

(iii) multiplication is associative;

(iv) addition and multiplication are connected by the *distributive laws*: $a(b+c) = ab + ac$ and $(b + c)a = ba + bc$ for every $a, b, c \in F$. ♣

So, *ring axioms* (i), (ii), and (iv) are the same as the corresponding field axioms, but (iii) does not contain the commutativity of multiplication and the requirements about identity and inverses. Note that the rules for addition are exactly the same as for fields or vector spaces.

As multiplication is not (necessarily) commutative, we have to impose two distributivity conditions. These are independent, e.g., the functions $\mathbf{R} \to \mathbf{R}$ with the usual addition and the composition as multiplication satisfy all ring axioms except one of the two distributive laws.

Similar to the field, the requirements about zero and negatives can be replaced by subtraction. So, we can summarize the notion of a ring as a set with addition, subtraction, and multiplication satisfying the usual operational identities except that multiplication is not necessarily commutative.

A ring is *commutative* if multiplication is commutative; it has an *identity* if multiplication has an identity. So, a *commutative field* is a commutative ring with identity where every non-zero element has an inverse.

In a ring R, every $a \in R$ satisfies $0a = a0 = 0$ (see Exercise A.6.4). This implies that if a ring has at least two elements, then the zero and the identity must differ, and the 0 can have neither left, nor right inverses.

In many rings a product of two non-zero elements can be zero. This leads to the notion of *zero divisors*.

Definition A.6.2. In a ring, an element $a \neq 0$ is a *left zero divisor* if $ab = 0$ for some element $b \neq 0$.

Similarly, $a \neq 0$ is a *right zero divisor* if $ca = 0$ for some $c \neq 0$. ♣

Theorem A.6.3. *If R is a ring with identity and $a \in R$ has a left inverse, then a cannot be a left zero divisor.* ♣

Of course, left can be replaced by right.

Proof. Let e be the identity and let d be a left inverse of a. We have to show that $ab = 0$ implies $b = 0$. Multiplying $0 = ab$ by d from the left we obtain $0 = d0 = d(ab) = (da)b = eb = b$. ∎

The converse of Theorem A.6.3 is false, e.g., the ring of integers is free of zero divisors, whereas only 1 and -1 have an inverse.

An important corollary of Theorem A.6.3 is that every field is zero divisor free.

The commutative rings with identity and without zero divisors are called *integral domains*.

Examples of rings

E1 Every field is a ring (even the non-commutative ones).

E2 The sets below with the usual addition and multiplication form an integral domain: (A) integers; (B) real numbers $a + b\sqrt{2}$ where a and b are integers; (C) Gaussian integers, i.e., complex numbers $a + bi$ where a and b are integers; (D) polynomials with real coefficients; (E) polynomials with integer coefficients.

E3 The even numbers or the polynomials with constant term 0 form (with the usual operations) a commutative ring without zero divisors that has no identity.

E4 The sequences of real numbers with the element-wise addition and multiplication, or the functions $\mathbf{R} \to \mathbf{R}$ with the usual addition and multiplication of functions form a commutative ring with identity having zero divisors.

E5 The modulo m residue classes with addition and multiplication defined using the representatives of the classes form a commutative ring with identity. In another interpretation, we can consider this as the ring of modulo m remainders $0, 1, 2, \dots, m-1$ where addition and multiplication mean the remainders of the sum and product of the elements (see Example E4 in Section A.4). We denote this ring by $\mathbf{Z}_m$. It is commutative and 1 is an identity. A remainder/residue class has an inverse if and only if it is coprime to m. For a residue class this means that every element in it is coprime to m, we call such a class a *reduced* residue class modulo m. A non-zero remainder/residue class is a zero divisor if and only if it is not coprime to m. For a residue class this means that no element in it is coprime to m. The ring $\mathbf{Z}_m$ is a field if and only if m is a prime.

E6 The ring of $n \times n$ (e.g., real) matrices is discussed in Section 2.2 in detail. It has an identity, but is not commutative for $n > 1$. A matrix has an inverse exactly if its determinant is not 0, all other non-zero matrices are both left and right zero divisors.

E7 Consider all subsets of a set S and define addition as the symmetric difference and multiplication as the intersection: $A \oplus B = (A \setminus B) \cup (B \setminus A)$ and $A \odot B = A \cap B$. We get a commutative ring with identity. Only the identity has an inverse and all other non-zero elements are zero divisors.

Exercises A.6

A.6.1. Verify that Examples E1–E7 define rings having the properties stated there.

A.6.2. (a) Which elements have an inverse in the rings of Example E2?
(b) Which elements have an inverse and which are zero divisors in the rings of Example E4?
(c) Find the inverse of the remainder/residue class 37 in $\mathbf{Z}_{100}$.

A.6.3. Find all rings among the examples of Exercises A.5.1–A.5.3 that do not form a field. Describe the left and right zero divisors in each. In the rings with identity, determine which elements have an inverse.

A.6.4. Prove $0a = a0 = 0$ for every element a in a ring.

A.6.5. (a) Verify that the ring in Example E7 is commutative, and every element equals both its negative and its square.
(b) Is there some connection in general between the three properties in (a) for an arbitrary ring?

A.6.6. How can we characterize the elements c in a ring which make possible cancellation from the left: $ca = cb \Rightarrow a = b$? What does this mean in particular in a field, in the ring of integers, and in $\mathbf{Z}_m$?

A.6.7. Let c and d denote elements of a ring. True or false?
(a) If c is a right zero divisor, then $cd = 0$ or cd is a right zero divisor.
(b) If cd is a right zero divisor, then so is c.
**(c) If at least one of c and d is a right zero divisor, then $cd = 0$ or cd is a right zero divisor.
(d) If cd is a right zero divisor, then so is at least one of c and d.
(e) If both c and d are right zero divisors, then $c + d = 0$ or $c + d$ is a right zero divisor.
(f) If $c + d$ is a right zero divisor, then so is at least one of c and d.

A.6.8. Prove that if a finite ring of at least two elements is zero divisor free, then it is a (possibly not commutative) field.

Remark. By a (hard) theorem by Wedderburn, every finite field is commutative, so the option in the parentheses cannot hold.

*A.6.9. Show that if a ring has exactly one left identity, then this is a two-sided identity.

A.6.10. The *subring* of a ring R is a subset $S \subseteq R$ that is a ring with (the restriction of) the addition and multiplication in R. E.g., the even numbers form a subring in the ring of integers. Verify that the zero of the subring must be the same as the zero of the original ring.

A.6.11. Let R be a ring with identity and let S be a non-zero subring in R. True or false?
(a) S necessarily has an identity.
(b) If S has an identity, then this must be the same as the identity in R.

(c) If R is zero divisor free and S has an identity, then this must be the same as the identity in R.

(d) If S is zero divisor free and S has an identity, then this must be the same as the identity in R.

*A.6.12. Prove that if in a not necessarily commutative ring of at least two elements the equation $xb = a$ has a solution for every $b \neq 0$ and a, then the ring is a not necessarily commutative field.

A.7. Polynomials

In this section we sketch (occasionally slightly sloppily) the most important facts about polynomials over (commutative) fields.

1. Polynomials

Skipping the difficulties of a precise introduction, a *polynomial* over a field F is a formal expression $\alpha_0 + \alpha_1 x + \cdots + \alpha_n x^n$ where the coefficients α_i are from F. The polynomials $f = \alpha_0 + \alpha_1 x + \cdots + \alpha_n x^n$ and $g = \beta_0 + \beta_1 x + \cdots + \beta_k x^k$ are equal if apart from eventual terms with 0 coefficient, the corresponding coefficients are equal: if (say) $k \geq n$, then $\alpha_0 = \beta_0, \alpha_1 = \beta_1, \ldots, \alpha_n = \beta_n, \beta_{n+1} = \cdots = \beta_k = 0$.

2. Polynomial functions

A polynomial in itself is not a function, but every polynomial induces a *polynomial function* in a natural way: the function belonging to the polynomial $\alpha_0 + \alpha_1 x + \cdots + \alpha_n x^n$ is the $F \to F$ map $\gamma \mapsto \alpha_0 + \alpha_1 \gamma + \cdots + \alpha_n \gamma^n$. So, we can think of a polynomial as a formula for the polynomial function. The notions of polynomials and polynomial functions cannot be identified since certain polynomial functions can be defined with various such formulas. E.g., the distinct polynomials x and x^p over the field F_p of the modulo p residue classes create the same function as $a^p \equiv a \pmod{p}$ for every a by Fermat's Little Theorem. It can be proven that such anomalies occur only for *finite* fields but there they are typical (see Exercise A.7.1). For infinite fields the connection between polynomials and polynomial functions is bijective.

In general, we make no distinction in notation and wording between a polynomial and the corresponding polynomial function; we denote both by (say) f, or by $f(x)$ if we want to accentuate the variable, and use mostly the word polynomial for a polynomial function. So, e.g., the value assumed by a polynomial means the value assumed by the corresponding polynomial function. We follow this common terminology in the other parts of the book, but in this section we consistently distinguish the two notions in wording to make the roles absolutely clear-cut.

3. Operations

We add two polynomials by adding the corresponding terms, so the sum of the two polynomials f and g in part 1 is

$$f + g = (\alpha_0 + \beta_0) + (\alpha_1 + \beta_1)x + \cdots + (\alpha_n + \beta_n)x^n + \beta_{n+1}x^{n+1} + \cdots + \beta_k x^k.$$

The product is formed by multiplying each term in the first polynomial with each term in the second one, adding these products, and uniting the terms with the same exponent, so the product of f and g is

$$fg = \alpha_0\beta_0 + (\alpha_0\beta_1 + \alpha_1\beta_0)x + \cdots + \Big(\sum_{i+j=m} \alpha_i\beta_j\Big)x^m + \cdots + \alpha_n\beta_k x^{n+k}.$$

A natural explanation for the above definitions of addition and multiplication is that we consider polynomials as (formal) sums and perform the operations accordingly. These operations relate well to the usual addition and multiplication of polynomial functions: the map assigning to each polynomial the polynomial function induced by it preserves both addition and multiplication.

4. The ring of polynomials

The polynomials over a field F with the two operations defined in part 3 form a commutative and zero divisor free ring with identity. We denote this ring by $F[x]$. Its zero is the 0 polynomial with all coefficients 0 (of F), the identity is the constant polynomial 1. Only the non-zero constant polynomials have an inverse.

Also the polynomial functions form a ring with the usual addition and multiplication of functions. This ring is commutative and has an identity, but it contains zero divisors if F is a finite field (see Exercise A.7.3). If F is an infinite field, then the (natural) map from the polynomials to the polynomial functions is bijective and preserves the operations. So, if F is infinite, then the rings of polynomials and polynomial functions are *isomorphic* (hence neither the polynomial functions contain zero divisors in this case).

5. Degree

The *degree* of a polynomial $f = \alpha_0 + \alpha_1 x + \cdots + \alpha_n x^n$ is the non-negative integer n if $\alpha_n \neq 0$. We denote the degree of f by $\deg f$. Apart from the 0 polynomial, every polynomial has a degree. It is straightforward from the definitions of the operations and degree that if none of f, g, and $f+g$ is the 0 polynomial, then $\deg(fg) = \deg f + \deg g$ and $\deg(f+g) \leq \max(\deg f, \deg g)$. The coefficient of x^n in a polynomial f of degree n is called the *leading coefficient* of f. If the leading coefficient is 1, then we say that f is *monic*.

Note that the degree belongs to a (non-zero) polynomial and not to a polynomial function: e.g., over the field F_p the polynomial x has degree 1, the polynomial x^p has degree p whereas they induce the same polynomial function. (Over an infinite field it causes no error to speak about the degree of a polynomial function as there is a bijection between polynomials and polynomial functions.)

6. Roots

An element $\gamma \in F$ is a *root* (or *zero*) of a polynomial *function* if the value assigned to γ is $0 (\in F)$. An important connection is that γ is a root of the *polynomial function* belonging to the polynomial f if and only if the *polynomial* f has a decomposition

$f = (x - \gamma)g$ with some polynomial g. This makes it possible to speak about the root of a polynomial, too. The linear polynomial $x - \gamma$ is called a *root factor* of the polynomial f.

7. Multiplicity

We say that the *multiplicity* of a root $\gamma \in F$ is (exactly) k in a non-zero polynomial(!) f if $f = (x - \gamma)^k g$ where γ is not a root of the polynomial g. If the multiplicity is at least 2, then we speak about a *multiple root* (or *repeated root*).

Similar to the degree, multiplicity refers to polynomials and not to polynomial functions. Using the previous example, over the field F_p the multiplicity of 0 is 1 in the polynomial x and is p in the polynomial x^p but the two polynomials induce the same polynomial function.

8. The number of roots

Every element in F is a root of the zero polynomial but any other polynomial f has at most $\deg f$ roots. (This theorem is no more valid for *non-commutative* fields; see Exercise 5.6.22.)

The *fundamental theorem of algebra* states that over the *complex* field every non-constant polynomial has a (complex) root. This implies that over the complex field every non-zero polynomial f has *exactly* $\deg f$ roots if roots are counted with multiplicity, or equivalently, f is the product of linear polynomials.

The multiplicities of a *complex* root and of its conjugate in a polynomial with *real* coefficients are the same. This and the fundamental theorem of algebra imply that every polynomial with real coefficients is the product of polynomials with real coefficients of degree 1 or 2, and if the degree is odd, then it must have a *real* root.

9. Finding the roots

The only root of a linear polynomial $\alpha_0 + \alpha_1 x$ (where $\alpha_1 \neq 0$) is $-\alpha_0/\alpha_1$ for any field F. To find the roots of a quadratic polynomial we can use the quadratic formula for the real, complex, and many other fields.

Over the real or complex field, there exist similar universal formulas for polynomials of third and fourth degrees. These formulas express the roots of the polynomial by applying the four basic operations and taking roots of positive integer exponents finitely many times starting from the coefficients. For polynomials of fifth or higher degree there exists no such general algebraic method; in fact we can find concrete polynomials such that their roots cannot be obtained from the coefficients as described above.

The rational root test is a simple algorithm to determine the rational roots of a polynomial with rational coefficients. Multiplying the polynomial with the least common multiple of the denominators of the coefficients we obtain a polynomial with integer coefficients $a_0 + a_1 x + \cdots + a_n x^n$ having the same roots as the original polynomial. We may assume $a_n \neq 0$, and dividing by a suitable power of x if necessary (this makes no difference for the non-zero roots) we can achieve $a_0 \neq 0$. If a rational number

c/d (where c and d are coprime integers) is a root of this polynomial, then necessarily $c \mid \alpha_0$ and $d \mid \alpha_n$. We select the actual (non-zero) rational roots by checking these finitely many rational numbers.

The roots of a polynomial over a field F_p, or equivalently, the solutions of a congruence modulo a prime p can be found by checking the finitely many elements of the field at the worst. If $\deg f \geq p$, then we can reduce f to a polynomial g of degree at most $p-1$: g is the remainder in the division algorithm of f by $x^p - x$, i.e., we obtain g by replacing x^p in f by x as long as this is possible. This g is a polynomial of degree not greater than $p-1$ (or $g = 0$) and induces the same polynomial function as f, so their roots are the same, too.

10. The derivative of a polynomial

Over an arbitrary(!) commutative field F we define the *derivative* of a polynomial $f = \alpha_0 + \alpha_1 x + \cdots + \alpha_n x^n$ by the formal derivation rules: $f' = \alpha_1 + \cdots + n\alpha_n x^{n-1}$ where $j(\alpha_j x^{j-1})$ means an addition j times in $F[x]$.

An easy calculation verifies that the derivatives of sums, products, and powers obey the usual derivation rules.

The derivative is closely connected to the multiplicity of roots: If the multiplicity of a root γ is exactly $k(\geq 1)$ in f, then it is at least $k-1$ in f'. The latter multiplicity is not always exactly $k-1$. E.g., over F_2 the derivative of $f = x^5 + x = x(x+1)^4$ is $f' = x^4 + 1 = (x+1)^4$, and the multiplicity of 1 is exactly 4 in both. Even $f' = 0$ can occur, e.g., the derivative of $f = x^p$ over the field F_p is $f' = px^{p-1} = 0$. If, however, adding any non-zero element in F to itself arbitrarily many times, we never get zero, i.e., $\alpha + \alpha + \cdots + \alpha = 0 \Rightarrow \alpha = 0$, then the above theorem is valid in the stronger form that the multiplicity of γ is exactly $k-1$ in f'.

11. Relation between roots and coefficients: the Vieta formulas

If a polynomial f of degree n has exactly n roots counted with multiplicity, let these be $\gamma_1, \ldots, \gamma_n$, then

$$f = \alpha_0 + \alpha_1 x + \cdots + \alpha_n x^n = \alpha_n \prod_{j=1}^{n} (x - \gamma_j)$$

implies $\sigma_m = (-1)^m \alpha_{n-m}/\alpha_n$, $m = 1, 2, \ldots, n$, where σ_m is the sum of all $\binom{n}{m}$ products of m factors formed from the roots γ_j. The special cases $m = 1$ and $m = n$ give that the sum of the roots γ_i is $-\alpha_{n-1}/\alpha_n$, and their product is $(-1)^n \alpha_0/\alpha_n$.

12. Number theory of polynomials

Divisibility and all other number theoretic notions are defined in $F[x]$ exactly as in any commutative ring with identity and without zero divisors.

The *units* are the polynomials dividing every polynomial, i.e., having a (multiplicative) inverse. These are exactly the non-zero constant polynomials. Two polynomials are associates if they differ only in a unit factor.

A polynomial is *irreducible* if it is not a unit and it cannot be factored into a product except if one of the factors is a unit. In other words, it is divisible only by units and its associates. A polynomial different from 0 and units is called *reducible* if it is not irreducible.

The long division provides a *division algorithm* in $F[x]$: For any polynomials $g \neq 0$ and f there exist polynomials q and r satisfying $f = gh + r$ and $\deg r < \deg g$ or $r = 0$ (q and r are unique). This means that $F[x]$ is a *Euclidean domain*, hence *unique prime factorization* is valid: every non-zero and non-unit polynomial is the product of finitely many irreducible factors, and this representation is unique apart from associates and the order of the factors: if $f = s_1 \cdot \dots \cdot s_m = t_1 \cdot \dots \cdot t_k$ where every s_i and t_j is irreducible, then $m = k$ and the polynomials s_i and t_j can be ordered into associate pairs.

The *greatest common divisor* or *gcd* of two polynomials f and g is a polynomial dividing both f and g and being a multiple of all their common divisors. It is denoted by (f, g) or $\gcd(f, g)$. Similar to the integers, its existence follows from the *Euclidean algorithm* consisting of a series of division algorithms (see after Theorem A.2.2). The Euclidean algorithm serves as a practical procedure to determine the gcd and implies that the gcd is the sum of suitable multiples of the two polynomials: for any $f, g \in F[x]$ there exist $u, v \in F[x]$ satisfying $(f, g) = fu + gv$ (u and v are not unique). We infer from the definition that the gcd is unique apart from associates: If $d = (f, g)$, then all associates of d but no other polynomials possess this property.

A polynomial is a *prime* if it differs from zero and units and it cannot divide the product of two polynomials unless it divides at least one of the factors. A polynomial is irreducible if and only if it is a prime; this can be verified using the greatest common divisor.

13. Irreducible polynomials

We always have to specify the field over which we consider a given polynomial as, e.g., a polynomial with rational coefficients is also a polynomial with real or complex coefficients, and it can happen that it is irreducible over the rationals but reducible over the reals. (The expressions "f is irreducible over F" and "f is irreducible in $F[x]$" are synonyms.)

A linear polynomial is irreducible over any field. Also the converse is true for $\mathbf{C}[x]$ by the fundamental theorem of algebra. This implies that in $\mathbf{R}[x]$ the only other irreducible polynomials are the ones of degree two having no real roots.

There is a much bigger diversity of irreducible polynomials in $\mathbf{Q}[x]$. A useful sufficient (but by no means necessary) condition for irreducibility is the *criterion by Schönemann and Eisenstein*: If $f = a_0 + a_1x + \dots + a_nx^n$ has *integer* coefficients and each of $a_0, a_1, \dots, a_{n-1}$ is divisible by a prime p, but a_n is not divisible by p, and a_0 is not divisible by p^2, then f is irreducible over the *rational* field. This makes possible the construction of irreducible polynomials of any degree in $\mathbf{Q}[x]$.

To determine whether a given polynomial with rational coefficients is irreducible, we multiply the polynomial by the least common multiple of the denominators of its coefficients, and divide the new polynomial by the greatest common divisor of its coefficients. This does not affect the irreducibility over **Q** as we multiplied only by a constant, i.e., a unit. The resulting polynomial has integer coefficients that are coprime; we call such a polynomial *primitive*. The two *Gauss's lemmas* have both theoretical and practical significance:

I. The product of two primitive polynomials is primitive.

II. If a polynomial $F \neq 0$ with *integer* coefficients is the product of polynomials g and h with *rational* coefficients: $F = gh$, then F can be written as $F = GH$ where G and H have *integer* coefficients and are constant multiples of g and h (so $\deg G = \deg g$ and $\deg H = \deg h$).

This reduces the question of reducibility over **Q** to the question of factoring a polynomial with integer coefficients into the product of two non-constant polynomials with integer coefficients.

An important class of irreducible polynomials over **Q** are the *cyclotomic polynomials*. The mth cyclotomic polynomial Φ_m is monic (i.e., has a leading coefficient 1) and its roots are the complex primitive mth roots of unity. So the degree of Φ_m is $\varphi(m)$. E.g., $\Phi_4 = x^2 + 1$, $\Phi_{11} = x^{10} + x^9 + \cdots + 1$. Φ_m has integer coefficients; this can be verified by induction using the formula $x^m - 1 = \prod_{d|m} \Phi_d$. If m is a prime or a prime power, then its irreducibility can be proven applying the criterion by Schönemann and Eisenstein after a linear substitution. For an arbitrary m, the proof is much more difficult.

14. Polynomials with integer coefficients

Most properties of polynomials over a commutative field remain valid if the coefficients are taken from any commutative ring with identity and without zero divisors (see Exercise A.7.2).

Perhaps the simplest and most important case is the investigation of polynomials with integer coefficients. We discuss here only the number theoretic aspects of $\mathbf{Z}[x]$. In contrast to the polynomials over a commutative field, we emphasize two essential differences: the only units are ± 1, and there is no division algorithm.

Let us take a closer look at the division algorithm. It is easy to see that if we want to perform a division algorithm $f = gq + r$, e.g., for $f = x$ and $g = 2$, then there exists no remainder r having a smaller degree than g, and $r = 0$ does not work either. It could be, however, that a division algorithm is still possible for some other Euclidean function instead of the degree. We show that this is not the case. The division algorithm is the basis for the Euclidean algorithm which implies that the greatest common divisor of x and 2, i.e., the polynomial 1 can be written in the form $1 = xu + 2v$ with suitable polynomials u and v with *integer* coefficients. But this is impossible as the constant term on the right-hand side is even.

Though the division algorithm fails, unique prime factorization is valid in $\mathbf{Z}[x]$. As a consequence of Gauss's lemma II, $f \in \mathbf{Z}[x]$ is irreducible over **Z** if and only if either f is a constant polynomial p where p is a prime, or f is a primitive polynomial irreducible over **Q**. This means that the irreducibility in $\mathbf{Z}[x]$ can be characterized by the irreducibility in $\mathbf{Q}[x]$ and $\mathbf{Z}$, so the unique prime factorization theorem in the latter two implies its validity also in $\mathbf{Z}[x]$.

Remark. Finally we call attention to *interpolation polynomials* in Theorem 3.2.4 and related Exercises 3.2.7–3.2.15.

Exercises A.7

F denotes a commutative field.

A.7.1. Let F be an *arbitrary* finite field.
- (a) Prove the existence of distinct polynomials over F inducing the same polynomial function.
- (b) (Continued) Moreover, *every* polynomal function comes from *infinitely* many polynomials.

A.7.2. Verify that the statements below are true also for polynomials with coefficients from a commutative zero divisor free ring with identity (instead of a commutative field).
- I. The degree of the product of two polynomials is the sum of degrees of the factors.
- II. γ is a root of the function belonging to the polynomial f if and only if $x - \gamma$ is a factor of f.
- III. A polynomial of degree n has at most n roots.

Which statements remain valid for commutative rings with identity containing zero divisors?

A.7.3. Let F be an arbitrary *finite* field.
- (a) Check that the $F \to F$ polynomial functions form a ring under the usual functions addition and multiplication.
- (b) Show that this ring always contains zero divisors.
- *(c) Determine the number of zero divisors.

A.7.4. Verify for every infinite commutative field F the existence of a function $F \to F$ that is not a polynomial function (cf. with Exercise 3.2.14).

A.7.5. A polynomial G with integer coefficients has degree ten and 11 divides $G(n)$ for every integer n. Prove that every coefficient of G is divisible by 11.

A.7.6.
- (a) Which subsets of the real numbers occur as a range of a polynomial function with real coefficients?
- (b) Which subsets of the complex numbers occur as a range of a polynomial function with complex coefficients?

A.7.7. Given a non-linear polynomial $f = a_0 + a_1x + \cdots + a_nx^n$ with integer coefficients satisfying $a_0 \neq 0, a_n \neq 0$, we investigate the effect of modifying a coefficient on the existence of a rational root. Verify:
- (a) We can replace a_0 by infinitely many integers for the resulting polynomial to have a rational root.
- (b) We can replace a_n by infinitely many integers for the resulting polynomial to have a rational root.

(c) For any fixed $1 \leq i \leq n-1$ we can replace a_i by at least one but at most finitely many integers for the resulting polynomial to have a rational root.

(d) For any fixed $1 \leq i \leq n-1$ we can replace a_i by infinitely many *rational* numbers for the resulting polynomial to have a rational root.

(e) (Now we modify two coefficients.) For any fixed $i \neq j$ we can replace a_i and a_j by infinitely many pairs of integers for the resulting polynomial to have a rational root.

*(f) For any fixed i we can replace a_i by infinitely many integers for the resulting polynomial *not* to have a rational root.

A.7.8. Show that the polynomial $1 + x + x^2/2 + x^3/6 + \cdots + x^n/(n!)$ has n distinct complex roots.

A.7.9. Devise an algorithm to determine the exact roots of polynomials of degree five with complex coefficients having a multiple root.

A.7.10. Prove that a polynomial irreducible over $\mathbf{Q}$ cannot have a complex multiple root.

A.7.11. Show that a polynomial f with complex coefficients has a multiple root if and only if f and f' are not coprime, i.e., $(f, f') \neq 1$.

A.7.12. Describe all polynomials with complex coefficients divisible by their derivatives.

A.7.13. Let F be a polynomial with integer coefficients of degree 28. True or false?

(a) F must have a divisor of degree 17 in $\mathbf{C}[x]$.

(b) F must have a divisor of degree 17 in $\mathbf{R}[x]$.

(c) F must have a divisor of degree 18 in $\mathbf{R}[x]$.

(d) F must have a divisor of degree 18 in $\mathbf{Q}[x]$.

A.7.14. (a) Let every coefficient of f and g be integers. Then f and g can be considered as polynomials over $\mathbf{Z}$, $\mathbf{Q}$, $\mathbf{C}$, or F_2. So $f \mid g$ may have four different meanings. What is the relation between these divisibilities?

(b) What can we assert if every coefficient of f and g is 0 or 1?

A.7.15. Prove $x^2 + x + 1 \mid x^{3m} + x^{3n+1} + x^{3k+2}$ for any non-negative integers m, n, and k.

A.7.16. Determine the greatest common divisor of polynomials $x^n - 1$ and $x^k - 1$.

A.7.17. Is there a polynomial of degree ten with real coefficients yielding the same remainder at the long division by the polynomials $x^5 + 2$ and $2x^6 + 3x + 1$?

A.7.18. We investigate a linear "Diophantine" equation for polynomials: for given polynomials $f, g, h \in F[x]$ we want to find polynomials $u, v \in F[x]$ satisfying $fu + gv = h$. What is a necessary and sufficient condition for solvability, what is the number of solutions, and how can we obtain all solutions?

A.7.19. Let f be a polynomial of degree $n \geq 1$ with rational coefficients. Determine all values of n for which the following statements are true.

I. If f is irreducible over $\mathbf{Q}$, then f has no rational roots.

II. If f has no rational roots, then f is irreducible over $\mathbf{Q}$.

A.7.20. Factor $x^4 + 1$ into the product of irreducible polynomials over

(a) $\mathbf{C}$; (b) $\mathbf{R}$; (c) $\mathbf{Q}$; (d) F_2; (e) F_3.

A.7.21. List infinitely many positive integers c such that $1^4 + c$, $2^4 + c$, $3^4 + c$, ..., $n^4 + c, \dots$ are all composite numbers.

A.7.22. Which polynomials are irreducible over $\mathbf{Q}$?

(a) $x^2 + 2500$; (b) $x^4 + 2500$; (c) $x^4 + 3000$; (d) $x^4 + 100000$.

*A.7.23. Prove that if $a_1, \dots, a_k$ are distinct integers, then $(x - a_1) \cdot \dots \cdot (x - a_k) - 1$ is irreducible over $\mathbf{Q}$.

A.7.24. How can we obtain Φ_{2m} from Φ_m (where Φ_j denotes the jth cyclotomic polynomial)?

*A.7.25. Verify

$$\Phi_m = \frac{(x^m - 1) \cdot \prod_{p,q}(x^{m/pq} - 1) \cdot \dots}{\prod_p (x^{m/p} - 1) \cdot \prod_{p,q,r}(x^{m/pqr} - 1) \cdot \dots}$$

where $p, q, r, \dots$ denote the distinct prime factors of m.

A.7.26. (a) Find the roots of $f = x^{4k} + x^{3k} + x^{2k} + x^k + 1$.

(b) For which values of k is f a cyclotomic polynomial?

*A.7.27. Compute the product of the lengths of all the $n-1$ sides and diagonals starting from a vertex of a regular n-gon inscribed into a unit circle.

A.7.28. Assume that the polynomial $f = x^n + \alpha_{n-1}x^{n-1} + \dots + \alpha_0$ with real coefficients satisfies $\alpha_{n-1}^2 - 2\alpha_{n-2} < 0$. Prove that f has a non-real complex root.

A.7.29. How can we recognize from the coefficients that the three complex roots of a polynomial $ax^3 + bx^2 + cx + d \in \mathbf{C}[x]$ $(a \neq 0)$ form an arithmetic progression?

A.7.30. A polynomial with integer coefficients has leading coefficient 1 and all its complex roots have absolute value less than 1. Determine the coefficients of the polynomial.

A.8. Groups

Definition A.8.1. A non-empty set G is a *group* if an associative operation with identity is defined on G and every element has an inverse. ♣

We arrive at the same notion if — as seen for fields and rings, based on Theorem A.4.7 — we replace the requirements for an identity and inverses by the existence of inverse operations from both sides.

In general, we denote the identity element by e and the inverse of an element g by g^{-1}. If the operation is commutative, then G is a *commutative* or *Abelian* group.

Examples of groups

E1 Any ring (and so any field) is a commutative group under addition. Thus the integers, the even, rational, real, or complex numbers, the quaternions (see Example E5 in Section 5.6), the appropriate matrices, polynomials, functions, the modulo m residue classes, etc. form an Abelian group under the usual addition.

E2 The non-zero elements of any (commutative or non-commutative) field form a group under multiplication. Accordingly, the **non-zero(!)** rational, real, or complex numbers, the non-zero quaternions, the non-zero residue classes modulo a prime p, etc. form a group under the usual multiplication.

E3 The invertible elements in a ring with identity form a group. A special case is just Example E2 for fields. Two other important groups are provided by the rings of square matrices and the modulo m residue classes: the $n \times n$ invertible matrices over a field F and the reduced residue classes modulo m form a group under the usual multiplication.

E4 The complex numbers have many subsets that are groups under multiplication: e.g., the nth roots of unity, all roots of unity, the complex numbers of absolute value 1, etc.

E5 The bijections of the set $\{1, 2, \ldots, n\}$ form a group under composition. This is the *symmetric group* of degree n and is denoted by S_n. So S_n consists of all *permutations* of the elements $1, 2, \ldots, n$, $|S_n| = n!$.

E6 All congruences or similitudes of the plane or space (i.e., the geometric transformations preserving distance or proportions, resp.) form a group under composition. The same holds for certain special congruences, e.g., all rotations around a given point or all translations in the plane, etc. (under composition).

E7 The *symmetry group* of a figure in the plane or space consists of all congruences carrying the figure onto itself; the operation is composition. For example, the elements of the symmetry group of a regular n-gon are the reflections through the n axes of symmetry and the rotations around the center by angles $2k\pi/n$, $k = 0, 1, \ldots, n-1$. The rotation by angle 0 is the identity. This group is called the *dihedral group* of degree n and is denoted by D_n, $|D_n| = 2n$. Let t be the rotation by $2\pi/n$ and let f be a reflection over an axis of symmetry. Then the elements of D_n have a unique representation as $f^i t^j$, $0 \le i \le 1$, $0 \le j \le n-1$ and the rules of multiplication are $f^2 = t^n = e$ and $tf = ft^{n-1}$.

In a group, we denote the exponentiation for *integer exponents* by the usual rules: if n is a positive integer, then g^n is a product of n copies of g; $g^0 = e$; and $g^{-n} = (g^n)^{-1}$. (Exponentiation with rational, real, etc. exponents generally makes no sense in a group.) The laws $a^m a^k = a^{m+k}$ and $(a^m)^k = a^{mk}$ are valid in any group (where a is an arbitrary element of the group and the exponents k and m can be arbitrary integers including zero and negative integers, too). However, $(ab)^m = a^m b^m$ is false in general as multiplication is not necessarily commutative.

Definition A.8.2. The *order* of an element g in a group G is the *smallest positive* integer n satisfying $g^n = e$ (where e is the identity in the group). If there is no such n, then the order of g is infinite. ♣

We denote the order of g by $o(g)$. For example, in D_n, $o(f) = 2, o(t) = n$; in the additive group of integers the order of every non-zero element is infinite.

We obtain important special cases for the two multiplicative groups of the non-zero complex numbers and the reduced residue classes modulo m. In the latter, the notion of order is exactly the same as seen at the congruences (Definition A.2.7). In the first group, the roots of unity are the elements of finite order (Definition A.3.6).

In any group, two powers of an element of order n are equal if and only if the exponents are congruent modulo n. If the order of an element is infinite, then the equality of its two powers implies the equality of the exponents. So, the number of distinct powers of an element is equal to its order.

Definition A.8.3. A group is *cyclic* if it consists of all integer powers of a single element. Such an element is called a *generator* of the group. ♣

The cyclic group generated by g is denoted by $\langle g\rangle$. As stated at the order, $|\langle g\rangle| = o(g)$. So if $o(g) = n$, then $\langle g\rangle = \{e, g, g^2, \ldots, g^{n-1}\}$ (and these elements are all distinct). If $o(g) = \infty$, then all powers of g are distinct (including the ones with negative integer exponents, too).

The integers and the modulo m residue classes form cyclic groups under addition with 1 as a generator. The nth complex roots of unity and the non-zero residue classes modulo a prime p form cyclic groups under multiplication with the *primitive nth roots of unity* and the modulo *p primitive roots* as generators.

Since any two powers of an element commute, every cyclic group is necessarily commutative. Thus, no non-commutative group can be cyclic.

Non-cyclic commutative groups are, e.g., the rational, real, or complex numbers under addition, the non-zero rational, real, or complex numbers under multiplication, the symmetry group of a rectangle, an at least two-dimensional vector space over a (finite or infinite) field under addition, the reduced residue classes modulo 15 under multiplication, etc. In general, the multiplicative group of the reduced residue classes modulo m is cyclic if and only if $m = 2, 4, p^k$, or $2p^k$ where p is an odd prime and $k \geq 1$.

Definition A.8.4. A subset $H \subseteq G$ is a *subgroup* in G if H is a group under the (restriction of the) operation of G. ♣

A non-empty subset H is a subgroup in G if and only if H is closed for the operation and inverses in G (i.e., $r, s \in H \Rightarrow rs \in H, r^{-1} \in H$).

Examples. Some subgroups in the additive group of polynomials with real coefficients are the polynomials with integer coefficients; the ones of degree at most five and the 0 polynomial; the multiples of $x^2 + 1$, etc. In any group G, the entire group itself and the subset consisting solely of e are the *trivial subgroups* (we shall denote the latter simply by e instead of $\{e\}$). All powers (with integer exponents) of an element form a subgroup, this is the cyclic subgroup generated by the given element.

Definition A.8.5. Let H be a subgroup in G and let $g \in G$ be an arbitrary element. The set $gH = \{gh \mid h \in H\}$ is called a *left coset* of H. ♣

Any two left cosets of H are either disjoint, or coincide, and the union of all left cosets is G. The number of (distinct) left cosets is the *index* of H in G and is denoted by $|G : H|$.

We can define the *right* cosets of H similarly. These are generally not the same as the left cosets, but the number of the left and right cosets is the same, i.e., $|G : H|$.

Let G be a finite group. Since the size of every (say left) coset is $|H|$, we have $|G| = |H| \cdot |G : H|$. An immediate corollary is the next theorem.

Theorem A.8.6 (Lagrange's Theorem). *The size of a subgroup of a finite group divides the size of the group.* ♣

The number of elements in a group is sometimes called the order of the group.

As the order of an element is equal to the size of the cyclic subgroup it generates, Lagrange's Theorem implies $o(g) \bigm| |G|$ for any element g of G. This is equivalent to $g^{|G|} = e$ by the properties of the order. If G is the multiplicative group of the reduced residue classes modulo m, we obtain the Euler–Fermat Theorem A.2.6.

Finally, we mention that similar to other algebraic structures, two groups G_1 and G_2 are *isomorphic* if there exists a bijection between them preserving the operation(!). Notation: $G_1 \cong G_2$. For example, the additive group of the real numbers and the multiplicative group of the positive real numbers are isomorphic since the map $x \mapsto 2^x$ is a bijection preserving the operation. Two cyclic groups are isomorphic if and only if they are of the same size.

Exercises A.8

G denotes a group.

*A.8.1. How many elements are in the symmetry groups of the five Platonic solids?

A.8.2. Verify that G is commutative if and only if $(ab)^2 = a^2b^2$ for every $a, b \in G$. Is a similar statement true for fourth powers instead of squares, i.e., if $(ab)^4 = a^4b^4$ holds for every $a, b \in G$?

A.8.3. Prove $o(ab) \bigm| [o(a), o(b)]$ for commutative groups.

A.8.4. Can the order of the product of two elements of finite order be infinite?

A.8.5. True or false?
(a) If $|G| < \infty$, then the order of every element in G is finite.
(b) If the order of every element in G is finite, then $|G| < \infty$.
(c) If $2 \bigm| |G|$, then G contains an element of order 2.
(d) If $4 \bigm| |G|$, then G contains an element of order 4.

A.8.6. Prove $s \mid \varphi(k^s - 1)$ for every $k, s > 1$ where $\varphi(n)$ is Euler's function φ.

*A.8.7. (a) What is the product of all elements in a finite Abelian group?
(b) Let p be a prime. What is the remainder of $(p-1)!$ modulo p?

A.8.8. Find the isomorphic groups.
(a) The reduced residue classes modulo 15 under multiplication.
(b) The reduced residue classes modulo 16 under multiplication.
(c) The reduced residue classes modulo 24 under multiplication.
(d) The non-reduced residue classes modulo 16 under addition.

(e) The 8th complex roots of unity under multiplication.

(f) The quaternions $\pm 1, \pm i, \pm j, \pm k$ under multiplication (see Example E5 in Section 5.6).

(g) The matrices $\begin{pmatrix} 1 & c \\ 0 & \pm 1 \end{pmatrix}$ over the modulo 4 residue classes under multiplication.

(h) F_2^3 under addition.

(i) The symmetry group of a square.

(j) The symmetry group of a brick.

*A.8.9. (a) Prove that if $|G_1| = |G_2|$ and every non-identity element has order 2, then G_1 and G_2 are isomorphic.

(b) The similar statement is false for order 3.

A.8.10. (a) Which groups have only trivial subgroups?

(b) Which groups have finitely many subgroups?

A.8.11. Show an example of a group that is the union of three proper subgroups. Can such a group have an odd number of elements?

*A.8.12. How many subgroups are (a) in a cyclic group of n elements; (b) in D_n?

*A.8.13. Verify that a non-empty subset C in a group is a left coset of some subgroup if and only if $a, b, c \in C \Rightarrow ab^{-1}c \in C$.

A.9. Ideals and Factor Rings

Definition A.9.1. A non-empty subset $I \subseteq R$ of a ring R is an *ideal* of R if

(i) I is closed under addition and taking negatives (in R), i.e.

$$i, j \in I \Rightarrow i + j \in I, -i \in I;$$

(ii) multiplying any element of I by an element of R from either side, we get an element of I, i.e.,

$$i \in I, r \in R \Rightarrow ri \in I, ir \in I. \clubsuit$$

So an ideal I is a subring where the product of an element in I and an element outside I is in I (see the definition of a subring in Exercise A.6.10).

Examples. The multiples of a given m form an ideal in the ring of integers; the polynomials divisible by a given polynomial g form an ideal in the ring of polynomials with (say) real coefficients; the polynomials with an even constant term form an ideal in the ring of polynomials with integer coefficients. The constant polynomials or the polynomials with integer coefficients (form a subring but) do *not* form an ideal in the ring of polynomials with real coefficients. A field has only the two *trivial ideals*, itself and the subset containing only the 0.

The simplest but also most important ideals are the ones generated by a single element, these are the *principal ideals*. For convenience, we restrict ourselves to commutative rings with identity.

Definition A.9.2. Let R be a commutative ring with identity and let $a \in R$. The set $\{ra \mid r \in R\}$ is called the *principal ideal* generated by a and is denoted by (a). ♣

So, the principal ideal (a) generated by a consists of the multiples of a. The phrase "ideal generated by a" is justified by the following theorem.

Theorem A.9.3. *The subset $(a) = \{ra \mid r \in R\}$ of a commutative ring R with identity satisfies:*

(i) *(a) is an ideal in R;*

(ii) *$a \in (a)$;*

(iii) *if I is an ideal in R and $a \in I$, then necessarily $(a) \subseteq I$.* ♣

So the principal ideal (a) is the *tightest* ideal containing a.

The third example after Definition A.9.1 is not a principal ideal (see Exercise A.9.9(c)), but the first two are principal ideals generated by m and g, resp. More generally, we get the following theorem.

Theorem A.9.4. *Every ideal in the ring $\mathbf{Z}$ of integers and in the ring $F[x]$ of polynomials over a commutative field F is a principal ideal.* ♣

The statement of Theorem A.9.4 is true in every commutative and zero divisor free ring with identity that satisfies a division algorithm, or in other words, every *Euclidean domain* is a *principal ideal domain.*

All this suggests that ideals are strongly connected to number theory. In fact, they do originate from number theory: Kummer introduced the notion of "ideal numbers" in his investigations of Fermat's Last Theorem. We discuss a few number-theoretic relations in Exercise A.9.10.

Now we turn to the construction of factor rings. This notion is a generalization of the ring $\mathbf{Z}_m$ of the modulo m residue classes.

We saw that the multiples of m form an ideal I in the ring $\mathbf{Z}$ of integers. The modulo m residue class containing an integer c is the set $c+I=\{c+i \mid i\in I\}=\{c+km \mid k\in\mathbf{Z}\}$, and c is a representative of this residue class. We define addition and multiplication of residue classes with the help of representatives, i.e., $(c + I) + (d + I) = (c + d) + I$ and $(c + I)(d + I) = cd + I$. We have to confirm that these functions define an operation for the *residue classes*, i.e., the resulting class is *unique*, it does *not depend* on which *representatives* were chosen from the two original classes $c + I$ and $d + I$. Performing and analyzing the proof, we see that this uniqueness is guaranteed by I being an ideal. This gives rise to the following generalization.

Theorem A.9.5. *Let I be an ideal in R. Then the (distinct) cosets $c + I = \{c + i \mid i \in I\}$ form a ring under the addition and multiplication*

$$(c + I) + (d + I) = c + d + I, \qquad (c + I)(d + I) = cd + I.$$

This ring is called the factor ring of R mod *I and is denoted by R/I.* ♣

The zero element of the factor ring is the coset $0 + I$, i.e., the ideal I itself.

Note that the coset $c + I$ is just a coset of the additive subgroup of I in the additive group of R (we do not have to distinguish left or right cosets as addition is commutative in a ring). Accordingly, the distinct cosets $\bmod I$ constitute a partition of R into disjoint subsets.

We now study factor rings of a polynomial ring $F[x]$ over a commutative field F. Every ideal in $F[x]$ is a principal ideal by Theorem A.9.4. Let $I = (g)$ where $g \neq 0$. Then a coset is formed by the polynomials giving the same remainder upon division by g. So, similar to the integers, every coset can be uniquely characterized by this common remainder, i.e., a polynomial of degree at most $\deg g - 1$ (including the 0 polynomial representing I itself). In the factor ring $F[x]/(g)$ we perform the computations with these remainders, so, e.g., when multiplying two cosets we essentially multiply the two remainders and take the remainder of the product modulo g (exactly as, e.g., modulo 15 the product of 7 and 6 is 12).

As an example, consider the factor ring $\mathbf{R}[x]/(x^2 + 1)$. Every coset has a unique representative polynomial $a+bx$ of degree at most one (with real coefficients) identified as a remainder upon division by $x^2 + 1$. Accordingly addition and multiplication are defined by

$$(a + bx) + (c + dx) = (a + c) + (b + d)x$$

and

$$\begin{aligned}(a + bx)(c + dx) &= ac + (ad + bc)x + bdx^2 \\ &= ac + (ad + bc)x - bd + bd(x^2 + 1) \\ &= (ac - bd) + (ad + bc)x.\end{aligned}$$

These are exactly the same rules as for the complex numbers $a + bi$. So the factor ring $\mathbf{R}[x]/(x^2 + 1)$ is a field isomorphic to $\mathbf{C}$ (we denote this fact as $\mathbf{R}[x]/(x^2 + 1) \cong \mathbf{C}$).

The next theorem characterizes in general which factor rings $F[x]/(g)$ form a field.

Theorem A.9.6. *Let F be a commutative zero divisor free ring with identity and let $g \in F[x]$. The factor ring $F[x]/(g)$ is a field if and only if g is irreducible over F.* ♣

Exercises A.9

A.9.1. Consider an arbitrary additive Abelian group and define the product of any two elements to be zero. Prove that we obtain a ring. What are the ideals? (Such rings are called rings of square zero or sometimes just zero rings, but some literature keeps this latter name only for the one-element ring consisting of 0 alone.)

A.9.2. Determine all integers $m > 1$ such that the zero divisors and 0 form an ideal in the ring $\mathbf{Z}_m$ of the modulo m residue classes.

A.9.3. (a) Let R be a ring without zero divisors. Prove that the intersection of arbitrarily (but finitely) many non-zero ideals is a non-zero ideal.
(b) Show an example of a ring where the intersection of two non-zero ideals can be zero.
(c) Show an example of a ring with zero divisors where the intersection of finitely many non-zero ideals cannot be zero.

A.9.4. (a) Verify that a field has only trivial ideals.
(b) Assume that the ring R is commutative, has only trivial ideals, and not all products ab are zero (i.e., R is not a ring of square zero). Prove that R is a field.
(c) Verify that commutativity plays an important role in (b) by showing that the ring of matrices $F^{n\times n}$ has only trivial ideals.

A.9.5. (a) Prove that every ideal in $\mathbf{Z}_m$ is a principal ideal.
*(b) Let $k \mid m$. What is a necessary and sufficient condition for the principal ideal (k) (of $\mathbf{Z}_m$) to be a ring isomorphic to $\mathbf{Z}_{m/k}$?
*(c) Verify that the factor ring $\mathbf{Z}_m/(k)$ is isomorphic to $\mathbf{Z}_k$ for any $k \mid m$.

A.9.6. Prove Theorems A.9.3–A.9.6.

A.9.7. Generalizing the notion of a principal ideal we introduce the *finitely generated ideals*. For a commutative ring R with identity, we define the ideal generated by $a_1, \dots, a_k \in R$ as $(a_1, \dots, a_k) = \left\{\sum_{i=1}^{k} r_i a_i \mid r_i \in R\right\}$. Formulate and prove the analogue of Theorem A.9.3 for $(a_1, \dots, a_k)$.

A.9.8. Denote the ring in Example E7 of Section A.6 by R_S.
(a) Characterize the principal ideals of R_S.
(b) Prove that every ideal is a principal ideal in R_S if $|S| < \infty$.
(c) Verify that if $|S| = \infty$, then all finite subsets of S form an ideal in R_S that is not a principal ideal, moreover it cannot be generated by finitely many elements.
(d) Let $A \subseteq S$. Show that the factor ring $R_S/(A)$ is isomorphic to $R_{S\setminus A}$.

A.9.9. Find a simpler form for the ideals below generated by two elements. Which are principal ideals? Describe the factor rings mod each ideal.
(a) $(30, 42)$ in the ring $\mathbf{Z}$.
(b) $(30, 42)$ in the ring $\mathbf{Z}_{100}$.
(c) $(2, x)$ in the ring $\mathbf{Z}[x]$.

A.9.10. Let R be a commutative ring with identity free of zero divisors and define divisibility, units, and greatest common divisor as usual: a unit is an element dividing every element and the greatest common divisor of two elements is a common divisor that is a multiple of every common divisor. The parentheses will be reserved for the generated ideals and a, b, d denote the elements of the ring R. So (a, b) is the ideal generated by a and b, and not the gcd of a and b [though the two notions are closely related, see parts (c)–(e)]. Prove:
(a) $a \mid b \iff (b) \subseteq (a)$.
(b) $(a) = (b) \iff$ a and b are associates (i.e., they differ only in a unit factor).

(c) If $(a, b) = (d)$, then d is a greatest common divisor of a and b.
(d) In the ring $\mathbf{Z}$ of integers or in the ring $F[x]$ of polynomials over a commutative field F, the converse of part (c) is also true.
(e) In the ring $\mathbf{Z}[x]$ of polynomials with integer coefficients the converse of part (c) is false.

A.9.11. Let R be the ring of 2×2 matrices with integer entries. Prove that the matrices with even entries form an ideal I. Determine $|R/I|$. Which well-known ring is isomorphic to the factor ring R/I?

A.9.12. Let R be the usual ring of all real functions and define f as $f(x) = x$, if $x \geq 5$, and $f(x) = 0$, if $x < 5$.
(a) Which functions are in the principal ideal (f)?
(b) Verify that the factor ring $R/(f)$ is isomorphic to R.

A.9.13. Let I be a non-trivial ideal in a ring R. True or false?
(a) If R is commutative, then so is R/I.
(b) If R/I is commutative, then so is R.
(c) If R has an identity, then so does R/I.
(d) If R/I has an identity, then so does R.
(e) If R is zero divisor free, then so is R/I.
(f) If R/I is zero divisor free, then so is R.

A.9.14. Which factor rings form a field?
(a) $\mathbf{R}[x]/(x^2)$;
(b) $\mathbf{C}[x]/(x^2 + 1)$;
(c) $\mathbf{Q}[x]/(x^6 - 2)$;
(d) $F_2[x]/(x^4 + x + 1)$.

A.10. Field Extensions

Similar to the other parts of the chapter, field always means a *commutative* field in this section.

Definition A.10.1. The field M is an *extension* of the field L if L is a *subfield* in M, i.e., $L \subseteq M$ and the operations in L are (the restrictions of) the operations in M. ♣

The usual notation is $M \mid L$, but as this might be misleading and does not reflect properly the relation of L and M, we rather apply another widely accepted notation $M : L$.

If M is an extension of L, then M is a vector space over L under the naturally arising operations from the field structure: we add two “vectors” in M as the elements of the field M, and we multiply a “vector” in M by a “scalar” of L by multiplying these two elements in the field M.

We introduce a special name and notation for the *dimension* of M as a vector space over the field L.

Definition A.10.2. If M is an extension of L, then the dimension of M as a vector space over L is the *degree* of the extension and is denoted by $\deg(M : L)$. ♣

Examples. $\deg(\mathbf{C} : \mathbf{R}) = 2$, $\deg(\mathbf{R} : \mathbf{Q}) = \infty$.

A basic theorem states that in a chain of consecutive extensions the degrees get multiplied.

Theorem A.10.3 (Tower Theorem). *If $L \subseteq M \subseteq N$ is a chain of extensions, then*

$$\deg(N : L) = \deg(N : M) \cdot \deg(M : L). \clubsuit$$

The simplest but also most important extensions are generated by a single element. We define it for arbitrary fields, but it is highly recommended to understand it through the special case $L = \mathbf{Q}$, $M = \mathbf{C}$.

In an extension $M : L$ we shall denote the elements of L and M by lower case and uppercase Greek letters, respectively.

Let L be a subfield of M and $\Theta \in M$. The *simple extension* of L by Θ is the set obtained applying the field operations and inverses (in M) to Θ and the elements of L. This means that we take all possible elements $\alpha_0 + \alpha_1\Theta + \cdots + \alpha_n\Theta^n \in M$ where n is an *arbitrary* non-negative integer and $\alpha_i \in L$, and we form the quotients of such expressions. Observe that $\alpha_0 + \alpha_1\Theta + \cdots + \alpha_n\Theta^n = g(\Theta) \in M$ with the polynomial $g = \alpha_0 + \alpha_1 x + \cdots + \alpha_n x^n \in L[x]$. So, the quotients in question are the elements $g(\Theta)/h(\Theta) \in M$ where g and h are arbitrary polynomials in $L[x]$ and $h(\Theta) \neq 0$. These quotients constitute the *tightest* subfield of M containing Θ and L. We formulate all this precisely in the next definition and theorem.

Definition A.10.4. Let L be a subfield of M and let $\Theta \in M$ be an arbitrary element. Then

$$\left\{ \frac{g(\Theta)}{h(\Theta)} \,\middle|\, g, h \in L[x],\ h(\Theta) \neq 0 \right\},$$

i.e.,

$$\left\{ \frac{\sum_{i=0}^{n} \alpha_i \Theta^i}{\sum_{j=0}^{k} \beta_j \Theta^j} \,\middle|\, \alpha_i, \beta_j \in L,\ \sum_{j=0}^{k} \beta_j \Theta^j \neq 0, \quad n, k = 0, 1, 2, \ldots \right\}$$

is called a *simple extension* of L by Θ and is denoted by $L(\Theta)$. ♣

Theorem A.10.5. *$L(\Theta)$ is the tightest subfield of M containing the element Θ and the field L, i.e.,*

(i) *$L(\Theta)$ is a subfield of M;*

(ii) *$\Theta \in L(\Theta)$, $L \subseteq L(\Theta)$;*

(iii) *if F is a subfield of M and $\Theta \in F$, $L \subseteq F$, then necessarily $L(\Theta) \subseteq F$.* ♣

We can write the elements of $L(\Theta)$ in a simpler form in some cases. We need the notion of an *algebraic element* for this characterization.

Definition A.10.6. Let L be a subfield of M. Then $\Theta \in M$ is an *algebraic element* over the field L if it is a root of some non-zero polynomial $f \in L[x]$, i.e., $f(\Theta) = 0$. ♣

In the above-mentioned special case $L = \mathbf{Q}$ and $M = \mathbf{C}$ this is the notion of an *algebraic number*: a complex number Θ is an algebraic number if it is a root of some non-zero polynomial with rational coefficients. So, e.g., $\sqrt{2}$, $\sqrt[3]{5}$, or $i\sqrt[7]{10}$ are algebraic numbers as they are roots of x^2-2, x^3-5, and $x^{14}+100$, resp. The algebraic numbers form a subfield of the complex numbers (see Exercise A.10.6).

The non-algebraic complex numbers are called *transcendental*, e.g., π, e (the base of the natural logarithm), $\log_{10} 2$, and $\sin 1$ (the angle measured in radians) are transcendental. There are only countably many algebraic numbers, so "most" complex numbers are transcendental. But it is generally very hard to determine whether a given complex number is algebraic or transcendental. This question is unsolved, e.g., for $e+\pi$, moreover we do not even know whether this number is rational or irrational.

An algebraic element $\Theta \in M$ is a root of many polynomials $f \in L[x]$, as any multiple of such a polynomial has this property. The ones of minimal degree play an important role.

Definition A.10.7. A *minimal polynomial* of a $\Theta \in M$ algebraic element over L is a polynomial in $L[x]$ of minimal degree having Θ among its roots. We denote the minimal polynomial of Θ by m_Θ. ♣

Theorem A.10.8.

(i) *m_Θ is unique apart from a constant multiple (in L).*

(ii) *Let $f \in L[x]$. Then $f(\Theta)=0 \iff m_\Theta \mid f$.*

(iii) *m_Θ is irreducible over L.*

(iv) *If f is irreducible over L and $f(\Theta)=0$, then $f=m_\Theta$.* ♣

Note that m_Θ can denote any of the minimal polynomials of Θ. This causes no problem as these polynomials differ only in a constant factor by (i). If we want to make the notation unique, then we can choose, e.g., the monic version of the minimal polynomial. In this case, the conclusion of (iv) should be modified to $f=\gamma m_\Theta$ where γ is any constant in L.

Definition A.10.9. The *degree* of a $\Theta \in M$ algebraic element over L is the degree of its minimal polynomial: $\deg \Theta = \deg m_\Theta$. ♣

In our previous examples for $L=\mathbf{Q}, M=\mathbf{C}$ the polynomials listed at the algebraic numbers were irreducible over $\mathbf{Q}$, so the monic minimal polynomials of $\sqrt{2}$, $\sqrt[3]{5}$, and $i\sqrt[7]{10}$ are x^2-2, x^3-5, and $x^{14}+100$, and the degrees of the three numbers are 2, 3, and 14, resp. (For $i\sqrt[7]{10}$, we can verify this by Theorems A.10.3 and A.10.11 below.) The algebraic numbers of first degree are the roots of the linear polynomials with rational coefficients, so these are the rational numbers themselves. More generally, the algebraic elements of first degree over L are just the elements of L.

Now we show that if Θ is algebraic, then the elements of the extension $L(\Theta)$ can be written in a much simpler form than described in Definition A.10.4.

As an example, consider the extension $\mathbf{Q}(\sqrt{2})$ of the rational field by $\sqrt{2}$. This is the set F of the numbers of the form $\alpha_0+\alpha_1\sqrt{2}$ where $\alpha_i \in \mathbf{Q}$ (see Example E3 in Section A.5), since F is a field containing $\sqrt{2}$ and the rational numbers and is obviously the smallest set satisfying these properties. This means that compared to the form in

Definition A.10.4, we do not need division and the powers of $\sqrt{2}$ with an exponent greater than one. Similarly, considering the extension $\mathbf{Q}(\sqrt[3]{5})$, we need just the powers of $\sqrt[3]{5}$ with exponents at most two, as the third and higher powers can be expressed with these and suitable rational numbers.

In the general case, we have the following theorem.

Theorem A.10.10. *If $\Theta \in M$ is an algebraic element of degree n over the field L, then the elements of $L(\Theta)$ have a unique representation as $\alpha_0 + \alpha_1\Theta + \cdots + \alpha_{n-1}\Theta^{n-1}$ where $\alpha_i \in L$.* ♣

In another formulation, the elements $1, \Theta, \dots, \Theta^{n-1}$ form a basis in $L(\Theta)$ considered as a vector space over L. So the dimension of this vector space, i.e., the degree of the extension $L(\Theta) : L$ is n. We state this important fact as a separate theorem.

Theorem A.10.11. *If $\Theta \in M$ is an algebraic element over the field L, then* $\deg\big(L(\Theta) : L\big) = \deg \Theta$. ♣

We can add to Theorems A.10.10–A.10.11 that if Θ is a non-algebraic (i.e., *transcendental*) element over L, then the elements of $L(\Theta)$ *cannot* be represented in a simpler form than described in Definition A.10.4 and the degree of the extension $L(\Theta) : L$ is *infinite*.

Another approach to characterize the field $L(\Theta)$ is the next theorem.

Theorem A.10.12. *If $\Theta \in M$ is an algebraic element over the field L, then the field $L(\Theta)$ is isomorphic to the factor ring $L[x]/(m_\Theta)$.* ♣

For example, the extension of the real numbers by i yields the complex numbers: $\mathbf{C} = \mathbf{R}(i)$. The minimal polynomial of i is $m_i = x^2+1$, so, by Theorem A.10.12, $\mathbf{C} = \mathbf{R}(i)$ is isomorphic to the factor ring $\mathbf{R}[x]/(x^2+1)$, a result established already in Section A.9.

By Theorem A.10.12, we can handle the elements of $L(\Theta)$ as remainders of polynomials upon division by the polynomial m_Θ.

If Θ is *transcendental* over L, then $L(\Theta)$ is a field isomorphic to the *algebraic fractions* over L, i.e., to the field of formal quotients of polynomials in $L[x]$.

Finally, we note that Theorem A.10.12 makes possible the construction of $L(\Theta)$ even if no larger field M is given; see Exercise A.10.18.

Exercises A.10

A.10.1. Assume that the degree of the extension $M : L$ is finite, $\deg(M : L) = n$. Prove that every $\Theta \in M$ is algebraic over L and $\deg \Theta \le n$, moreover $\deg \Theta \mid n$.

A.10.2. What causes the difference that whereas the minimal polynomial of an algebraic element is always irreducible, the minimal polynomial of a linear transformation (see Section 6.3) can be reducible over the field in question?

A.10.3. Prove Theorems A.10.3, 5, 8, 10, and 12.

Exercises A.10.4–A.10.17 deal with algebraic numbers (i.e., with the algebraic elements over $\mathbf{Q}$ among the complex numbers).

A.10.4. Show that a polynomial $f \neq 0$ with *complex* coefficients has a multiple $g \neq 0$ among the polynomials with *rational* coefficients if and only if every (complex) root of f is an algebraic number.

A.10.5. Verify that all nth roots of algebraic numbers with positive integer exponents n are algebraic numbers.

*A.10.6. Prove that the algebraic numbers form a subfield of the complex numbers.

A.10.7. What can we say about the sum of two complex numbers if (a) one is algebraic and the other is transcendental; (b) both are transcendental? (Possible answers: algebraic for sure; transcendental for sure; both can occur.)

A.10.8. (a) Denote the sum of two complex numbers by S and their product by P. What can we say about the two original numbers (being algebraic or transcendental) if
   (i) S is algebraic and P is transcendental;
   (ii) S is transcendental and P is algebraic;
   (iii) both S and P are transcendental;
   (iv) both S and P are algebraic?
(b) Investigate the analogous problem for rational and irrational numbers instead of algebraic and transcendental ones.

A.10.9. Consider a complex number $z(\neq 0)$ of algebraic form $z = a+bi$ and of trigonometric form $z = r(\cos\varphi + i\sin\varphi)$. Show that z is algebraic if and only if
(a) both a and b are algebraic;
(b) both r and $\cos\varphi$ are algebraic.

A.10.10. Determine whether the complex numbers below are algebraic or transcendental, and determine the degree of the algebraic ones:
(a) $\sqrt[100]{3000}$;
(b) $\sqrt{2} + \sqrt{3}$;
(c) $e + \pi i$;
(d) $\pi^{1000} + 3\pi^9 + 7$;
(e) $\sqrt[3]{2} + \sqrt[3]{4}$;
(f) $\cos 20°$;
(g) a 101st root of unity;
(h) a 6th root of unity;
(i) a primitive nth root of unity;
*(j) $\cos 1°$.

*A.10.11. Are there algebraic numbers on the unit circle besides the roots of unity?

*A.10.12. Find all algebraic numbers of odd degree on the unit circle.

A.10.13. Let Θ be an algebraic number of degree k. What are the possible values of $\deg(\Theta^2)$?

A.10.14. Verify (a) $\mathbf{Q}(\sqrt{8}) = \mathbf{Q}(\sqrt{18})$; (b) $\mathbf{Q}(\sqrt[9]{2}) \cap \mathbf{Q}(\sqrt[6]{2}) = \mathbf{Q}(\sqrt[3]{2})$.

A.10.15. Prove the existence of a polynomial f with *rational* coefficients satisfying $f(1 + 3\sqrt[7]{25} + 11\sqrt[7]{125} + 1000\sqrt[7]{625}) = \sqrt[7]{5}$.

***A.10.16.** Verify $\mathbf{Q}(z) \cap \mathbf{R} = \mathbf{Q}(\mathrm{Re}\, z)$ if $|z| = 1$.

***A.10.17.** Show that if every coefficient of a polynomial $f \in \mathbf{C}[x]$ is an algebraic number, then every (complex) root of f is an algebraic number.

Remark. Let A denote the field of algebraic numbers. Another formulation of the statement is that every non-constant polynomial $f \in A[x]$ has a root in A, or equivalently, the number of roots of f in A counted with multiplicity is equal to the degree of f. This means that similar to the complex field, the analogue of the fundamental theory of algebra also holds for the field of algebraic numbers. Fields having this property are called *algebraically closed* fields.

*A.10.18. Let L be an arbitrary (commutative) field and let f be an irreducible polynomial over L. Construct a field M satisfying the requirements below:

(i) M has a subfield L^* isomorphic to L;

(ii) if the coefficients of $f^* \in L^*[x]$ are obtained from the coefficients of f via the isomorphism $L \to L^*$, then f^* has a root $\Theta \in M$;

(iii) $M = L^*(\Theta)$.

Remark. This makes it possible to construct an extension of L by a not yet existing(!) root of an irreducible polynomial even if we have no larger field containing L.

A.11. Finite Fields

We used the field F_p of the residue classes modulo p where p is a prime in various parts of the book. Some other finite fields occurred, too: Exercises A.5.2(b) and A.9.14(d) displayed fields of 9 and 16 elements.

In this section we describe and characterize the finite fields from various aspects. We use the notation M for a general finite field, Greek uppercase letters for its elements, and 0 and 1 for its zero and identity.

We note that the commutative law for multiplication in a finite field follows from the other field axioms; this is a famous (and hard) theorem by Wedderburn.

1. Size

We start the systematic discussion by giving the possible number of elements in a finite field.

Theorem A.11.1.

(i) *The size of a finite field can only be a power of a prime (including the first powers, i.e., the primes themselves).*

(ii) *For every prime power p^k there exists exactly one finite field M up to isomorphism satisfying $|M| = p^k$.* ♣

Thus no field can have, e.g., 100 elements, but there exists a field of 81 elements. The theorem implies that if two finite fields have the same size, then they are isomorphic. So, if p is a prime, then the residue classes mod p provide the only field of p elements.

2. Addition

The next theorem describes the structure of the additive group in a finite field.

Theorem A.11.2. *Let $|M| = p^k$. Then, concerning addition, M is isomorphic to the additive group of the k-dimensional vector space F_p^k over F_p.* ♣

So, the elements of M are k-dimensional vectors where every component is a residue class modulo p, and we perform addition accordingly. This implies that adding p times any $\Theta \in M$, we always get 0. In other words, the non-zero elements in the additive group of M have (additive) order p.

3. Multiplication

The multiplicative structure of a finite field is also very nice.

Theorem A.11.3. *The non-zero elements of a finite field form a cyclic group for multiplication.* ♣

This means that there exists a generator element $\Delta \in M$ so that all non-zero elements in M appear among its powers. Then the multiplicative order of Δ is $o(\Delta) = |M| - 1 = p^k - 1$, so the non-zero elements in M have a unique representation as Δ^j where $j = 0, 1, 2, \ldots, p^k - 2$.

By a general result for cyclic groups, a field of p^k elements has (not just one, but) $\varphi(p^k - 1)$ generators (where φ is Euler's function); these are called *primitive elements* of the field. If $|M| = p$, i.e., $M = F_p$, then we use also the term *primitive root* modulo p.

4. Vector space

An important fact is that every finite field contains a field of type F_p. This is the subfield generated by the identity 1 and consists of the elements $1, 1 + 1, \ldots, 1 + 1 + \cdots + 1$. If $|M| = p^k$, then the additive order of 1 is p by Theorem A.11.2, so there are p such (distinct) elements. It can be easily verified that these form a field isomorphic to F_p.

Thus, we can consider M as an *extension* of the field F_p. Then M is a k-dimensional *vector space* over F_p, so the isomorphism stated in Theorem A.11.2 holds not just for the additive groups but for the vector spaces, too.

We summarize this in the next theorem.

Theorem A.11.4. *If the finite field M has $|M| = p^k$ elements, then M contains a subfield isomorphic to F_p and M is a k-dimensional vector space over this subfield.* ♣

5. Field extension

We discuss some properties of the extension $M : F_p$ (we rely strongly on Section A.10).

Since $\deg(M : F_p) = k$, Exercise A.10.1 implies that every $\Theta \in M$ is an *algebraic* element over F_p and $\deg \Theta \mid k$.

We show that $M : F_p$ is a *simple* extension. Let Δ be a generator element of the multiplicative group of M. Then the tightest subfield containing Δ is M since we obtain every non-zero element using just multiplication. So $F_p(\Delta) = M$.

It follows that $k = \deg(M : F_p) = \deg\big(F_p(\Delta) : F_p\big) = \deg m_\Delta$, so the degree of Δ over the field F_p is exactly k.

Then $1, \Delta, \Delta^2, \dots, \Delta^{k-1}$ is a basis of (the vector space) M over F_p (here 1 denotes the common identity element of the fields F_p and M).

Of course, besides the generators of the multiplicative group, i.e., the primitive elements, other $\Theta \in M$ (can) satisfy $M = F_p(\Theta)$; this holds exactly if the degree of Θ is k (over F_p).

We summarize this in the following theorem.

Theorem A.11.5. *Let M be a finite field, $|M| = p^k$, let Δ be a generator of the multiplicative group, and let F_p be the subfield of p elements. Then*

(i) $M = F_p(\Delta)$;

(ii) $1, \Delta, \Delta^2, \dots, \Delta^{k-1}$ *is a basis in M considered as a vector space over F_p;*

(iii) $\deg \Theta \mid k$ *for every* $\Theta \in M$;

(iv) $\Theta \in M$, $\deg \Theta = k \iff M = F_p(\Theta)$. ♣

6. Factor ring

The representation $M = F_p(\Delta)$ implies that the field M is isomorphic to the factor ring $F_p[x]/(m_\Delta)$ by Theorem A.10.12. Combining this with Theorem A.9.6 we obtain a useful characterization of finite fields.

Theorem A.11.6. *We obtain the finite field of p^k elements as the factor ring $F_p[x]/(f)$ where f is an irreducible polynomial of degree k over F_p.* ♣

So the elements of a field of size p^k can be considered as polynomials of degree not greater than $k-1$ in $F_p[x]$ and the operations are performed as with remainders under the division by the irreducible polynomial f.

Theorems A.11.1(ii) and A.11.6 imply that there exist irreducible polynomials of degree k over F_p for any k. The same holds over any finite field; moreover we can determine the number of irreducible polynomials of a given degree (see Exercise A.11.12(b) and the remark at the hint).

The minimal polynomials of the *primitive elements* are called *primitive polynomials* (over F_p). (This notion has no connection to the primitive polynomials over the integers occurring in Gauss's lemma; see part 13 in Section A.7.)

7. An example

We illustrate the theory by constructing a field M of 125 elements.

We need an irreducible polynomial of third degree over the field F_5. A polynomial of third degree is irreducible exactly if it has no roots in the given field (this is false for higher degrees, see Exercise A.7.19). We find that, e.g., $f = x^3 + x + 1$ is irreducible since $f(u) \neq 0$ holds for all the five elements $u \in F_5$.

By Theorem A.11.6, the elements of M are polynomials $\alpha_0 + \alpha_1 x + \alpha_2 x^2$ where $\alpha_i \in F_5$ and we add and multiply them as remainders modulo $x^3 + x + 1$.

E.g., the sum and product of $3 + 4x$ and $3 + x^2$ are $(3+4x)+(3+x^2) = 1+4x+x^2$ and

$$(3 + 4x)(3 + x^2) = 4 + 2x + 3x^2 + 4x^3 = 4 + 2x + 3x^2 + 4(-1 - x) = 3x + 3x^2.$$

We can denote the elements of M by vectors in F_5^3. Choosing the natural basis $1, x, x^2$, the above operations are

$$\begin{pmatrix}3\\4\\0\end{pmatrix} + \begin{pmatrix}3\\0\\1\end{pmatrix} = \begin{pmatrix}1\\4\\1\end{pmatrix} \quad \text{and} \quad \begin{pmatrix}3\\4\\0\end{pmatrix}\begin{pmatrix}3\\0\\1\end{pmatrix} = \begin{pmatrix}0\\3\\3\end{pmatrix}.$$

This notation makes addition easy but concerning multiplication we have to think in polynomials and use the reduction modulo the minimal polynomial.

How can we obtain a generator of the multiplicative group of the field? The remainder x (or, using the second notation, the vector $\begin{pmatrix}0\\1\\0\end{pmatrix}$) can serve as a generator if its (multiplicative) order is $|M| - 1 = 124$. By Lagrange's theorem, the 124th power of every element in the group is the identity, so we have to check whether some smaller (positive) power equals 1. Since $124 = 2^2 \cdot 31$, it is enough to compute the 4th and 62nd powers: if none of these is the identity, then the order does not divide them, so it must be 124.

Using $x^3 = -1 - x$, we get $x^4 = -x - x^2 \neq 1$, so $o(x) \nmid 4$.

To check the 62nd power, the quickest way is to use

$$1 = x^{62} = (x^{31})^2 \iff x^{31} = \pm 1 \iff x^{32} = \pm x$$

since a field is zero divisor free and $x \neq 0$. Squaring $x^4 = -(x + x^2)$ we obtain

$$x^8 = x^2 + 2x^3 + x^4 = x^2 - (2 + 2x) - (x + x^2) = -(2 + 3x).$$

We raise this to the fourth power applying the binomial theorem:

$$\begin{aligned} x^{32} = (2 + 3x)^4 &= 2^4 + 4 \cdot 2^3 \cdot 3x + 6 \cdot 2^2 \cdot 3^2 x^2 + 4 \cdot 2 \cdot 3^3 x^3 + 3^4 x^4 \\ &= 1 + x + x^2 + x^3 + x^4 = 1 + x + x^2 - (1 + x) - (x + x^2) = -x. \end{aligned}$$

This implies $x^{62} = 1$ (in fact $o(x) = 62$), so x is not a primitive element.

Then $(x^m)^{62} = 1$ for any m; therefore the powers of x can neither be primitive elements. Besides them the (altogether two) elements of order four are not primitive. Hence, each of the remaining $124 - 62 - 2 = 60$ elements generates the multiplicative group of the field.

Another option is to find some other irreducible polynomial of third degree that is primitive over F_5 and use this one to form the factor ring characterizing the field of 125 elements. E.g., $x^3 + 3x + 2$ is primitive, so x (as a residue modulo $x^3 + 3x + 2$) is a generator of the multiplicative group.

Exercises A.11

A.11.1. Show that in a field of p^k elements one can raise a sum to the pth power by terms, i.e., $(\Theta + \Psi)^p = \Theta^p + \Psi^p$.

A.11.2. Compute the sum and product of all non-zero elements in a finite field.

A.11.3. Determine the number of roots of the polynomial $x^m - 1$ in the field of p^k elements.

A.11.4. How many pairs of elements $\Theta \neq \Psi$ are the cube roots of each other in the field of 13^k elements?

A.11.5. (a) Assume that in a ring free of zero divisors adding some non-zero element with itself certain number of times we obtain 0. Prove that there exists a unique prime p such that adding any element of the ring with itself p times we get 0.

(b) Find an *infinite* field with this property.

Remark. This prime p is called the *characteristic* of the ring (free of zero divisors). So, the characteristic of a field of p^k elements is p. If there is no such p, i.e., adding a non-zero element with itself arbitrarily many times we never get 0, then the characteristic is 0 (or infinite, in an older terminology). E.g., $\mathbf{Q}$, $\mathbf{R}$, and $\mathbf{C}$ have characteristic 0. If a ring contains zero divisors, then there is no such uniform minimal number of terms in general, e.g., in $\mathbf{Z}_6$, we have to add 1 six times with itself to get 0, whereas $2+2+2=0$ and $3+3=0$.

A.11.6. Construct a field of (a) 169; (b) 81 elements.

A.11.7. Show that the field M of p^k elements consists of the roots of the polynomial $x^{p^k} - x$, i.e., $x^{p^k} - x = \prod_{\Theta \in M}(x - \Theta)$.

***A.11.8.** Let f be an irreducible polynomial over F_p. Verify: $f \mid x^{p^k} - x \iff \deg f \mid k$.

A.11.9. Prove that the field of p^k elements contains a subfield of p^n elements if and only if $n \mid k$.

A.11.10. Prove that two (distinct) subfields of a finite field cannot have the same number of elements.

*A.11.11. Let $f = x^k + \alpha_{k-1}x^{k-1} + \dots + \alpha_1 x + \alpha_0$ be a primitive polynomial over F_p and

$$A = \begin{pmatrix} 0 & 0 & 0 & \dots & 0 & -\alpha_0 \\ 1 & 0 & 0 & \dots & 0 & -\alpha_1 \\ 0 & 1 & 0 & \dots & 0 & -\alpha_2 \\ \vdots & \vdots & \vdots & & \vdots & \vdots \\ 0 & 0 & 0 & \dots & 1 & -\alpha_{k-1} \end{pmatrix} \in F_p^{k \times k}.$$

Show that the (distinct) powers of A and the zero matrix form a field of p^k elements under the usual matrix addition and multiplication.

***A.11.12.** Determine the number of monic (a) primitive; (b) irreducible polynomials of degree k over F_p.

A.11.13. Consider the field M of p^3 elements as a three-dimensional vector space over F_p. Let us call *points* and *lines* the one- and two-dimensional subspaces of M, resp. A point is on a line if the corresponding larger subspace contains the smaller one. Show that we obtain a (non-degenerate) finite projective plane of order p (see Exercise 9.5.11):

(a) Any two lines have exactly one point in common and any two points determine exactly one line.

(b) There are $p^2 + p + 1$ points and $p^2 + p + 1$ lines.

(c) Every line consists of $p + 1$ points and every point lies on exactly $p + 1$ lines.

Remarks. The same construction works for a prime power $n = p^m$ instead of $n = p$: the finite field of p^{3m} elements should be considered as a vector space over its subfield of p^m elements. As mentioned in Exercise 9.5.11, it is an old unsolved problem whether there exist finite projective planes where n is not a prime power.

We can illuminate the geometric relation better if we disregard the zero vector in the one-dimensional subspaces defining the points. Then the vectors characterizing a point are just scalar multiples of each other. This can be interpreted as characterizing a point by *homogeneous coordinates* since we get the same point if we multiply the three coordinates with the same non-zero scalar. This is a perfect analogy to the projective plane over the real numbers. This real projective plane is obtained by extending the Euclidean plane with intersection points of parallel lines in every given direction called *ideal points*. The ideal points form an ideal line. An ordinary point in the Euclidean plane having coordinates (a_1, a_2) corresponds to the triples of homogeneous coordinates $(\alpha_1, \alpha_2, \alpha_3)$ where $\alpha_1/\alpha_3 = a_1$ and $\alpha_2/\alpha_3 = a_2$.

A.11.14. Let M be as in the previous exercise, but now both the points and the lines are the one-dimensional subspaces. A point lies on a line if the two one-dimensional subspaces corresponding to the point and the line are perpendicular. Verify the statements (a)–(c) of the previous exercise.

Index

We generally indicate the first occurrence only. The data include the typical notation (if any), the serial number of the definition, theorem, etc. explaining the notion or denomination, and finally the page number in parentheses.

D2.1.6 means Definition 2.1.6, and letters T, L, E instead of D refer to the theorem, lemma, and exercise with the given number. P1.3.3 stands for the proof of Theorem 1.3.3, 4.1.E3 denotes Example 3 in Section 4.1, 5.7 means Section 5.7, and A.7/2 refers to Subsection 2 in Section A.7.

We add a sign "−" or "+" to the number of definition, theorem, etc., if the notion is introduced not in the given definition, theorem, etc., but just before or after it, without a new serial number. E.g., DA.10.6+ indicates at the phrase "transcendental number" that it is explained *after* Definition A.10.6.

In some cases, we include an important theorem besides the definition. E.g., for "subspace" we refer both to Definition 4.2.1 and Theorem 4.2.2.

For information about notation used in the book, please consult part "Technical details" in the Introduction. The phrasing of a definition or theorem is closed by a ♣ sign, the end of a proof is denoted by ■. We add that as mentioned in another part of the Introduction, exercises marked with one or two asterisks are considered hard or extra hard by our judgement, and the boldface serial number indicates that a detailed solution can be found at `bookstore.ams.org/amstext-66`.

Selected Published Titles in This Series

66 **Róbert Freud,** Linear Algebra, 2024
65 **Yitzhak Katznelson and Yonatan Katznelson,** An Introduction to Real Analysis, 2024
64 **Johannes A. Buchmann,** Introduction to Quantum Algorithms, 2024
63 **Bettina Richmond and Thomas Richmond,** A Discrete Transition to Advanced Mathematics, Second Edition, 2023
62 **Scott A. Taylor,** Introduction to Mathematics, 2023
61 **Bennett Chow,** Introduction to Proof Through Number Theory, 2023
60 **Lorenzo Sadun,** The Six Pillars of Calculus: Biology Edition, 2023
59 **Oscar Gonzalez,** Topics in Applied Mathematics and Modeling, 2023
58 **Sebastian M. Cioabă and Werner Linde,** A Bridge to Advanced Mathematics, 2023
57 **Meighan I. Dillon,** Linear Algebra, 2023
56 **Lorenzo Sadun,** The Six Pillars of Calculus: Business Edition, 2023
55 **Joseph H. Silverman,** Abstract Algebra, 2022
54 **Rustum Choksi,** Partial Differential Equations, 2022
53 **Louis-Pierre Arguin,** A First Course in Stochastic Calculus, 2022
52 **Michael E. Taylor,** Introduction to Differential Equations, Second Edition, 2022
51 **James R. King,** Geometry Transformed, 2021
50 **James P. Keener,** Biology in Time and Space, 2021
49 **Carl G. Wagner,** A First Course in Enumerative Combinatorics, 2020
48 **Róbert Freud and Edit Gyarmati,** Number Theory, 2020
47 **Michael E. Taylor,** Introduction to Analysis in One Variable, 2020
46 **Michael E. Taylor,** Introduction to Analysis in Several Variables, 2020
45 **Michael E. Taylor,** Linear Algebra, 2020
44 **Alejandro Uribe A. and Daniel A. Visscher,** Explorations in Analysis, Topology, and Dynamics, 2020
43 **Allan Bickle,** Fundamentals of Graph Theory, 2020
42 **Steven H. Weintraub,** Linear Algebra for the Young Mathematician, 2019
41 **William J. Terrell,** A Passage to Modern Analysis, 2019
40 **Heiko Knospe,** A Course in Cryptography, 2019
39 **Andrew D. Hwang,** Sets, Groups, and Mappings, 2019
38 **Mark Bridger,** Real Analysis, 2019
37 **Mike Mesterton-Gibbons,** An Introduction to Game-Theoretic Modelling, Third Edition, 2019
36 **Cesar E. Silva,** Invitation to Real Analysis, 2019
35 **Álvaro Lozano-Robledo,** Number Theory and Geometry, 2019
34 **C. Herbert Clemens,** Two-Dimensional Geometries, 2019
33 **Brad G. Osgood,** Lectures on the Fourier Transform and Its Applications, 2019
32 **John M. Erdman,** A Problems Based Course in Advanced Calculus, 2018
31 **Benjamin Hutz,** An Experimental Introduction to Number Theory, 2018
30 **Steven J. Miller,** Mathematics of Optimization: How to do Things Faster, 2017
29 **Tom L. Lindstrøm,** Spaces, 2017
28 **Randall Pruim,** Foundations and Applications of Statistics: An Introduction Using R, Second Edition, 2018
27 **Shahriar Shahriari,** Algebra in Action, 2017
26 **Tamara J. Lakins,** The Tools of Mathematical Reasoning, 2016

For a complete list of titles in this series, visit the
AMS Bookstore at **www.ams.org/bookstore/amstextseries/**.